London Mathematical Society Lecture Note Series: 323

Poisson Geometry, Deformation Quantisation and Group Representations

Edited by

SIMONE GUTT
Université Libre de Bruxelles and Université de Metz

JOHN RAWNSLEY
University of Warwick

DANIEL STERNHEIMER
Université de Bourgogne

CAMBRIDGE
UNIVERSITY PRESS

CAMBRIDGE UNIVERSITY PRESS
Cambridge, New York, Melbourne, Madrid, Cape Town, Singapore, São Paulo

Cambridge University Press
The Edinburgh Building, Cambridge CB2 2RU, UK

Published in the United States of America by Cambridge University Press, New York

www.cambridge.org
Information on this title: www.cambridge.org/9780521615051

First published 2005

Printed in the United Kingdom at the University Press, Cambridge

A catalogue record for this book is available from the British Library

Library of Congress Cataloguing in Publication data

ISBN-13 978-0-521-61505-1 paperback
ISBN-10 0-521-61505-4 paperback

LONDON MATHEMATICAL SOCIETY LECTURE NOTE SERIES

Managing Editor: Professor N.J. Hitchin, Mathematical Institute,
University of Oxford, 24–29 St. Giles, Oxford OX1 3LB, United Kingdom

The title below are available from booksellers, or from Cambridge University Press at www.cambridge.org

161 Lectures on block theory, BURKHARD KÜLSHAMMER
163 Topics in varieties of group representations, S.M. VOVSI
164 Quasi-symmetric designs, M.S. SHRIKANDE & S.S. SANE
166 Surveys in combinatorics, 1991, A.D. KEEDWELL (ed)
168 Representations of algebras, H. TACHIKAWA & S. BRENNER (eds)
169 Boolean function complexity, M.S. PATERSON (ed)
170 Manifolds with singularities and the Adams-Novikov spectral sequence, B. BOTVINNIK
171 Squares, A.R. RAJWADE
172 Algebraic varieties, GEORGE R. KEMPF
173 Discrete groups and geometry, W.J. HARVEY & C. MACLACHLAN (eds)
174 Lectures on mechanics, J.E. MARSDEN
175 Adams memorial symposium on algebraic topology 1, N. RAY & G. WALKER (eds)
176 Adams memorial symposium on algebraic topology 2, N. RAY & G. WALKER (eds)
177 Applications of categories in computer science, M. FOURMAN, P. JOHNSTONE & A. PITTS (eds)
178 Lower K- and L-theory, A. RANICKI
179 Complex projective geometry, G. ELLINGSRUD *et al*
180 Lectures on ergodic theory and Pesin theory on compact manifolds, M. POLLICOTT
181 Geometric group theory I, G.A. NIBLO & M.A. ROLLER (eds)
182 Geometric group theory II, G.A. NIBLO & M.A. ROLLER (eds)
183 Shintani Zeta Functions, A. YUKIE
184 Arithmetical functions, W. SCHWARZ & J. SPILKER
185 Representations of solvable groups. O. MANZ & T.R. WOLF
186 Complexity: knots, colourings and counting, D.J.A. WELSH
187 Surveys in combinatorics, 1993, K. WALKER (ed)
188 Local analysis for the odd order theorem, H. BENDER & G. GLAUBERMAN
189 Locally presentable and accessible categories, J. ADAMEK & J. ROSICKY
190 Polynomial inwariants of finite groups, D.J. BENSON
191 Finite geometry and combinatorics, F. DE CLERCK *et al*
192 Symplectic geometry, D. SALAMON (ed)
194 Independent random variables and rearrangment invariant spaces, M. BRAVERMAN
195 Arithmetic of blowup algebras, WOLMER VASCONCELOS
196 Microlocal analysis for differential operators, A. GRIGIS & J. SJÖSTRAND
197 Two-dimensional homotopy and combinatorial group theory, C. HOG-ANGELONI *et al*
198 The algebraic characterization of geometric 4-manifolds, J.A. HILLMAN
199 Invariant potential theory in rhe unit ball of C^n, MANFRED STOLL
200 The Grothendieck theory of dessins d'enfant, L. SCHNEPS (ed)
201 Singularities, JEAN-PAUL BRASSELET (ed)
202 The technique of pseudodifferential operators, H.O. CORDES
203 Hochschild cohomology of von Neumann algebras, A. SINCLAIR & R. SMITH
204 Combinatorial and geometric group theory, A.J. DUNCAN, N.D. GILBERT & J. HOWIE (eds)
205 Ergodic theory and its connections with harmonic analysis, K. PETERSEN & I. SALAMA (eds)
207 Groups of Lie type and their geometries, W.M. KANTOR & L. DI MARTINO (eds)
208 Vector bundles in algebraic geometry, N.J. HITCHIN, P. NEWSTEAD & W.M. OXBURY (eds)
209 Arithmetic of diagonal hypersutrfaces over finite fields, Q. GOUVÉA & N. YUI
210 Hilbert C*-modules, E.C. LANCE
211 Groups 93 Galway / St Andrews I, CM. CAMPBELL *et al* (eds)
212 Groups 93 Galway / St Andrews II, C.M. CAMPBELL *et al* (eds)
214 Generalised Euler-Jacobi inversion formula and asymptotics beyond all orders, V. KOWALENKO *el al*
215 Number theory 1992–93, S. DAVID (ed)
216 Stochastic partial differential equations, A. ETHERIDGE (ed)
217 Quadratic forms with applications to algebraic geometry and topology, A. PFISTER
218 Surveys in combinatorics, 1995, PETER ROWLINSON
220 Algebraic set theory, A. JOYAL & I. MOERDIJK
221 Harmonic approximation, S.J. GARDINER
222 Advances in linear logic, J.-Y. GIRARD. Y. LAFONT & L. REGNIER (eds)
223 Analytic semigroups and semilinear initial boundary value problems, KAZUAKI TAIRA
224 Comrutability, enumerability, unsolvability, S.B. COOPER, T.A. SLAMAN & S.S. WAINER (eds)
225 A mathematical introduction to string theory, S. ALBEVERIO *et al*
226 Novikov conjectures, index theorems and rigidity I. S. FERRY, A. RANICKI & J. ROSENBERG (eds)
227 Novikov conjectures, index theorems and rigidity II, S. FERRY, A. RANICK1 & J. ROSENBERG (eds)
228 Ergodic theory of Z_d actions, M. POLLICOTT & K. SCHMIDT (eds)
229 Ergodicity for infinite dimensional systems, G. DA PRATO & J. ZABCZYK
230 Prolegomena to a middlebrow arithmetic of curves of genus 2, J.W.S. CASSELS & E.V. FLYNN
231 Semigmup theory and its applications. K.H. HOFMANN & M.W. MISLOVE (eds)
232 The descriptive set theory of Polish group actions, H. BECKER & A.S. KECHRIS
233 Finite fields and applications, S. COHEN & H. NEDERREITER (eds)
234 Introduction to subfactors, V. JONES & V.S. SUNDER
235 Number theory 1993–94. S. DAVID (ed)

236 The James forest, H. FETTER & B. GAMBOA DE BUEN
237 Sieve methods. exponential sums, and their applications in number theory, G.R.H. GREAVES *et al*
238 Representation theory and algebraic geometry, A. MARTSINKOVSKY & G. TODOROV (eds)
240 Stable groups, FRANK O. WAGNER
241 Surveys in combinatorics, 1997. R.A. BAILEY (ed)
242 Geometric Galois actions I, L. SCHNEPS & P. LOCHAK (eds)
243 Geometric Galois actions II, L. SCHNEPS & P. LOCHAK (eds)
244 Model theory of groups and automrphism groups, D. EVANS (ed)
245 Geometry, combinatorial designs and related structures, J.W.P HIRSCHFELD *et al*
246 p-Automorphisms of Finite-groups, E. KHUKHRO
247 Analytic number theory, Y. MOTOHASHI (ed)
248 Tame topology and o-minimal structures, LOU VAN DEN DRIES
249 The atlas of finite gmups: ten years on, ROBERT CURTIS & ROBERT WILSON (eds)
250 Characters and blocks of finite groups. G. NAVARRO
251 Groner bases and applications, B. BUCHBERGER & E WINKLER (eds)
252 Geometry and cohomology in group theory, P. KROPHOLLER, G. NIBLO, R. STÖHR (eds)
253 The q-Schur algebra, S. DONKIN
254 Galois representations in arithmetic algebraic geometry, A.J. SCHOLL & R.L. TAYLOR (eds)
255 Symmetries and integrability of difference equations, P.A. CLARKSON & F.W. NIJHOFF (eds)
256 Aspects of Galois theory, HELMUT VÖLKLBIN *et al*
257 An introduction to nancommutative differential geometty and its physical applications 2ed, J. MADORE
258 Sets and proofs, S.B. COOPER & J. TRUSS (eds)
259 Models and computability, S.B. COOPER & J. TRUSS (eds)
260 Groups St Andrews 1997 in Bath, I, CM. CAMPBELL *et al*
261 Groups St Andrews 1997 in Bath, II, C.M. CAMPBELL *et al*
262 Analysis and logic, C.W. HENSON, J. IOVINO, A.S. KECHRIS & E. ODELL
263 Singularity theoy, BILL BRUCE & DAVID MOND (eds)
264 New trends in algebraic geometry, K. HULEK, F. CATANESE, C. PETERS & M. REID (eds)
265 Elliptic curves in cryptography, I. BLAKE, G. SEROUSSI & N. SMART
267 Surveys in combinatorics, 1999, J.D. LAMB & D.A. PREECE (eds)
268 Spectral asymptatics in the semi-classical limit, M. DIMASSI & J. SJÖSTRAND
269 Ergodic theory and topological dynamics, M.B. BEKKA & M. MAYER
270 Analysis on Lie groups, N.T. VAROPOULOS & S. MUSTAPHA
271 Singular perturbations of differential operators, S. ALBEVERIO & P. KURASOV
272 Character theory for the odd order theorem, T. PETERFALVI
273 Spectral theory geometry, E.B. DAVIES & Y. SAFAROV (eds)
274 The Mandlebrot set, theme and variations, TAN LEI (ed)
275 Descriptive set theory and dynamical systems, M. FOREMAN *et al*
276 Singularities of plane curves, E. CASAS-ALVERO
277 Computatianal and geometric aspects of modern algebra, M.D. ATKINSON *et al*
278 Glbal attractors in abstract parabolic problems, J.W. CHOLEWA & T. DLOTKO
279 Topics in symbolic dynamics and applications, F. BLANCHARD, A. MAASS & A. NOGUEIRA (eds)
280 Chamctw and automorphism groups of compact Riemann surfaces, THOMAS BREUER
281 Explicit birational geometry of 3-folds, ALESSIO CORTI & MILES REID (eds)
282 Auslander-Buchweitz approximations of equivariant modules, M. HASHIMOTO
283 Nonlinear elasticity, Y. FU & R.W. OGDEN (eds)
284 Foundations of computational mathematics, R. DEVORE, A. ISERLES & E. SÜLI (eds)
285 Rational points on curves over finite fields. H. NIEDERREITER & C. XING
286 Clifford algebras and spinors 2ed, P. LOUNESTO
287 Topics on Riemann surfaces and Fuchsian groups, E. BUJALANCE, A.F. COSTA & E. MARTINEZ (eds)
288 Surveys in cambinatorics, 2001, J. HIRSCHFELD (ed)
289 Aspects of Sobolev-type inequalities. L. SALOFF-COSTE
290 Quantum groups and Lie theory, A. PRESSLEY (ed)
291 Tits buildings and the model theory of groups, K. TENT (ed)
292 A quantum groups primer, S. MAJID.
293 Second order partial diffetcntial equations in Hilbett spaces, G. DA PRATO & I. ZABCZYK
294 Imwduction to the theory of operator spaces, G. PISIER
295 Geometry and integrability. LIONEL MASON & YAVUZ NUTKU (eds)
296 Lectures on invariant theory, IGOR DOLGACHEV
297 The homotopy category of simply connected 4-manifolds, H.-J. BAUES
299 Kleinian groups and hyperbolic 3-manifolds, Y. KOMORI, V. MARKOVIC, & C. SERIES (eds)
300 Introduction to Möbius differential geometry, UDO HERTRICH-JEROMIN
301 Stable modules and the D(2)-problem, F.E.A. JOHNSON
302 Diicrete and continuous nonlinear Schrödinger systems, M.J. ABLOWITZ, B. PRINARI, & A.D. TRUBATCH
303 Number theory and algebraic geometry, MILES REID & ALEXEI SKOROBOOATOV (eds)
304 Groups St Andrews 2001 in Oxford Vol. I, COLIN CAMPBELL, EDMUND ROBERTSON
305 Groups St Andrews 2001 in Oxford Vol. 2. C.M. CAMPBELL, E.F. ROBERTSON & G.C. SMITH (eds)
307 Surveys in combinatorics 2003, C.D. WENSLEY (ed)
308 Topology, Geometry and Quantum Field Theory, U. TILLMANN (ed)
309 Corings and comodules, TOMASZ BRZEZINSKI & ROBERT WISBAUER
310 Topics in dynamics and ergodic theory, SERGEY BEZUGLYI & SERGIY KOLYADA (eds)
311 Groups, T.W. MÜLLER
312 Foundations of computational mathematics, Minneapolis 2002, FELIPE CUCKER *et al* (eds)
313 Transcendantal Aspects of Algebraic Cycles, S. MÜLLER-STACH & C. PETERS (eds)
314 Spectral Generalizations of Line Graphs, D CVETKOVIC, P. ROWLINSON & S. SIMIC
315 Structured Ring Spectra, A. BAKER & B. RICHTER (eds)

Contents

Preface

This volume is devoted to the lecture courses given to the EuroSchool PQR2003 on "Poisson Geometry, Deformation Quantisation and Group Representations" held at the Université Libre de Bruxelles from the 13th to the 17th of June, 2003.

The EuroSchool was followed by a EuroConference from the 18th to the 22nd June with a large intersection of the audiences. These linked EuroSchool and EuroConference were made possible by the generous support of the European Commission through its High Level Scientific Conferences programme, contract HPCF-CT-2002-00180, the Fonds National belge de la Recherche Scientifique, the Communauté Française de Belgique, and the Université Libre de Bruxelles. The proceedings of the Euroconference appear separately in a volume of Letters in Mathematical Physics.

The idea to organise this meeting was born soon after the Conferences Moshe Flato in Dijon, in 1999 and 2000. We thought that Moshe would have appreciated having a new meeting centred around Poisson Geometry, Deformation Quantisation and Group Representations and which would incidentally celebrate important birthdays of four of our friends who have made fundamental contributions to this area: Boris Fedosov, Wilfried Schmid, Daniel Sternheimer, and Alan Weinstein. We particularly thank Alan, Daniel and Wilfried who accepted this proposal and helped with organising it.

There were four series of four one-hour lectures each given by Alberto Cattaneo on *Formality and Star Products*; Ieke Moerdijk and Janez Mrčun on *Lie Groupoids and Lie Algebroids*; Wilfried Schmid on *Geometric Methods in Representation Theory*; Alan Weinstein on *Morita Equivalence in Poisson Geometry*; and a two-hour broad presentation given by Daniel Sternheimer on *Deformation theory: A powerful tool in physics modelling.* We are extremely grateful to the lecturers and their co-authors for agreeing to write up and publish their lectures, to the London Mathematical Society for publishing them in their Lecture Notes series and to Cambridge University Press for their editorial support.

The editors wish to thank all those who took part in both meetings, with special thanks to the members of the scientific committee and to the lecturers. We also wish to thank all those who helped with the practical organisation, in particular Christine and Luc Lemaire, Monique Parker and Isabelle Renders.

Simone Gutt
Université Libre de Bruxelles & Université de Metz

John Rawnsley
University of Warwick

Daniel Sternheimer
Université de Bourgogne

PART ONE

Poisson geometry and Morita equivalence

Henrique Bursztyn[1]
henrique@math.toronto.edu

Department of Mathematics
University of Toronto
Toronto, Ontario M5S 3G3
Canada

Alan Weinstein[2]
alanw@math.berkeley.edu

Department of Mathematics
University of California
Berkeley CA, 94720-3840 USA
and
Mathematical Sciences Research Institute
17 Gauss Way
Berkeley, CA 94720-5070 USA

[1] research partially supported by DAAD
[2] research partially supported by NSF grant DMS-0204100
MSC2000 Subject Classification Numbers: 53D17 (Primary), 58H05 16D90 (secondary)
Keywords: Picard group, Morita equivalence, Poisson manifold, symplectic groupoid, bimodule

1
Introduction

Poisson geometry is a "transitional" subject between noncommutative algebra and differential geometry (which could be seen as the study of a very special class of commutative algebras). The physical counterpart to this transition is the correspondence principle linking quantum to classical mechanics.

The main purpose of these notes is to present an aspect of Poisson geometry which is inherited from the noncommutative side: the notion of Morita equivalence, including the "self-equivalences" known as Picard groups.

In algebra, the importance of Morita equivalence lies in the fact that Morita equivalent algebras have, by definition, equivalent categories of modules. From this it follows that many other invariants, such as cohomology and deformation theory, are shared by all Morita equivalent algebras. In addition, one can sometimes understand the representation theory of a given algebra by analyzing that of a simpler representative of its Morita equivalence class. In Poisson geometry, the role of "modules" is played by Poisson maps from symplectic manifolds to a given Poisson manifold. The simplest such maps are the inclusions of symplectic leaves, and indeed the structure of the leaf space is a Morita invariant. (We will see that this leaf space sometimes has a more rigid structure than one might expect.)

The main theorem of algebraic Morita theory is that Morita equivalences are implemented by bimodules. The same thing turns out to be true in Poisson geometry, with the proper geometric definition of "bimodule".

Here is a brief outline of what follows this introduction.

Chapter 2 is an introduction to Poisson geometry and some of its recent generalizations, including Dirac geometry and "twisted" Poisson geometry in the presence of a "background" closed 3-form. Both of these generalizations are used simultaneously to get a geometric understanding of new notions of symmetry of growing importance in mathematical physics, especially with background 3-forms arising throughout string theory (in the guise of the more familiar closed 2-forms on spaces of curves).

In Chapter 3, we review various flavors of the algebraic theory of Morita equivalence in a way which transfers easily to the geometric case. In fact, some of our examples come from geometry: algebras of smooth functions. Others come from the quantum side: operator algebras.

Chapter 4 is the heart of these notes, a presentation of the geometric Morita theory of Poisson manifolds and the closely related Morita theory of symplectic groupoids. We arrive at this theory via the Morita theory of Lie groupoids in general.

In Chapter 5, we attempt to remedy a defect in the theory of Chapter 4. Poisson manifolds with equivalent (even isomorphic) representation categories may not be Morita equivalent. We introduce refined versions of the representation category (some of which are not really categories!) which do determine the Morita equivalence class. Much of the material in this chapter is new and has not yet appeared in print. (Some of it is based on discussions which came after the PQR Euroschool where this course was presented.)

Along the way, we comment on a pervasive problem in the geometric theory. Many constructions involve forming the leaf space of a foliation, but these leaf spaces are not always manifolds. We make some remarks about the use of differentiable stacks as a language for admitting pathological leaf spaces into the world of smooth geometry.

Acknowledgements:

We would like to thank all the organizers and participants at the Euroschool on Poisson Geometry, Deformation Quantization, and Representation Theory for the opportunity to present this short course, and for their feedback at the time of the School. We also thank Stefan Waldmann for his comments on the manuscript.

H.B. thanks Freiburg University for its hospitality while part of this work was being done.

2
Poisson geometry and some generalizations

2.1 Poisson manifolds

Let P be a smooth manifold. A **Poisson structure** on P is an $\mathbb{R}$-bilinear Lie bracket $\{\cdot,\cdot\}$ on $C^\infty(P)$ satisfying the Leibniz rule

$$\{f, gh\} = \{f, g\}h + g\{f, h\}, \quad \text{for all } f, g, h \in C^\infty(P). \tag{1}$$

A **Poisson algebra** is a commutative associative algebra which is also a Lie algebra so that the associative multiplication and the Lie bracket are related by (1).

For a function $f \in C^\infty(P)$, the derivation $X_f = \{f, \cdot\}$ is called the **hamiltonian vector field** of f. If $X_f = 0$, we call f a **Casimir function** (see Remark 2.4). It follows from (1) that there exists a bivector field $\Pi \in \mathcal{X}^2(P) = \Gamma(\bigwedge^2 TP)$ such that

$$\{f, g\} = \Pi(df, dg);$$

the Jacobi identity for $\{\cdot,\cdot\}$ is equivalent to the condition $[\Pi, \Pi] = 0$, where $[\cdot,\cdot]$ is the Schouten- Nijenhuis bracket, see e.g. [85].

In local coordinates $(x_1, \cdots, x_n)$, the tensor Π is determined by the matrix

$$\Pi_{ij}(x) = \{x_i, x_j\}. \tag{2}$$

If this matrix is invertible at each x, then Π is called nondegenerate or **symplectic**. In this case, the local matrices $(\omega_{ij}) = (-\Pi_{ij})^{-1}$ define a global 2-form $\omega \in \Omega^2(P) = \Gamma(\bigwedge^2 T^*P)$, and the condition $[\Pi, \Pi] = 0$ is equivalent to $d\omega = 0$.

Example 2.1 (*Constant Poisson structures*)
Let $P = \mathbb{R}^n$, and suppose that the $\Pi_{ij}(x)$ are constant. By a linear

change of coordinates, one can find new coordinates

$$(q_1, \cdots, q_k, p_1, \cdots, p_k, e_1, \cdots, e_l), \quad 2k + l = n,$$

so that

$$\Pi = \sum_i \frac{\partial}{\partial q_i} \wedge \frac{\partial}{\partial p_i}.$$

In terms of the bracket, we have

$$\{f, g\} = \sum_i \left(\frac{\partial f}{\partial q_i} \frac{\partial g}{\partial p_i} - \frac{\partial f}{\partial p_i} \frac{\partial g}{\partial q_i} \right)$$

which is the original Poisson bracket in mechanics. In this example, all the coordinates e_i are Casimirs.

Example 2.2 (*Poisson structures on* $\mathbb{R}^2$)
Any smooth function $f : \mathbb{R}^2 \to \mathbb{R}$ defines a Poisson structure in $\mathbb{R}^2 = \{(x_1, x_2)\}$ by

$$\{x_1, x_2\} := f(x_1, x_2),$$

and every Poisson structure on $\mathbb{R}^2$ has this form.

Example 2.3 (*Lie-Poisson structures*)
An important class of Poisson structures are the linear ones. If P is a (finite-dimensional) vector space V considered as a manifold, with linear coordinates $(x_1, \cdots, x_n)$, a linear Poisson structure is determined by constants c_{ij}^k satisfying

$$\{x_i, x_j\} = \sum_{k=1}^n c_{ij}^k x_k. \tag{3}$$

(We may assume that $c_{ij}^k = -c_{ji}^k$.) Such Poisson structures are usually called **Lie-Poisson structures**, since the Jacobi identity for the Poisson bracket implies that the c_{ij}^k are the structure constants of a Lie algebra $\mathfrak{g}$, which may be identified in a natural way with V^*. (Also, these Poisson structures were originally introduced by Lie [56] himself.) Note that we may also identify V with $\mathfrak{g}^*$. Conversely, any Lie algebra $\mathfrak{g}$ with structure constants c_{ij}^k defines by (3) a linear Poisson structure on $\mathfrak{g}^*$.

Remark 2.4 (*Casimir functions*)
Deformation quantization of the Lie-Poisson structure on $\mathfrak{g}^*$, see e.g. [10, 45], leads to the universal enveloping algebra $U(\mathfrak{g})$. Elements of the center of $U(\mathfrak{g})$ are known as Casimir elements (or Casimir operators,

when a representation of $\mathfrak{g}$ is extended to a representation of $U(\mathfrak{g})$). These correspond to the center of the Poisson algebra of functions on $\mathfrak{g}^*$, hence, by extension, the designation "Casimir functions" for the center of any Poisson algebra.

2.2 Dirac structures

We now introduce a simultaneous generalization of Poisson structures and closed 2-forms. (We will often refer to closed 2-forms as **presymplectic**.)

Each 2-form ω on P corresponds to a bundle map

$$\widetilde{\omega} : TP \to T^*P, \quad \widetilde{\omega}(v)(u) = \omega(v, u). \tag{4}$$

Similarly, for a bivector field $\Pi \in \mathcal{X}^2(P)$, we define the bundle map

$$\widetilde{\Pi} : T^*P \to TP, \quad \beta(\widetilde{\Pi}(\alpha)) = \Pi(\alpha, \beta). \tag{5}$$

The matrix representing $\widetilde{\Pi}$ in the bases (dx_i) and $(\partial/\partial x_i)$ corresponding to local coordinates induced by coordinates $(x_1, \ldots, x_n)$ on P is, up to a sign, just (2). So bivector fields (or 2-forms) are nondegenerate if and only if the associated bundle maps are invertible.

By using the maps in (4) and (5), we can describe both closed 2-forms and Poisson bivector fields as subbundles of $TP \oplus T^*P$: we simply consider the graphs

$$L_\omega := \mathrm{graph}(\widetilde{\omega}), \quad \text{and} \quad L_\Pi := \mathrm{graph}(\widetilde{\Pi}).$$

To see which subbundles of $TP \oplus T^*P$ are of this form, we introduce the following canonical structure on $TP \oplus T^*P$:

1) The symmetric bilinear form $\langle \cdot, \cdot \rangle_+ : TP \oplus T^*P \to \mathbb{R}$,

$$\langle (X, \alpha), (Y, \beta) \rangle_+ := \alpha(Y) + \beta(X). \tag{6}$$

2) The bracket $[\![\cdot , \cdot]\!] : \Gamma(TP \oplus T^*P) \times \Gamma(TP \oplus T^*P) \to \Gamma(TP \oplus T^*P)$,

$$[\![(X, \alpha), (Y, \beta)]\!] := ([X, Y], \mathcal{L}_X \beta - i_Y d\alpha). \tag{7}$$

Remark 2.5 (*Courant bracket*)

The bracket (7) is the non-skew-symmetric version, introduced in [57] (see also [80]), of T. Courant's original bracket [27]. The bundle $TP \oplus T^*P$ together with the brackets (6) and (7) is an example of a **Courant algebroid** [57].

Using the brackets (6) and (7), we have the following result [27]:

Proposition 2.6 *A subbundle* $L \subset TP \oplus T^*P$ *is of the form* $L_\Pi = \mathrm{graph}(\widetilde{\Pi})$ *(resp.* $L_\omega = \mathrm{graph}(\widetilde{\omega})$*) for a bivector field* Π *(resp. 2-form* ω*) if and only if*

i) $TP \cap L = \{0\}$ *(resp.* $L \cap T^*P = \{0\}$*) at all points of* P*;*
ii) L *is maximal isotropic with respect to* $\langle \cdot, \cdot \rangle_+$*;*

furthermore, $[\Pi, \Pi] = 0$ *(resp.* $d\omega = 0$*) if and only if*

iii) $\Gamma(L)$ *is closed under the Courant bracket* (7).

Recall that L being isotropic with respect to $\langle \cdot, \cdot \rangle_+$ means that, at each point of P,

$$\langle (X, \alpha), (Y, \beta) \rangle_+ = 0$$

whenever $(X, \alpha), (Y, \beta) \in L$. Maximality is equivalent to the dimension condition $\mathrm{rank}(L) = \dim(P)$.

A **Dirac structure** on P is a subbundle $L \subset TP \oplus T^*P$ which is maximal isotropic with respect to $\langle \cdot, \cdot \rangle_+$ and whose sections are closed under the Courant bracket (7); in other words, a Dirac structure satisfies conditions *ii)* and *iii)* of Prop. 2.6 but is not necessarily the graph associated to a bivector field or 2-form.

If L satisfies only *ii)*, it is called an **almost Dirac structure**, and we refer to *iii)* as the **integrability condition** of a Dirac structure. The next example illustrates these notions in another situation.

Example 2.7 (*Regular foliations*)

Let $F \subseteq TP$ be a subbundle, and let $F^\circ \subset T^*P$ be its annihilator. Then $L = F \oplus F^\circ$ is an almost Dirac structure; it is a Dirac structure if and only if F satisfies the Frobenius condition

$$[\Gamma(F), \Gamma(F)] \subset \Gamma(F).$$

So regular foliations are examples of Dirac structures.

Example 2.8 (*Vector Dirac structures*)

If V is a finite-dimensional real vector space, then a **vector Dirac structure** on V is a subspace $L \subset V \oplus V^*$ which is maximal isotropic with respect to the symmetric pairing (6).[1]

[1] Vector Dirac structures are sometimes called "linear Dirac structures," but we will eschew this name to avoid confusion with linear (i.e. Lie-) Poisson structures. (See Example 2.3)

Let L be a vector Dirac structure on V. Let $\mathrm{pr}_1 : V \oplus V^* \to V$ and $\mathrm{pr}_2 : V \oplus V^* \to V^*$ be the canonical projections, and consider the subspace

$$R := \mathrm{pr}_1(L) \subseteq V.$$

Then L induces a skew-symmetric bilinear form θ on R defined by

$$\theta(X, Y) := \alpha(Y), \tag{8}$$

where $X, Y \in R$ and $\alpha \in V^*$ is such that $(X, \alpha) \in L$.

Exercise
Show that θ is well defined, i.e., (8) is independent of the choice of α.

Conversely, any pair (R, θ), where $R \subseteq V$ is a subspace and θ is a skew-symmetric bilinear form on R, defines a vector Dirac structure by

$$L := \{(X, \alpha),\ \ X \in R,\ \alpha \in V^* \text{ with } \alpha|_R = i_X\theta\}. \tag{9}$$

Exercise
Check that L defined in (9) is a vector Dirac structure on V with associated subspace R and bilinear form θ.

Example 2.8 indicates a simple way in which vector Dirac structures can be restricted to subspaces.

Example 2.9 (*Restriction of Dirac structures to subspaces*)
Let L be a vector Dirac structure on V, let $W \subseteq V$ be a subspace, and consider the pair (R, θ) associated with L. Then W inherits the vector Dirac structure L_W from L defined by the pair

$$R_W := R \cap W, \ \text{ and } \ \theta_W := \iota^*\theta,$$

where $\iota : W \hookrightarrow V$ is the inclusion map.

Exercise
Show that there is a canonical isomorphism

$$L_W \cong \frac{L \cap (W \oplus V^*)}{L \cap W^\circ}. \tag{10}$$

Let (P, L) be a Dirac manifold, and let $\iota : N \hookrightarrow P$ be a submanifold. The construction in Example 2.9, when applied to $T_xN \subseteq T_xP$ for all $x \in P$, defines a maximal isotropic "subbundle" $L_N \subset TN \oplus T^*N$. The problem is that L_N may not be a continuous family of subspaces. When L_N *is* a continuous family, it is a smooth bundle which then automatically satisfies the integrability condition [27, Cor. 3.1.4], so L_N defines a Dirac structure on N.

The next example is a special case of this construction and is one of the original motivations for the study of Dirac structures; it illustrates the connection between Dirac structures and "constraint submanifolds" in classical mechanics.

Example 2.10 (*Momentum level sets*)
Let $J : P \to \mathfrak{g}^*$ be the momentum [2] map for a hamiltonian action of a Lie group G on a Poisson manifold P [59]. Let $\mu \in \mathfrak{g}^*$ be a regular value for J, let G_μ be the isotropy group at μ with respect to the coadjoint action, and consider

$$Q = J^{-1}(\mu) \hookrightarrow P.$$

At each point $x \in Q$, we have a vector Dirac structure on T_xQ given by

$$(L_Q)_x := \frac{L_x \cap (T_xQ \oplus T_x^*P)}{L_x \cap T_xQ^\circ}. \tag{11}$$

To show that L_Q defines a smooth bundle, it suffices to verify that $L_x \cap T_xQ^\circ$ has constant dimension. (Indeed, if this is the case, then $L_x \cap (T_xQ \oplus T_x^*P)$ has constant dimension as well, since the quotient $L_x \cap (T_xQ \oplus T_x^*P)/L_x \cap T_xQ^\circ$ has constant dimension, and this insures that all bundles are smooth.) A direct computation shows that $L_x \cap T_xQ^\circ$ has constant dimension if and only if the stabilizer groups of the G_μ-action on Q have constant dimension, which happens whenever the G_μ-orbits on Q have constant dimension (for instance, when the action of G_μ on Q is locally free). In this case, L_Q is a Dirac structure on Q.

We will revisit this example in Section 2.7.

Remark 2.11 (*Complex Dirac structures and generalized complex geometry*)
Using the natural extensions of the symmetric form (6) and the Courant bracket (7) to $(TP \oplus T^*P) \otimes \mathbb{C}$, one can define a **complex Dirac structure** on a manifold P to be a maximal isotropic *complex* subbundle $L \subset (TP \oplus T^*P) \otimes \mathbb{C}$ whose sections are closed under the

[2] The term "moment" is frequently used instead of "momentum" in this context. In this paper, we will follow the convention, introduced in [61], that "moment" is used only in connection with groupoid actions. As we will see (e.g. in Example 4.16), many momentum maps, even for "exotic" theories, are moment maps as well.

Courant bracket. If a complex Dirac structure L satisfies the condition

$$L \cap \overline{L} = \{0\} \tag{12}$$

at all points of P (here $\overline{L}$ is the complex conjugate of L), then it is called a **generalized complex structure**; such structures were introduced in [43, 46] as a common generalization of complex and symplectic structures.

To see how complex structures fit into this picture, note that an almost complex structure $J : TP \to TP$ defines a maximal isotropic subbundle $L_J \subset (TP \oplus T^*P) \otimes \mathbb{C}$ as the i-eigenbundle of the map

$$(TP \oplus T^*P) \otimes \mathbb{C} \to (TP \oplus T^*P) \otimes \mathbb{C}, \quad (X, \alpha) \mapsto (-J(X), J^*(\alpha)).$$

The bundle L_J completely characterizes J, and satisfies (12); moreover L_J satisfies the integrability condition of a Dirac structure if and only if J is a complex structure.

Similarly, a symplectic structure ω on P can be seen as a generalized complex structure through the bundle $L_{\omega,\mathbb{C}}$, defined as the i-eigenbundle of the map

$$(TP \oplus T^*P) \otimes \mathbb{C} \to (TP \oplus T^*P) \otimes \mathbb{C}, \quad (X, \alpha) \mapsto (\widetilde{\omega}(X), -\widetilde{\omega}^{-1}(\alpha)).$$

Note that, by (12), a generalized complex structure is never the complexification of a real Dirac structure. In particular, for a symplectic structure ω, $L_{\omega,\mathbb{C}}$ is *not* the complexification of the real Dirac structure L_ω of Proposition 2.6.

2.3 Twisted structures

A "background" *closed* 3-form $\phi \in \Omega^3(P)$ can be used to "twist" the geometry of P [48, 69], leading to a modified notion of Dirac structure [80], and in particular of Poisson structure. The key point is to use ϕ to alter the ordinary Courant bracket (7) as follows:

$$[\![(X, \alpha), (Y, \beta)]\!]_\phi := ([X, Y], \mathcal{L}_X \beta - i_Y d\alpha + \phi(X, Y, \cdot)). \tag{13}$$

We now simply repeat the definitions in Section 2.2 replacing (7) by the **ϕ-twisted Courant bracket** (13).

A **ϕ-twisted Dirac structure** on P is a subbundle $L \subset TP \oplus T^*P$ which is maximal isotropic with respect to $\langle \cdot, \cdot \rangle_+$ (6) and for which

$$[\![\Gamma(L), \Gamma(L)]\!]_\phi \subseteq \Gamma(L). \tag{14}$$

With this new integrability condition, one can check that the graph of a bivector field Π is a ϕ-twisted Dirac structure if and only if

$$\frac{1}{2}[\Pi, \Pi] = \wedge^3 \widetilde{\Pi}(\phi);$$

such bivector fields are called **ϕ-twisted Poisson structures**. Similarly, the graph of a 2-form ω is a ϕ-twisted Dirac structure if and only if

$$d\omega + \phi = 0,$$

in which case ω is called a **ϕ-twisted presymplectic structure**.

Remark 2.12 (*Terminology*)
The term "twisted Dirac structure" and its cousins represent a certain abuse of terminology, since it is not the Dirac (or Poisson, etc.) structure which is twisted, but rather the *notion* of Dirac structure. Nevertheless, we have chosen to stick to this terminology, rather than the alternative "Dirac structure with background" [50], because it is consistent with such existing terms as "twisted sheaf", and because the alternative terms lead to some awkward constructions.

Example 2.13 (*Cartan-Dirac structures on Lie groups*)
Let G be a Lie group whose Lie algebra $\mathfrak{g}$ is equipped with a nondegenerate adjoint-invariant symmetric bilinear form $(\cdot,\cdot)_{\mathfrak{g}}$, which we use to identify TG and T^*G. In $TG \oplus TG \sim TG \oplus T^*G$, we consider the maximal isotropic subbundle

$$L_G := \{(v_r - v_l, \frac{1}{2}(v_r + v_l)),\ \ v \in \mathfrak{g}\}, \tag{15}$$

where v_r and v_l are the right and left invariant vector fields corresponding to v. One can show that L_G is a ϕ^G-twisted Dirac structure, where ϕ^G is the bi-invariant Cartan 3-form on G, defined on Lie algebra elements by

$$\phi^G(u, v, w) = \frac{1}{2}(u, [v, w])_{\mathfrak{g}}.$$

We call L_G the **Cartan-Dirac structure** on G associated with $(\cdot,\cdot)_{\mathfrak{g}}$. Note that L_G is of the form L_Π only at points g for which $\mathrm{Ad}_g + 1$ is invertible, see also Example 2.19.

These Dirac structures are closely related to the theory of quasi-hamiltonian spaces and group-valued momentum maps [3, 15, 97], as well as to quasi-Poisson manifolds [2, 14].

2.4 Symplectic leaves and local structure of Poisson manifolds

If Π is a symplectic Poisson structure on P, then Darboux's theorem asserts that, around each point of P, one can find coordinates $(q_1, \cdots, q_k, p_1, \cdots, p_k)$ such that

$$\Pi = \sum_i \frac{\partial}{\partial q_i} \wedge \frac{\partial}{\partial p_i}.$$

The corresponding symplectic form ω is

$$\omega = \sum_i dq_i \wedge dp_i.$$

In general, the image of the bundle map (5), $\widetilde{\Pi}(T^*P) \subseteq TP$, defines an integrable singular distribution on P; in other words, P is a disjoint union of "leaves" $\mathcal{O}$ satisfying $T_x\mathcal{O} = \widetilde{\Pi}(T_x^*P)$ for all $x \in P$. The leaf $\mathcal{O}$ through x can be described as the points which can be reached from x through piecewise hamiltonian paths.

If $\widetilde{\Pi}$ has locally constant rank, we call the Poisson structure Π **regular**, in which case it defines a foliation of P in the ordinary sense. Note that this is always the case on an open dense subset of P, called the **regular part**.

The local structure of a Poisson manifold (P, Π) around a regular point is given by the Lie-Darboux theorem: If Π has constant rank k around a given point, then there exist coordinates $(q_1, \ldots, q_k, p_1, \ldots, p_k, e_1, \ldots, e_l)$ such that

$$\{q_i, p_j\} = \delta_{ij}, \text{ and } \{q_i, q_j\} = \{p_i, p_j\} = \{q_i, e_j\} = \{p_i, e_j\} = 0.$$

Thus, the local structure of a regular Poisson manifold is determined by that of the vector Poisson structures on any of its tangent spaces (in a given connected component).

In the general case, we have the local splitting theorem [88]:

Theorem 2.14 *Around any point x_0 in a Poisson manifold P, there exist coordinates*

$$(q_1, \ldots, q_k, p_1, \ldots, p_k, e_1, \ldots, e_l), \quad (q, p, e)(x_0) = (0, 0, 0),$$

such that

$$\Pi = \sum_{i=1}^{k} \frac{\partial}{\partial q_i} \wedge \frac{\partial}{\partial p_i} + \frac{1}{2} \sum_{i,j=1}^{l} \eta_{ij}(e) \frac{\partial}{\partial e_i} \wedge \frac{\partial}{\partial e_j}$$

and $\eta_{ij}(0) = 0$.

The splitting of Theorem 2.14 has a symplectic factor associated with the coordinates (q_i, p_i) and a totally degenerate factor (i.e., with all Poisson brackets vanishing at $e = 0$) associated with the coordinates (e_j). The symplectic factor may be identified with an open subset of the leaf $\mathcal{O}$ through x_0; patching them together defines a symplectic structure on each leaf of the foliation determined by Π. So Π canonically defines a singular foliation of P by **symplectic leaves**. The totally degenerate factor in the local splitting is well-defined up to isomorphism. Its isomorphism class is the same at all points in a given symplectic leaf, so one refers to the totally degenerate factor as the **transverse structure** to Π along the leaf.

Example 2.15 (*Symplectic leaves of Poisson structures on* $\mathbb{R}^2$)
Let $f : \mathbb{R}^2 \to \mathbb{R}$ be a smooth function, and let us consider the Poisson structure on $\mathbb{R}^2 = \{(x_1, x_2)\}$ defined by

$$\{x_1, x_2\} := f(x_1, x_2).$$

The connected components of the set where $f(x_1, x_2) \neq 0$ are the 2-dimensional symplectic leaves; in the set where f vanishes, each point is a symplectic leaf.

Example 2.16 (*Symplectic leaves of Lie-Poisson structures*)
Let us consider $\mathfrak{g}^*$, the dual of the Lie algebra $\mathfrak{g}$, equipped with its Lie-Poisson structure, see Example 2.3. The symplectic leaves are just the coadjoint orbits for any connected group with Lie algebra $\mathfrak{g}$. Since $\{0\}$ is always an orbit, a Lie-Poisson structure is not regular unless $\mathfrak{g}$ is abelian.

Exercise
Describe the symplectic leaves in the duals of $\mathfrak{su}(2)$, $\mathfrak{sl}(2, \mathbb{R})$ and $\mathfrak{a}(1)$ (nonabelian 2-dimensional Lie algebra).

Remark 2.17 (*Linearization problem*)
By linearizing at x_0 the functions η_{ij} in Theorem 2.14, we can write

$$\{e_i, e_j\} = \sum_k c_{ij}^k e_k + O(e^2), \tag{16}$$

and it turns out that c_{ij}^k define a Lie-Poisson structure on the normal space to the symplectic leaf at x_0. The **linearization problem** consists of determining whether one can choose suitable "transverse" coordinates $(e_1, \ldots, e_l)$ with respect to which $O(e^2)$ in (16) vanishes. For example, if the Lie algebra structure on the conormal bundle to a symplectic leaf determined by the linearization of Π at a point x_0 is semi-simple and of compact type, then Π is linearizable around x_0 through a smooth change of coordinates. The first proof of this theorem, due to Conn [25], used many estimates and a "hard" implicit function theorem. A "soft" proof, using only the sort of averaging usually associated with compact group actions (but for groupoids instead of groups), has recently been announced by Crainic and Fernandes [31]. There is also a "semilocal" problem of linearization in the neighborhood of an entire symplectic leaf. This problem was first addressed by Vorobjev [86], with further developments by Davis and Wade [32]. For overviews of linearization and more general normal form questions, we refer to the article of Fernandes and Monnier [38] and the forthcoming book of Dufour and Zung [35].

A local normal form for Dirac structures was found by Courant [27] under a strong regularity assumption. The general case has been studied recently by Dufour and Wade in [34].

2.5 Presymplectic leaves and Poisson quotients of Dirac manifolds

Let $\mathrm{pr}_1 : TP \oplus T^*P \to TP$ and $\mathrm{pr}_2 : TP \oplus T^*P \to T^*P$ be the canonical projections. If $L \subset TP \oplus T^*P$ is a (twisted) Dirac structure on P, then

$$\mathrm{pr}_1(L) \subseteq TP \tag{17}$$

defines a singular distribution on P. Note that if $L = L_\Pi$ for a Poisson structure Π, then $\mathrm{pr}_1(L) = \widetilde{\Pi}(T^*P)$, so this distribution coincides with the one defined by Π, see Section 2.4. It turns out that the integrability condition for (twisted) Dirac structures guarantees that (17) is integrable in general, so a (twisted) Dirac structure L on P determines a decomposition of P into leaves $\mathcal{O}$ satisfying

$$T_x\mathcal{O} = \mathrm{pr}_1(L)_x$$

at all $x \in P$.

Just as leaves of foliations associated with Poisson structures carry symplectic forms, each leaf of a (twisted) Dirac manifold P is naturally equipped with a (twisted) presymplectic 2-form θ: at each $x \in P$, θ_x is the bilinear form defined in (8). These forms fit together into a smooth leafwise 2-form, which is nondegenerate on the leaves just when L is a (twisted) Poisson structure. If L is twisted by ϕ, then θ is twisted by the pull back of ϕ to each leaf.

Remark 2.18 (*Lie algebroids*)
The fact that $\mathrm{pr}_1(L) \subseteq TP$ is an integrable singular distribution is a consequence of a more general fact: the restriction of the Courant bracket $[\![\cdot,\cdot]\!]_\phi$ to $\Gamma(L)$ defines a Lie algebra bracket making $L \to P$ into a *Lie algebroid* with anchor $\mathrm{pr}_1|_L$, and the image of the anchor of any Lie algebroid is always an integrable distribution (its leaves are also called **orbits**). We refer to [20, 63] for more on Lie algebroids.

Example 2.19 (*Presymplectic leaves of Cartan-Dirac structures*)
Let L_G be a Cartan-Dirac structure on G with respect to $(\cdot,\cdot)_\mathfrak{g}$, see (15). Then the associated distribution on G is

$$\mathrm{pr}_1(L_G) = \{v_r - v_l,\ v \in \mathfrak{g}\}.$$

Since vector fields of the form $v_r - v_l$ are infinitesimal generators of the action of G on itself by conjugation, it follows that the twisted presymplectic leaves of L_G are the connected components of the conjugacy classes of G. With $v_G = v_r - v_l$, the corresponding twisted presymplectic forms can be written as

$$\theta_g(v_G, w_G) := \frac{1}{2}((\mathrm{Ad}_{g^{-1}} - \mathrm{Ad}_g)v, w)_\mathfrak{g}, \tag{18}$$

at $g \in G$. These 2-forms were introduced in [44] in the study of the symplectic structure of certain moduli spaces. They are analogous to the Kostant-Kirillov-Souriau symplectic forms on coadjoint orbits, although they are neither nondegenerate nor closed: θ_g is degenerate whenever $1 + \mathrm{Ad}_g$ is not invertible, and, on a conjugacy class $\iota : \mathcal{O} \hookrightarrow G$, $d\theta = -\iota^*\phi^G$.

Just as the symplectic forms along coadjoint orbits on the dual of a Lie algebra are associated with Lie-Poisson structures, the 2-forms (18) along conjugacy classes of a Lie group are associated with Cartan-Dirac structures.

For any ϕ-twisted Dirac structure L, the (topologically) closed family of subspaces $TP \cap L = \ker(\theta)$ in TP is called the **characteristic distribution** of L and is denoted by $\ker(L)$. It is always contained in $\mathrm{pr}_1(L)$. When $\ker(L)$ has constant fibre dimension, it is integrable if and only if

$$\phi(X, Y, Z) = 0 \quad \text{for all } X, Y \in \ker(\theta),\ Z \in \mathrm{pr}_1(L), \tag{19}$$

at each point of P. In this case, the leaves of the corresponding **charac-**

teristic foliation are the null spaces of the presymplectic forms along the leaves. On each leaf $\iota : \mathcal{O} \hookrightarrow P$, the 2-form θ is basic with respect to the characteristic foliation if and only if

$$\ker(\theta) \subseteq \ker(\iota^*\phi) \tag{20}$$

at all points of $\mathcal{O}$. In this case, forming the leaf space of this foliation (locally, or globally when the foliation is simple) produces a quotient manifold bearing a singular foliation by twisted symplectic leaves; it is in fact a twisted Poisson manifold. In particular, when $\phi = 0$, conditions (19) and (20) are satisfied, and the quotient is an ordinary Poisson manifold. Thus, Dirac manifolds can be regarded as "pre-Poisson" manifolds, since, in nice situations, they become Poisson manifolds after they are divided out by the characteristic foliation.

Functions which are annihilated by all tangent vectors in the characteristic distribution (equivalently, have differentials in the projection of L to T^*P) are called **admissible** [27]. For admissible f and g, one can define

$$\{f,g\} := \theta(X_f, X_g), \tag{21}$$

where X_f is any vector field such that $(X_f, df) \in L$. (Note that any two choices for X_f differ by a characteristic vector, so the bracket (21) is well defined.) If (20) holds, then the algebra of admissible functions is closed under this bracket, but it is not in general a Poisson algebra, due to the presence of ϕ. In particular, if the characteristic foliation is regular and simple, the admissible functions are just the functions on the (twisted) Poisson quotient.

Example 2.20 (*Nonintegrable characteristic distributions*)
Consider the presymplectic structure $x_1 dx_1 \wedge dx_2$ on $\mathbb{R}^2$. Its characteristic distribution consists of the zero subspace at points where $x_1 \neq 0$ and the entire tangent space at each point of the x_2 axis. Thus, the points off the axis are integral manifolds, while there are no integral manifolds through points on the axis. The only admissible functions are constants.

On the other hand, if a 2-form is not closed, then its kernel may have constant fibre dimension and still be nonintegrable. For example, the characteristic distribution of the 2-form $x_2 dx_1 \wedge dx_4 - dx_3 \wedge dx_4$ on $\mathbb{R}^4$ is spanned by $\partial/\partial x_1 + x_2 \partial/\partial x_3$ and $\partial/\partial x_2$. A direct computation shows that this 2-dimensional distribution does not satisfy the Frobenius condition, so it is not integrable.

Example 2.21 (*A nonreducible 2-form*)
The characteristic foliation of the 2-form $(x_3^2 + 1)dx_1 \wedge dx_2$ on $\mathbb{R}^3$ consists of lines parallel to the x_3-axis, so it is simple. However, the form is not basic with respect to this foliation.

We will say more about presymplectic leaves and quotient Poisson structures in Section 2.7.

2.6 Poisson maps

Although we shall see later that the following notion of morphism between Poisson manifolds is not the only one, it is certainly the most obvious one.

Let (P_1, Π_1) and (P_2, Π_2) be Poisson manifolds. A smooth map $\psi : P_1 \to P_2$ is a **Poisson map**[3] if $\psi^* : C^\infty(P_2) \to C^\infty(P_1)$ is a homomorphism of Poisson algebras, i.e.,

$$\{f, g\}_2 \circ \psi = \{f \circ \psi, g \circ \psi\}_1$$

for $f, g \in C^\infty(P_2)$. One can reformulate this condition in terms of Poisson bivectors or hamiltonian vector fields as follows. A map $\psi : P_1 \to P_2$ is a Poisson map if and only if either of the following two equivalent conditions hold:

i) $\psi_* \Pi_1 = \Pi_2$, i.e., Π_1 and Π_2 are ψ-related.
ii) For all $f \in C^\infty(P_2)$, $X_f = \psi_*(X_{\psi^* f})$.

It is clear by condition *ii*) that trajectories of $X_{\psi^* f}$ project to those of X_f if ψ is a Poisson map. This provides a way to "lift" some paths from P_2 to P_1. However, knowing that X_f is complete does not guarantee that $X_{\psi^* f}$ is complete. In order to assure that there are no "missing points" on the lifted trajectory on P_1, we define a Poisson map $\psi : P_1 \to P_2$ to be **complete** if for any $f \in C^\infty(P_2)$ such that X_f is complete, then $X_{\psi^* f}$ is also complete. Alternatively, one can replace the condition of X_f being complete by X_f (or f itself) having compact support. Note that there is no notion of completeness (or "missing point") for a Poisson manifold by itself, only for a Poisson manifold *relative* to another.

[3] Following Lie [56], when P_1 is symplectic, we call ψ a symplectic realization of P_2.

Remark 2.22 (*Cotangent paths*)
The path lifting alluded to above is best understood in terms of so-called cotangent paths [40, 91]. A **cotangent path** on a Poisson manifold P is a path α in T^*P such that $(\pi \circ \alpha)' = \widetilde{\Pi} \circ \alpha$, where π is the cotangent bundle projection. If $\psi : P_1 \to P_2$ is a Poisson map, then a cotangent path α_1 on P_1 is a **horizontal lift** of the cotangent path α_2 on P_2 if $\alpha_1(t) = \psi^*(\alpha_2(t))$ for all t. It turns out that a cotangent path on P_2 has at most one horizontal lift for each initial value of $\pi \circ \alpha_1$. Furthermore, the existence of horizontal lifts for all cotangent paths α_2 and all initial data consistent with the initial value of α_2 is equivalent to the completeness of the map ψ.

This path lifting property suggests that complete Poisson maps play the role of "coverings" in Poisson geometry. This idea is borne out by some of the examples below.

Example 2.23 (*Complete functions*)
Let us regard $\mathbb{R}$ as a Poisson manifold, equipped with the zero Poisson bracket. (This is the only possible Poisson structure on $\mathbb{R}$.) Then any map $f : P \to \mathbb{R}$ is a Poisson map, which is complete if and only if X_f is a complete vector field.

Observe that the notion of completeness singles out the subset of $C^\infty(P)$ consisting of complete functions, which is preserved under complete Poisson maps.

Exercise
For which Poisson manifolds is the set of complete functions closed under addition? (Hint: when are all functions complete?)

Example 2.24 (*Open subsets of symplectic manifolds*)
Let (P, Π) be a symplectic manifold, and let $U \subseteq P$ be an open subset. Then the inclusion map $U \hookrightarrow P$ is complete if and only if U is closed (hence a union of connected components). More generally, the image of a complete Poisson map is a union of symplectic leaves.

Example 2.24 suggests that (connected) symplectic manifolds are "minimal objects" among Poisson manifolds.

Exercise
The inclusion of every symplectic leaf in a Poisson manifold is a complete Poisson map.

Exercise

Let P_1 be a Poisson manifold, and let P_2 be symplectic. Then any Poisson map $\psi : P_1 \to P_2$ is a submersion. Furthermore, if P_2 is connected and ψ is complete, then ψ is surjective (assuming that P_1 is nonempty).

The previous exercise is the first step in establishing that complete Poisson maps with symplectic target must be fibrations. In fact, if P_1 is symplectic and $\dim(P_1) = \dim(P_2)$, then a complete Poisson map $\psi : P_1 \to P_2$ is a covering map. In general, a complete Poisson map $\psi : P_1 \to P_2$, where P_2 is *symplectic*, is a locally trivial symplectic fibration with a flat Ehresmann connection: the horizontal lift in T_xP_1 of a vector X in $T_{\psi(x)}P_2$ is defined as

$$\widetilde{\Pi}_1((T_x\psi)^*\widetilde{\Pi}_2^{-1}(X)).$$

The horizontal subspaces define a foliation whose leaves are coverings of P_2, and P_1 and ψ are completely determined, up to isomorphism, by the holonomy

$$\pi_1(P_2, x) \to \mathrm{Aut}(\psi^{-1}(x)),$$

see [20, Sec. 7.6] for details.

2.7 Dirac maps

To see how to define Dirac maps, we first reformulate the condition for a map $\psi : (P_1, \Pi_1) \to (P_2, \Pi_2)$ to be Poisson in terms of the bundles $L_{\Pi_1} = \mathrm{graph}(\widetilde{\Pi}_1)$ and $L_{\Pi_2} = \mathrm{graph}(\widetilde{\Pi}_2)$. First, note that ψ is a Poisson map if and only if, at each $x \in P_1$,

$$(\widetilde{\Pi}_2)_{\psi(x)} = T_x\psi \circ (\widetilde{\Pi}_1)_x \circ (T_x\psi)^*. \tag{22}$$

Now, using (22), it is not difficult to check that ψ being a Poisson map is equivalent to

$$L_{\Pi_2} = \{(T\psi(X), \beta) \,|\, X \in TP_1,\, \beta \in T^*P_2,\, (X, (T\psi)^*(\beta)) \in L_{\Pi_1}\}. \tag{23}$$

Similarly, if (P_1, ω_1) and (P_2, ω_2) are presymplectic manifolds, then a map $\psi : P_1 \to P_2$ satisfies $\psi^*\omega_2 = \omega_1$ if and only if L_{ω_1} and L_{ω_2} are related by

$$L_{\omega_1} = \{(X, (T\psi)^*(\beta)) \,|\, X \in TP_1,\, \beta \in T^*P_2,\, (T\psi(X), \beta) \in L_{\omega_2}\}. \tag{24}$$

Since Dirac structures simultaneously generalize Poisson structures and presymplectic forms, and conditions (23) and (24) both make sense for arbitrary Dirac subbundles, we have *two* possible definitions: If (P_1, L_1) and (P_2, L_2) are (possibly twisted) Dirac manifolds, then a map $\psi : P_1 \to P_2$ is a **forward Dirac map** if

$$L_2 = \{(T\psi(X), \beta) \,|\, X \in TP_1,\, \beta \in T^*P_2,\, (X, (T\psi)^*(\beta)) \in L_1\}, \quad (25)$$

and a **backward Dirac map** if

$$L_1 = \{(X, (T\psi)^*(\beta)) \,|\, X \in TP_1,\, \beta \in T^*P_2,\, (T\psi(X), \beta) \in L_2\}. \quad (26)$$

Regarding vector Dirac structures as *odd* (in the sense of super geometry) analogues of lagrangian subspaces, one can interpret formulas (25) and (26) via composition of canonical relations [87], see [16].

The expression on the right-hand side of (25) defines at each point of P_1 a way to push a Dirac structure forward, whereas (26) defines a pull-back operation. For this reason, we often write

$$L_2 = \psi_* L_1$$

when (25) holds, following the notation for ψ-related vector or bivector fields; similarly, we may write

$$L_1 = \psi^* L_2$$

instead of (26). This should explain the terminology "forward" and "backward".

Remark 2.25 (*Isotropic and coisotropic subspaces*)
The notions of isotropic and coisotropic subspaces, as well as much of the usual lagrangian/coisotropic calculus [87, 90] can be naturally extended to Dirac vector spaces . This yields an alternative characterization of forward (resp. backward) Dirac maps in terms of their graphs being coisotropic (resp. isotropic) subspaces of the suitable product Dirac space [84].

Note that the pointwise pull back $\psi^* L_2$ is always a well defined family of maximal isotropic subspaces in the fibres of $TP_1 \oplus T^*P_1$, though it may not be continuous, whereas $\psi_* L_1$ may not be well-defined at all.

Exercise
Consider a smooth map $f : P_1 \to P_2$, and let L_2 be a ϕ-twisted Dirac structure on P_2. Show that if f^*L_2 defines a smooth vector bundle, then its sections are automatically closed under the $f^*\phi$-twisted Courant bracket on P_1 (so that f^*L_2 is a $f^*\phi$-twisted Dirac structure).

If P_1 and P_2 are symplectic manifolds, then a map $\psi : P_1 \to P_2$ is forward Dirac if and only if it is a Poisson map, and backward Dirac if and only if it pulls back the symplectic form on P_2 to the one on P_1, in which case we call it a **symplectic map**.

The next example shows that forward Dirac maps need not be backward Dirac, and vice versa.

Example 2.26 *(Forward vs. backward Dirac maps)*

Consider $\mathbb{R}^2 = \{(q,p)\}$, equipped with the symplectic form $dq \wedge dp$, and $\mathbb{R}^4 = \{(q_1,p_1,q_2,p_2)\}$, with symplectic form $dq_1 \wedge dp_1 + dq_2 \wedge dp_2$. Then a simple computation shows that the inclusion

$$\mathbb{R}^2 \hookrightarrow \mathbb{R}^4, \qquad (q,p) \mapsto (q,p,0,0),$$

is a symplectic (i.e. backward Dirac) map, but it does not preserve Poisson brackets. On the other hand, the projection

$$\mathbb{R}^4 \to \mathbb{R}^2, \quad (q_1,p_1,q_2,p_2) \mapsto (q_1,p_1),$$

is a Poisson (i.e. forward Dirac) map, but it is not symplectic.

Example 2.27 (*Backward Dirac maps and restrictions*)

Let (P,L) be a (possibly twisted) Dirac manifold, and let $\iota : N \hookrightarrow P$ be a submanifold. Let $L_N \subset TN \oplus T^*N$ be the subbundle defined pointwise by the restriction of L to N, see (10), and suppose that L_N is smooth, so that it defines a Dirac structure on N. A direct computation shows that

$$L_N = \iota^* L,$$

hence the inclusion ι is a backward Dirac map.

The next exercise explains when the notions of forward and backward Dirac maps coincide.

Exercise

Let V_1 and V_2 be vector spaces, and let $f : V_1 \to V_2$ be a linear map.

1. Let L be a vector Dirac structure on V_1. Then $f^* f_* L = L$ if and only if $\ker(f) \subseteq \ker(L)$, where $\ker(L) = V \cap L$.
2. Let L be a vector Dirac structure on V_2. Then $f_* f^* L = L$ if and only if $f(V_1) \supseteq R$, where $R = \mathrm{pr}_1(L) \subseteq V_2$.

It follows that $f^* f_*(L) = L$ for all L if and only if f is injective, and $f_* f^*(L) = L$ for all L if and only if f is surjective.

In particular, the previous exercise shows that if P_1 and P_2 are symplectic manifolds, then a Poisson map $P_1 \to P_2$ is symplectic if and only if it is an immersion, and a symplectic map $P_1 \to P_2$ is Poisson if and only if the map is a submersion (compare with Example 2.26). Thus, the only maps which are both symplectic and Poisson are local diffeomorphisms.

Using the previous exercise, we find important examples of maps which are both forward and backward Dirac.

Example 2.28 (*Inclusion of presymplectic leaves*)
Let (P, L) be a twisted Dirac manifold. Let $(\mathcal{O}, \theta)$ be a presymplectic leaf, and let $\iota : \mathcal{O} \hookrightarrow P$ be the inclusion. We regard $\mathcal{O}$ as a Dirac manifold, with Dirac structure $L_\theta = \mathrm{graph}(\widetilde{\theta})$. Then it follows from the definition of θ that ι is a backward Dirac map. On the other hand, since

$$T\iota(T\mathcal{O}) = \mathrm{pr}_1(L)$$

at each point, $\iota_* L_\theta = \iota_* \iota^* L = L$, so ι is also a forward Dirac map.

Note that θ is completely determined by either of the conditions that the inclusion be forward or backward Dirac.

Example 2.29 (*Quotient Poisson structures*)
Let (P, L) be a Dirac manifold, and suppose that its characteristic foliation is regular and simple. According to the discussion in Section 2.5, the leaf space P_{red} has an induced Poisson structure Π_{red}. Using the definition of Π_{red}, one can directly show that the natural projection

$$\mathrm{pr} : P \longrightarrow P_{red}$$

is a forward Dirac map, i.e., $\mathrm{pr}_* L = \mathrm{graph}(\Pi_{red})$. But since

$$\ker(T\mathrm{pr}) = \ker(L),$$

the previous exercise implies that $\mathrm{pr}^* \mathrm{pr}_* L = L$, so pr is a backward Dirac map as well.

As in Example 2.28, Π_{red} is uniquely determined by either of the conditions that pr be backward or forward Dirac.

Example 2.29 has an important particular case, which illustrates the connection between Dirac geometry and the theory of hamiltonian actions.

Example 2.30 (*Poisson reduction*)
Suppose that $J : P \to \mathfrak{g}^*$ is the momentum map for a hamiltonian

action of a Lie group G on a Poisson manifold (P, Π). Let $\mu \in \mathfrak{g}^*$ be a regular value for J, let $Q = J^{-1}(\mu)$, and assume that the orbit space

$$P_{red} = Q/G_\mu$$

is a smooth manifold such that the projection $Q \to P_{red}$ is a surjective submersion. Following Examples 2.10 and 2.27, we know that Q has an induced Dirac structure L_Q with respect to which the inclusion $Q \hookrightarrow P$ is a backward Dirac map.

Exercise

Show that the G_μ-orbits on Q coincide with the characteristic foliation of L_Q.

Thus, by Example 2.29, P_{red} inherits a Poisson structure Π_{red} for which the projection $Q \to P_{red}$ is both backward and forward Dirac (and either one of these conditions defines Π_{red} uniquely).

3

Algebraic Morita equivalence

There is another notion of morphism between Poisson manifolds which, though it does *not* include all the Poisson maps, is more closely adapted to the "representation theory" of Poisson manifolds. It is based on an algebraic idea which we present first. (The impatient reader may skip to Chapter 4.)

3.1 Ring-theoretic Morita equivalence of algebras

Let $\mathcal{A}$ and $\mathcal{B}$ be unital algebras over a fixed ground ring k, and let ${}_{\mathcal{A}}\mathfrak{M}$ and ${}_{\mathcal{B}}\mathfrak{M}$ denote the categories of left modules over $\mathcal{A}$ and $\mathcal{B}$, respectively. We call $\mathcal{A}$ and $\mathcal{B}$ **Morita equivalent** [65] if they have equivalent categories of left modules, i.e., if there exist functors

$$\mathcal{F} : {}_{\mathcal{B}}\mathfrak{M} \longrightarrow {}_{\mathcal{A}}\mathfrak{M} \quad \text{and} \quad \widetilde{\mathcal{F}} : {}_{\mathcal{A}}\mathfrak{M} \longrightarrow {}_{\mathcal{B}}\mathfrak{M} \tag{27}$$

whose compositions are naturally equivalent to the identity functors:

$$\mathcal{F} \circ \widetilde{\mathcal{F}} \cong \mathrm{Id}_{{}_{\mathcal{A}}\mathfrak{M}}, \quad \text{and} \quad \widetilde{\mathcal{F}} \circ \mathcal{F} \cong \mathrm{Id}_{{}_{\mathcal{B}}\mathfrak{M}}.$$

One way to construct such functors between module categories is via bimodules: if ${}_{\mathcal{A}}\mathrm{X}_{\mathcal{B}}$ is an $(\mathcal{A}, \mathcal{B})$-bimodule (i.e., X is a k-module which is a left $\mathcal{A}$-module and a right $\mathcal{B}$-module, and these actions commute), then we define an associated functor $\mathcal{F}_{\mathrm{X}} : {}_{\mathcal{B}}\mathfrak{M} \to {}_{\mathcal{A}}\mathfrak{M}$ by setting, at the level of objects,

$$\mathcal{F}_{\mathrm{X}}({}_{\mathcal{B}}\mathrm{M}) := {}_{\mathcal{A}}\mathrm{X}_{\mathcal{B}} \otimes_{\mathcal{B}} {}_{\mathcal{B}}\mathrm{M} \tag{28}$$

where the $\mathcal{A}$-module structure on $\mathcal{F}_{\mathrm{X}}({}_{\mathcal{B}}\mathrm{M})$ is given by

$$a \cdot (x \otimes_{\mathcal{B}} m) = (ax) \otimes_{\mathcal{B}} m.$$

For a morphism $T : {}_{\mathcal{B}}\mathrm{M} \longrightarrow {}_{\mathcal{B}}\mathrm{M}'$, we define

$$\mathcal{F}_{\mathrm{X}}(T) : {}_{\mathcal{A}}\mathrm{X}_{\mathcal{B}} \otimes_{\mathcal{B}} {}_{\mathcal{B}}\mathrm{M} \longrightarrow {}_{\mathcal{A}}\mathrm{X}_{\mathcal{B}} \otimes_{\mathcal{B}} {}_{\mathcal{B}}\mathrm{M}', \quad \mathcal{F}_{\mathrm{X}}(T)(x \otimes_{\mathcal{B}} m) = x \otimes_{\mathcal{B}} T(m). \tag{29}$$

This way of producing functors turns out to be very general. In fact, as we will see in Theorem 3.1, any functor establishing an equivalence between categories of modules is naturally equivalent to a functor associated with a bimodule.

Exercise

Let X and X$'$ be $(\mathcal{A}, \mathcal{B})$-bimodules. Show that the associated functors $\mathcal{F}_{\mathrm{X}}$ and $\mathcal{F}_{\mathrm{X}'}$ are naturally equivalent if and only if the bimodules X and X$'$ are isomorphic.

It follows from the previous exercise that the functors $\mathcal{F}_{\mathrm{X}} : {}_{\mathcal{B}}\mathfrak{M} \to {}_{\mathcal{A}}\mathfrak{M}$, associated with an $(\mathcal{A}, \mathcal{B})$-bimodule X, and $\mathcal{F}_{\mathrm{Y}} : {}_{\mathcal{A}}\mathfrak{M} \to {}_{\mathcal{B}}\mathfrak{M}$, associated with a $(\mathcal{B}, \mathcal{A})$-bimodule Y, are inverses of one another if and only if

$${}_{\mathcal{A}}\mathrm{X}_{\mathcal{B}} \otimes_{\mathcal{B}} {}_{\mathcal{B}}\mathrm{Y}_{\mathcal{A}} \cong \mathcal{A} \quad \text{and} \quad {}_{\mathcal{B}}\mathrm{Y}_{\mathcal{A}} \otimes_{\mathcal{A}} {}_{\mathcal{A}}\mathrm{X}_{\mathcal{B}} \cong \mathcal{B}. \tag{30}$$

The isomorphisms in (30) are *bimodule* isomorphisms, and $\mathcal{A}$ and $\mathcal{B}$ are regarded as $(\mathcal{A}, \mathcal{A})$- and $(\mathcal{B}, \mathcal{B})$-bimodules, respectively, in the natural way (with respect to left and right multiplications). So Morita equivalence is equivalent to the existence of bimodules satisfying (30).

One can see Morita equivalence as the notion of isomorphism in an appropriate category. For that, we think of an *arbitrary* $(\mathcal{A}, \mathcal{B})$-bimodule as a "generalized morphism" between $\mathcal{B}$ and $\mathcal{A}$. Note that, if $\mathcal{A} \overset{q}{\leftarrow} \mathcal{B}$ is an ordinary algebra homomorphism, then we can use it to make $\mathcal{A}$ into an $(\mathcal{A}, \mathcal{B})$-bimodule by

$$a \cdot x \cdot b := axq(b), \quad a \in \mathcal{A},\ x \in \mathcal{A},\ b \in \mathcal{B}. \tag{31}$$

Since the tensor product

$${}_{\mathcal{A}}\mathrm{X}_{\mathcal{B}} \otimes_{\mathcal{B}} {}_{\mathcal{B}}\mathrm{Y}_{\mathcal{C}}$$

is an $(\mathcal{A}, \mathcal{C})$-bimodule, we can see it as a "composition" of bimodules. As this composition is only associative up to isomorphism, we consider the collection of *isomorphism classes* of $(\mathcal{A}, \mathcal{B})$-bimodules, denoted by $\mathrm{Bim}(\mathcal{A}, \mathcal{B})$. Then $\otimes_{\mathcal{B}}$ defines an associative composition

$$\mathrm{Bim}(\mathcal{A}, \mathcal{B}) \times \mathrm{Bim}(\mathcal{B}, \mathcal{C}) \to \mathrm{Bim}(\mathcal{A}, \mathcal{C}). \tag{32}$$

We define the category Alg to be that in which the objects are unital

k-algebras and the morphisms $\mathcal{A} \leftarrow \mathcal{B}$ are the isomorphism classes of $(\mathcal{A}, \mathcal{B})$-bimodules, with composition given by (32); the identities are the algebras themselves seen as bimodules in the usual way. Note that a bimodule ${}_{\mathcal{A}}\mathrm{X}_{\mathcal{B}}$ is invertible in Alg if and only if it satisfies (30) for some bimodule ${}_{\mathcal{B}}\mathrm{Y}_{\mathcal{A}}$, so the notion of isomorphism in Alg coincides with Morita equivalence.

This is part of Morita's theorem [65], see also [4].

Theorem 3.1 *Let $\mathcal{A}$ and $\mathcal{B}$ be unital k-algebras.*

1. *A functor $\mathcal{F} : {}_{\mathcal{B}}\mathfrak{M} \to {}_{\mathcal{A}}\mathfrak{M}$ is an equivalence of categories if and only if there exists an invertible $(\mathcal{A}, \mathcal{B})$-bimodule X such that $\mathcal{F} \cong \mathcal{F}_{\mathrm{X}}$.*
2. *A bimodule ${}_{\mathcal{A}}\mathrm{X}_{\mathcal{B}}$ is invertible if and only if it is finitely generated and projective as a left $\mathcal{A}$-module and as a right $\mathcal{B}$-module, and $\mathcal{A} \to \mathrm{End}_{\mathcal{B}}(\mathrm{X})$ and $\mathcal{B} \to \mathrm{End}_{\mathcal{A}}(\mathrm{X})$ are algebra isomorphisms.*

Example 3.2 (*Matrix algebras*)
A unital algebra $\mathcal{A}$ is Morita equivalent to the matrix algebra $M_n(\mathcal{A})$, for any $n \geq 1$, through the $(M_n(\mathcal{A}), \mathcal{A})$-bimodule $\mathcal{A}^n$.

The following is a geometric example.

Example 3.3 (*Endomorphism bundles*)
Let $\mathcal{A} = C^\infty(M)$ be the algebra of complex-valued functions on a manifold M. The Serre-Swan theorem asserts that any finitely generated projective module over $C^\infty(M)$ can be identified with the space of smooth sections $\Gamma(E)$ of a complex vector bundle $E \to M$. In fact, $C^\infty(M)$ is Morita equivalent to $\Gamma(\mathrm{End}(E))$ via the $(\Gamma(\mathrm{End}(E)), C^\infty(M))$-bimodule $\Gamma(E)$. When E is the trivial bundle $\mathbb{C}^n \times M \to M$, we recover the Morita equivalence of $C^\infty(M)$ and $M_n(C^\infty(M))$ in Example 3.2. The same conclusion holds if $\mathcal{A}$ is the algebra of complex-valued continuous functions on a compact Hausdorff space.

Morita equivalence preserves many algebraic properties besides categories of representations, including ideal structures, cohomology groups and deformation theories [4, 39]. Another important Morita invariant is the center $\mathcal{Z}(\mathcal{A})$ of a unital algebra $\mathcal{A}$. If X is an invertible $(\mathcal{A}, \mathcal{B})$-bimodule then, for each $b \in \mathcal{Z}(\mathcal{B})$, there is a unique $a = a(b) \in \mathcal{Z}(\mathcal{A})$ determined by the condition $ax = xb$ for all $x \in \mathrm{X}$. In this way, X defines an isomorphism

$$h_{\mathrm{X}} : \mathcal{Z}(\mathcal{A}) \leftarrow \mathcal{Z}(\mathcal{B}), \quad h_{\mathrm{X}}(b) = a(b). \tag{33}$$

The group of automorphisms of an object $\mathcal{A}$ in Alg is called its **Picard**

group, denoted by $\mathrm{Pic}(\mathcal{A})$. More generally, the invertible morphisms in Alg form a "large" groupoid, called the **Picard groupoid** [9], denoted by Pic. (Here, "large" refers to the fact that the collection of objects in Pic is not a set, though the collection of morphisms between any two of them is.) The set of morphisms from $\mathcal{B}$ to $\mathcal{A}$ are the Morita equivalences; we denote this set by $\mathrm{Pic}(\mathcal{A}, \mathcal{B})$. Of course $\mathrm{Pic}(\mathcal{A}, \mathcal{A}) = \mathrm{Pic}(\mathcal{A})$. The orbit of an object $\mathcal{A}$ in Pic is its Morita equivalence class, while its isotropy $\mathrm{Pic}(\mathcal{A})$ parametrizes the different ways $\mathcal{A}$ can be Morita equivalent to any other object in its orbit. It is clear from this picture that Picard groups of Morita equivalent algebras are isomorphic.

Let us investigate the difference between $\mathrm{Aut}(\mathcal{A})$, the group of ordinary algebra automorphisms of $\mathcal{A}$, and $\mathrm{Pic}(\mathcal{A})$. Since ordinary automorphisms of $\mathcal{A}$ can be seen as generalized ones, see (31), we obtain a group homomorphism

$$j : \mathrm{Aut}(\mathcal{A}) \to \mathrm{Pic}(\mathcal{A}). \tag{34}$$

A simple computation shows that $\ker(j) = \mathrm{InnAut}(\mathcal{A})$, the group of inner automorphisms of $\mathcal{A}$. So the outer automorphisms $\mathrm{OutAut}(\mathcal{A}) := \mathrm{Aut}(\mathcal{A})/\mathrm{InnAut}(\mathcal{A})$ sit inside $\mathrm{Pic}(\mathcal{A})$.

Exercise

Morita equivalent algebras have isomorphic Picard groups. Do they always have isomorphic groups of outer automorphisms? (Hint: consider the direct sum of two matrix algebras of the same or different sizes.)

On the other hand, (33) induces a group homomorphism

$$h : \mathrm{Pic}(\mathcal{A}) \to \mathrm{Aut}(\mathcal{Z}(\mathcal{A})), \tag{35}$$

whose kernel is denoted by $\mathrm{SPic}(\mathcal{A})$, the **static Picard group** of $\mathcal{A}$.

Remark 3.4 If $\mathcal{A}$ is commutative, then each invertible bimodule induces an automorphism of $\mathcal{A}$ by (33), and $\mathrm{SPic}(\mathcal{A})$ consists of those bimodules "fixing" $\mathcal{A}$, which motivates our terminology. Bimodules in $\mathrm{SPic}(\mathcal{A})$ can also be characterized by having equal left and right module structures, and $\mathrm{SPic}(\mathcal{A})$ is often referred to in the literature as the "commutative" Picard group of $\mathcal{A}$.

If $\mathcal{A}$ is commutative, then the composition

$$\mathrm{Aut}(\mathcal{A}) \xrightarrow{j} \mathrm{Pic}(\mathcal{A}) \xrightarrow{h} \mathrm{Aut}(\mathcal{A})$$

is the identity. As a result, we can write $\mathrm{Pic}(\mathcal{A})$ as a semi-direct product,

$$\mathrm{Pic}(\mathcal{A}) = \mathrm{Aut}(\mathcal{A}) \ltimes \mathrm{SPic}(\mathcal{A}). \tag{36}$$

The action of $\mathrm{Aut}(\mathcal{A})$ on $\mathrm{SPic}(\mathcal{A})$ is given by $X \overset{q}{\mapsto} {}_q X_q$, where the left and right $\mathcal{A}$-module structures on ${}_q X_q$ are $a \cdot x := q(a)x$ and $x \cdot b := xq(b)$. Although the orbits of *commutative* algebras in Pic are just their isomorphism classes in the ordinary sense, (36) illustrates that their isotropy groups in Pic may be bigger than their ordinary automorphism groups. The following is a geometric example.

Example 3.5 (*Picard groups of algebras of functions*)
Let $\mathcal{A} = C^\infty(M)$ be the algebra of smooth complex-valued functions on a manifold M. Using the Serre-Swan identification of smooth complex vector bundles over M with projective modules over $\mathcal{A}$, one can check that $\mathrm{SPic}(\mathcal{A})$ coincides with $\mathrm{Pic}(M)$, the group of isomorphism classes of complex line bundles on M, which is isomorphic to $H^2(M, \mathbb{Z})$ via the Chern class map. We then have a purely geometric description of $\mathrm{Pic}(\mathcal{A})$ as

$$\mathrm{Pic}(C^\infty(M)) = \mathrm{Diff}(M) \ltimes H^2(M, \mathbb{Z}), \tag{37}$$

where the action of $\mathrm{Diff}(M)$ on $H^2(M, \mathbb{Z})$ is given by pull back. In (37), we use the identification of algebra automorphisms of $\mathcal{A}$ with diffeomorphisms of M, see e.g. [67].

3.2 Strong Morita equivalence of C^*-algebras

The notion of Morita equivalence of unital algebras has been adapted to several other classes of algebras. An example is the notion of *strong Morita equivalence* of C^*-algebras, introduced by Rieffel in [73, 74].

A **C^*-algebra** $\mathcal{A}$ is a complex Banach algebra with an involution * such that

$$\|aa^*\| = \|a\|^2, \;\; a \subset \mathcal{A}.$$

Important examples are the algebra of complex-valued continuous functions on a locally compact Hausdorff space and $\mathcal{B}(\mathcal{H})$, the algebra of bounded operators on a Hilbert space $\mathcal{H}$.

The relevant category of modules over a C^*-algebra, to be preserved under strong Morita equivalence, is that of Hilbert spaces on which the C^*-algebra acts through bounded operators. More precisely, for a given C^*-algebra $\mathcal{A}$, we consider the category $\mathrm{Herm}(\mathcal{A})$ whose objects are pairs $(\mathcal{H}, \rho)$, where $\mathcal{H}$ is a Hilbert space and $\rho : \mathcal{A} \to \mathfrak{B}(\mathcal{H})$ is a *nondegenerate* *-homomorphism of algebras, and morphisms are bounded linear intertwiners. (Here "nondegenerate" means that $\rho(\mathcal{A})h = 0$ implies that $h = 0$, which is always satisfied if $\mathcal{A}$ is unital and ρ preserves the unit.)

Since we are now dealing with more elaborate modules, it is natural that a bimodule giving rise to a functor $\mathrm{Herm}(\mathcal{B}) \to \mathrm{Herm}(\mathcal{A})$ analogous to (28) should be equipped with extra structure. If $(\mathcal{H}, \rho) \in \mathrm{Herm}(\mathcal{B})$ and ${}_{\mathcal{A}}\mathrm{X}_{\mathcal{B}}$ is an $(\mathcal{A}, \mathcal{B})$-bimodule, the key observation is that if X is itself equipped with an inner product $\langle \cdot, \cdot \rangle_{\mathcal{B}}$ *with values in* $\mathcal{B}$, then the map ${}_{\mathcal{A}}\mathrm{X}_{\mathcal{B}} \otimes_{\mathcal{B}} \mathcal{H} \times {}_{\mathcal{A}}\mathrm{X}_{\mathcal{B}} \otimes_{\mathcal{B}} \mathcal{H} \to \mathbb{C}$ uniquely defined by

$$(x_1 \otimes h_1, x_2 \otimes h_2) \mapsto \langle h_1, \rho(\langle x_1, x_2 \rangle_{\mathcal{B}}) h_2 \rangle \tag{38}$$

is an inner product on ${}_{\mathcal{A}}\mathrm{X}_{\mathcal{B}} \otimes_{\mathcal{B}} \mathcal{H}$, which we can complete to obtain a Hilbert space $\mathcal{H}'$. Moreover, the natural $\mathcal{A}$-action on ${}_{\mathcal{A}}\mathrm{X}_{\mathcal{B}} \otimes_{\mathcal{B}} \mathcal{H}$ gives rise to a *-representation $\rho' : \mathcal{A} \to \mathfrak{B}(\mathcal{H}')$. These are the main ingredients of Rieffel's induction of representations [73].

More precisely, let X be a right $\mathcal{B}$-module. Then a $\mathcal{B}$-valued inner product $\langle \cdot, \cdot \rangle_{\mathcal{B}}$ on X is a $\mathbb{C}$-sesquilinear pairing $\mathrm{X} \times \mathrm{X} \to \mathcal{B}$ (linear in the second argument) such that, for all $x_1, x_2 \in \mathrm{X}$ and $b \in \mathcal{B}$, we have

$$\langle x_1, x_2 \rangle_{\mathcal{B}} = \langle x_2, x_1 \rangle_{\mathcal{B}}^*, \quad \langle x_1, x_2 b \rangle_{\mathcal{B}} = \langle x_1, x_2 \rangle_{\mathcal{B}} b,$$

and

$$\langle x_1, x_1 \rangle_{\mathcal{B}} > 0 \ \text{ if } x_1 \neq 0.$$

(Inner products on left modules are defined analogously, but linearity is required in the first argument). One can show that $\|x\|_{\mathcal{B}} := \|\langle x, x \rangle_{\mathcal{B}}\|^{1/2}$ is a norm in X. A (right) **Hilbert $\mathcal{B}$-module** is a (right) $\mathcal{B}$-module X together with a $\mathcal{B}$-valued inner product $\langle \cdot, \cdot \rangle_{\mathcal{B}}$ so that X is complete with respect to $\| \cdot \|_{\mathcal{B}}$. Just as for Hilbert spaces, we denote by $\mathfrak{B}_{\mathcal{B}}(\mathrm{X})$ the algebra of endomorphisms of X possessing an adjoint with respect to $\langle \cdot, \cdot \rangle_{\mathcal{B}}$.

Example 3.6 (*Hilbert spaces*)
If $\mathcal{B} = \mathbb{C}$, then Hilbert $\mathcal{B}$-modules are just ordinary Hilbert spaces. In this case, $\mathfrak{B}_{\mathbb{C}}(\mathrm{X})$ coincides with the algebra of bounded linear operators on X, see e.g. [72].

Example 3.7 (*Hermitian vector bundles*)
Suppose $\mathcal{B} = C(X)$, the algebra of complex-valued continuous functions on a compact Hausdorff space X. If $E \to X$ is a complex vector bundle equipped with a hermitian metric h, then $\Gamma(E)$ is a Hilbert $\mathcal{B}$-

module with respect to the $C(X)$-valued inner product

$$\langle e, f\rangle_{\mathcal{B}}(x) := h_x(e(x), f(x)).$$

To describe the most general Hilbert modules over $C(X)$, one needs Hilbert bundles, which recover Example 3.6 when X is a point.

Example 3.8 (*C^*-algebras*)
Any C^*-algebra $\mathcal{B}$ is a Hilbert $\mathcal{B}$-module with respect to the inner product $\langle b_1, b_2\rangle_{\mathcal{B}} = b_1^* b_2$.

As in the case of unital algebras, one can define, for C^*-algebras $\mathcal{A}$ and $\mathcal{B}$, a "generalized morphism" $\mathcal{A} \leftarrow \mathcal{B}$ as a right Hilbert $\mathcal{B}$-module X, with inner product $\langle \cdot, \cdot\rangle_{\mathcal{B}}$, together with a *nondegenerate* *-homomorphism $\mathcal{A} \to \mathfrak{B}_{\mathcal{B}}(\mathrm{X})$. We "compose" ${}_{\mathcal{A}}\mathrm{X}_{\mathcal{B}}$ and ${}_{\mathcal{B}}\mathrm{Y}_{\mathcal{C}}$ through a more elaborate tensor product: we consider the algebraic tensor product ${}_{\mathcal{A}}\mathrm{X}_{\mathcal{B}} \otimes_{\mathbb{C}} {}_{\mathcal{B}}\mathrm{Y}_{\mathcal{C}}$, equipped with the semi-positive $\mathcal{C}$-valued inner product uniquely defined by

$$(x_1 \otimes y_1, x_2 \otimes y_2) \mapsto \langle y_1, \langle x_1, x_2\rangle_{\mathcal{B}} y_2\rangle_{\mathcal{C}}. \tag{39}$$

The null space of this inner product coincides with the span of elements of the form $xb \otimes y - x \otimes by$ [51], so (39) induces a positive-definite $\mathcal{C}$-valued inner product on ${}_{\mathcal{A}}\mathrm{X}_{\mathcal{B}} \otimes_{\mathcal{B}} {}_{\mathcal{B}}\mathrm{Y}_{\mathcal{C}}$. The completion of this space with respect to $\|\cdot\|_{\mathcal{C}}$ yields a "generalized morphism" from $\mathcal{C}$ to $\mathcal{A}$ denoted by ${}_{\mathcal{A}}\mathrm{X}_{\mathcal{B}} \widehat{\otimes}_{\mathcal{B}\mathcal{B}} \mathrm{Y}_{\mathcal{C}}$, called the **Rieffel tensor product** of ${}_{\mathcal{A}}\mathrm{X}_{\mathcal{B}}$ and ${}_{\mathcal{B}}\mathrm{Y}_{\mathcal{C}}$.

An isomorphism between "generalized morphisms" is a bimodule isomorphism preserving inner products. Just as ordinary tensor products, Rieffel tensor products are associative up to natural isomorphisms. So one can define a category C^* whose objects are C^*-algebras and whose morphisms are isomorphism classes of "generalized morphisms", with composition given by Rieffel tensor product; the identities are the algebras themselves, regarded as bimodules in the usual way, and with the inner product of Example 3.8.

Two C^*-algebras are **strongly Morita equivalent** if they are isomorphic in C^*. As in the case of unital algebras, isomorphic C^*-algebras are necessarily strongly Morita equivalent.

Remark 3.9 (*Equivalence bimodules*)
The definition of strong Morita equivalence as isomorphism in C^* coincides with Rieffel's original definition in terms of equivalence bimodules (also called imprimitivity bimodules) [73, 74]. In fact, any "generalized morphism" ${}_{\mathcal{A}}\mathrm{X}_{\mathcal{B}}$ which is invertible in C^* can be endowed with an $\mathcal{A}$-

valued inner product, compatible with its $\mathcal{B}$-valued inner product in the appropriate way, making it into an equivalence bimodule, see [53] and references therein. Conversely, any equivalence bimodule is automatically invertible in C^*.

Example 3.10 (*Compact operators*)
A Hilbert space $\mathcal{H}$, seen as a bimodule for $\mathbb{C}$ and the C^*-algebra $\mathcal{K}(\mathcal{H})$ of compact operators on $\mathcal{H}$, defines a strong Morita equivalence.

Example 3.11 (*Endomorphism bundles*)
Analogously to Example 3.3, a hermitian vector bundle $E \to X$, where X is a compact Hausdorff space, defines a strong Morita equivalence between $\Gamma(\mathrm{End}(E))$ and $C(X)$.

Any "generalized morphism" ${}_{\mathcal{A}}\mathrm{X}_{\mathcal{B}}$ in C^* defines a functor

$$\mathcal{F}_{\mathrm{X}} : \mathrm{Herm}(\mathcal{B}) \to \mathrm{Herm}(\mathcal{A}),$$

similar to (28), but with Rieffel's tensor product replacing the ordinary one, i.e., on objects,

$$\mathcal{F}_{\mathrm{X}}(\mathcal{H}) := {}_{\mathcal{A}}\mathrm{X}_{\mathcal{B}}\widehat{\otimes}_{\mathcal{B}}\mathcal{H}. \tag{40}$$

Such a functor is called **Rieffel induction of representations** [73]. It follows that strongly Morita equivalent C^*-algebras have equivalent categories of representations, although, in this setting, the converse is not true [74] (see [11] for a different approach where a converse does hold).

Remark 3.12 (*Strong vs. ring-theoretic Morita equivalence*)
By regarding unital C^*-algebras simply as unital algebras over $\mathbb{C}$, one can compare strong and ring-theoretic Morita equivalences. It turns out that two unital C^*-algebras are strongly Morita equivalent if and only if they are Morita equivalent as unital $\mathbb{C}$-algebras [6]. However, the Picard groups associated to each notion are different in general, see [18]. In terms of Picard groupoids, this means that, over unital C^*-algebras, the Picard groupoids associated with ring-theoretic and strong Morita equivalences have the same orbits, but generally different isotropy groups.

A study of Picard groups associated with strong Morita equivalence, analogous to the discussion in Section 3.1, can be found in [12].

3.3 Morita equivalence of deformed algebras

Let (P, Π) be a Poisson manifold and $C^\infty(P)$ be its algebra of smooth complex-valued functions. The general idea of a **deformation quantization** of P "in the direction" of Π is that of a family $\star_\hbar$ of associative algebra structures on $C^\infty(P)$ satisfying the following two conditions:

i.) $f \star_\hbar g = f \cdot g + O(\hbar)$;
ii.) $\frac{1}{i\hbar}(f \star_\hbar g - g \star_\hbar f) \longrightarrow \{f, g\}$, when $\hbar \to 0$.

There are several versions of deformation quantization. We will consider

1. **Formal deformation quantization** [5]: In this case, $\star_\hbar$ is an associative product on $C^\infty(P)[[\hbar]]$, the space of formal power series with coefficients in $C^\infty(P)$. Here $\hbar$ is a formal parameter, and the "limit" in *ii.)* above is defined simply by setting $\hbar$ to 0. A formal deformation quantization is also called a **star product**. The contribution by Cattaneo and Indelicato [23] to this volume contains a thorough exposition of the theory of star products and its history.
2. **Rieffel's strict deformation quantization** [76]: In this setting, one starts with a dense Poisson subalgebra of $C_\infty(P)$, the C^*-algebra of continuous functions on P vanishing at infinity, and considers families of associative products $\star_\hbar$ on it, defined along with norms and involutions such that the completions form a continuous field of C^*-algebras. The parameter $\hbar$ belongs to a closed subset of $\mathbb{R}$ having 0 as a non-isolated point, and one can make analytical sense of the limit in *ii.)* above. Variations of Rieffel's notion of deformation quantization are discussed in [52].

Intuitively, one should regard a deformation quantization $\star_\hbar$ as a path in the "space of associative algebra structures" on $C^\infty(P)$ for which the Poisson structure Π is the "tangent vector" at $\hbar = 0$. From this perspective, a direct relationship between deformation quantization and Poisson geometry is more likely in the formal case.

A natural question is when two algebras obtained by deformation quantization are Morita equivalent. In the framework of formal deformation quantization, the first observation is that if two deformation quantizations $(C^\infty(P_1, \Pi_1)[[\hbar]], \star^1_\hbar)$ and $(C^\infty(P_2, \Pi_2)[[\hbar]], \star^2_\hbar)$ are Morita equivalent (as unital algebras over $\mathbb{C}[[\hbar]]$), then the underlying Poisson manifolds are isomorphic. So we can restrict ourselves to a fixed Poisson manifold. The following result is proven in [13, 17]:

Theorem 3.13 *Let P be symplectic. If $\mathrm{Pic}(P) \cong H^2(P,\mathbb{Z})$ has no torsion, then it acts freely on the set of equivalence classes of star products on P, and two star products are Morita equivalent if and only if their classes lie in the same $H^2(P,\mathbb{Z})$-orbit, up to symplectomorphism.*

Recall that two star products $\star_\hbar^1$ and $\star_\hbar^2$ are **equivalent** if there exists a family of differential operators $T_r : C^\infty(P) \to C^\infty(P)$, $r = 1, 2\ldots$, so that $T = \mathrm{Id} + \sum_{r=1}^\infty T_r \hbar^r$ is an algebra isomorphism

$$(C^\infty(P)[[\hbar]], \star_\hbar^1) \xrightarrow{\sim} (C^\infty(P)[[\hbar]], \star_\hbar^2).$$

Equivalence classes of star products on a symplectic manifold are parametrized by elements in

$$\frac{1}{i\hbar}[\omega] + H^2_{dR}(P)[[\hbar]], \tag{41}$$

where ω is the symplectic form on P and $H^2_{dR}(P)$ is the second de Rham cohomology group of P with complex coefficients [5, 37, 68], called **characteristic classes**. As shown in [17], the $\mathrm{Pic}(P)$-action of Theorem 3.13 is explicitly given in terms of these classes by

$$[\omega_\hbar] \mapsto [\omega_\hbar] + 2\pi i c_1(L), \tag{42}$$

where $[\omega_\hbar]$ is an element in (41) and $c_1(L)$ is the image of the Chern class of the line bundle L in $H^2_{dR}(P)$.

Remark 3.14 A version of Theorem 3.13 holds for arbitrary Poisson manifolds (P, Π), see [13, 47]. In this general setting, equivalence classes of star products are parametrized by classes of formal Poisson bivectors $\Pi_\hbar = \Pi + \hbar\Pi_1 + \cdots$ (see [49] or the exposition in [23]), and the $\mathrm{Pic}(P)$-action on them, classifying Morita equivalent deformation quantizations of P, is via gauge transformations (see Section 4.8).

In the framework of strict deformation quantization and the special case of tori, a classification result for Morita equivalence was obtained by Rieffel and Schwarz in [77] (see also [55, 82]). Let us consider $\mathbb{T}^n = \mathbb{R}^n/\mathbb{Z}^n$ equipped with a constant Poisson structure, represented by a skew-symmetric real matrix Π: if $(\theta_1, \ldots, \theta_n)$ are coordinates on T^n, then

$$\Pi_{ij} = \{\theta_i, \theta_j\}.$$

Via the Fourier transform, one can identify the algebra $C^\infty(\mathbb{T}^n)$ with the space $\mathcal{S}(\mathbb{Z}^n)$ of complex-valued functions on $\mathbb{Z}^n$ with rapid decay at infinity. Under this identification, the pointwise product of functions becomes the convolution on $\mathcal{S}(\mathbb{Z}^n)$,

$$\widehat{f} * \widehat{g}(n) = \sum_{k \in \mathbb{Z}^n} \widehat{f}(n)\widehat{g}(n-k),$$

$\widehat{f}, \widehat{g} \in \mathcal{S}(\mathbb{Z}^n)$. One can now use the matrix Π to "twist" the convolution and define a new product

$$\widehat{f} *_\hbar \widehat{g}(n) = \sum_{k \in \mathbb{Z}^n} \widehat{f}(n)\widehat{g}(n-k) e^{-\pi i \hbar \Pi(k, n-k)} \tag{43}$$

on $\mathcal{S}(\mathbb{Z}^n)$, which can be pulled back to a new product in $C^\infty(\mathbb{T}^n)$. Here $\hbar$ is a real parameter. If we set $\hbar = 1$, this defines the algebra $\mathcal{A}_\Pi^\infty$, which can be thought of as the "algebra of smooth functions on the quantum torus $\mathbb{T}^n_\Pi$". A suitable completion of $\mathcal{A}_\Pi^\infty$ defines a C^*-algebra $\mathcal{A}_\Pi$, which is then thought of as the "algebra of continuous functions on $\mathbb{T}^n_\Pi$". (Note that, with $\hbar = 1$, we are no longer really considering a deformation.)

Exercise

Show that 1 is a unit for $\mathcal{A}_\Pi$. Let $u_j = e^{2\pi i \theta_j}$. Show that $u_j *_1 \bar{u}_j = \bar{u}_j *_1 u_j = 1$ and

$$u_j *_1 u_k = e^{2\pi i \Pi_{jk}} u_k *_1 u_j. \tag{44}$$

The algebra $\mathcal{A}_\Pi$ can be alternatively described as the universal C^*-algebra generated by n unitary elements $u_1, \ldots, u_n$ subject to the commutation relations (44).

In this context, the question to be addressed is when skew-symmetric matrices Π and Π' correspond to Morita equivalent C^*-algebras $\mathcal{A}_\Pi$ and $\mathcal{A}_{\Pi'}$. Let $O(n, n|\mathbb{R})$ be the group of linear automorphisms of $\mathbb{R}^n \oplus \mathbb{R}^{n*}$ preserving the inner product (6). One can identify elements of $O(n, n|\mathbb{R})$ with matrices

$$g = \begin{pmatrix} A & B \\ C & D \end{pmatrix},$$

where A, B, C and D are $n \times n$ matrices satisfying

$$A^t C + C^t A = 0 = B^t D + D^t B, \quad \text{and} \quad A^t D + C^t B = 1.$$

The group $O(n, n|\mathbb{R})$ "acts" on the space of all $n \times n$ skew-symmetric matrices by

$$\Pi \mapsto g \cdot \Pi := (A\Pi + B)(C\Pi + D)^{-1}. \tag{45}$$

Note that this is not an honest action, since the formula above only makes sense when $(C\Pi + D)$ is invertible.

Let $SO(n, n|\mathbb{Z})$ be the subgroup of $O(n, n|\mathbb{R})$ consisting of matrices with integer coefficients and determinant 1. The main result of [77], as improved in [55, 82], is

Theorem 3.15 *If* Π *is a skew-symmetric matrix,* $g \in SO(n, n|\mathbb{Z})$ *and* $g \cdot \Pi$ *is defined, then* $\mathcal{A}_\Pi$ *and* $\mathcal{A}_{g \cdot \Pi}$ *are strongly Morita equivalent.*

Remark 3.16 (*Converse results*)

The converse of Theorem 3.15 holds for $n = 2$ [75], but not in general. In fact, for $n = 3$, one can find Π and Π', *not* in the same $SO(n, n|\mathbb{Z})$-orbit, for which $\mathcal{A}_\Pi$ and $\mathcal{A}_{\Pi'}$ are isomorphic (hence Morita equivalent) [77].

On the other hand, for *smooth* quantum tori, Theorem 3.15 and its converse hold with respect to a refined notion of Morita equivalence, called "complete Morita equivalence" [78], in which bimodules carry connections of constant curvature.

For the algebraic Morita equivalence of smooth quantum tori, see [36].

Remark 3.17 (*Dirac structures and quantum tori*)

In [77], the original version of Theorem 3.15 was proven under an additional hypothesis. Rieffel and Schwarz consider three types of generators of $SO(n, n|\mathbb{Z})$, and prove that their action preserves Morita equivalence. In order to show that $\mathcal{A}_\Pi$ and $\mathcal{A}_{g\Pi}$ are Morita equivalent for an arbitrary $g \in SO(n, n|\mathbb{Z})$ (for which $g\Pi$ is defined), they need to assume that g can be written as a product of generators $g_r \cdots g_1$ in such a way that each of the products $g_k \cdots g_1 \Pi$ is defined. The result in Theorem 3.15, without this assumption, is conjectured in [77], and it was proven by Li in [55].

A geometric way to circumvent the difficulties in the Rieffel-Schwarz proof, in which Dirac structures play a central role, appears in [82]. The key point is the observation that, even if $g \cdot \Pi$ is not defined as a skew-symmetric matrix, it is still a Dirac structure on $\mathbb{T}^n$. The authors develop a way to quantize constant Dirac structures on $\mathbb{T}^n$ by attaching to each one of them a Morita equivalence class of quantum tori. They extend the $SO(n, n|\mathbb{Z})$ action to Dirac structures and prove that the Morita equivalence classes of the corresponding quantum tori is unchanged under the action.

4
Geometric Morita equivalence

In this chapter, we introduce a purely geometric notion of Morita equivalence of Poisson manifolds. This notion leads inevitably to the consideration of Morita equivalence of symplectic groupoids, so we will make a digression into the Morita theory of general Lie groups and groupoids. We end the chapter with a discussion of gauge equivalence, a geometric equivalence which is close to Morita equivalence, but is also related to the algebraic Morita equivalence of star products, as discussed in Section 3.3.

4.1 Representations and tensor product

In order to define Morita equivalence in Poisson geometry, we need notions of "representations" of (or "modules" over) Poisson manifolds as well as their tensor products.

As we saw in Example 2.24, symplectic manifolds are in some sense "irreducible" among Poisson manifolds. If one thinks of Poisson manifolds as algebras, then symplectic manifolds could be thought of as "matrix algebras". Following this analogy, a representation of a Poisson manifold P should be a symplectic manifold S together with a Poisson map $J : S \to P$ which is complete. At the level of functions, we have a "representation" of $C^\infty(P)$ by $J^* : C^\infty(P) \to C^\infty(S)$. This notion of representation is also suggested by the theory of geometric quantization, in which symplectic manifolds become "vector spaces" on which their Poisson algebras "act asymptotically".

More precisely, we define a left [right] P**-module** to be a complete [anti-] symplectic realization $J : S \to P$. Our first example illustrates how modules over Lie-Poisson manifolds are related to hamiltonian actions.

Example 4.1 (*Modules over* $\mathfrak{g}^*$ *and hamiltonian actions*)
Let (S, Π_S) be a symplectic Poisson manifold, $\mathfrak{g}$ be a Lie algebra, and suppose that $J : S \to \mathfrak{g}^*$ is a symplectic realization of $\mathfrak{g}^*$. The map

$$\mathfrak{g} \to \mathcal{X}(S), \quad v \mapsto \widetilde{\Pi}_S(dJ_v), \tag{46}$$

where $J_v(x) := \langle J(x), v \rangle$, defines a $\mathfrak{g}$-action on S by hamiltonian vector fields for which J is the momentum map. On the other hand, the momentum map $J : S \to \mathfrak{g}^*$ for a hamiltonian $\mathfrak{g}$-action on S is a Poisson map, so we have a one-to-one correspondence between symplectic realizations of $\mathfrak{g}^*$ and hamiltonian $\mathfrak{g}$-manifolds.

A symplectic realization $J : S \to \mathfrak{g}^*$ is complete if and only if the associated infinitesimal hamiltonian action is by complete vector fields, in which case it can be integrated to a hamiltonian G-action, where G is the connected and simply-connected Lie group having $\mathfrak{g}$ as its Lie algebra. So $\mathfrak{g}^*$-modules are just the same thing as hamiltonian G-manifolds.

Remark 4.2 (*More general modules over* $\mathfrak{g}^*$)
The one-to-one correspondence in Example 4.1 extends to one between Poisson maps into $\mathfrak{g}^*$ (from any Poisson manifold, not necessarily symplectic) and hamiltonian $\mathfrak{g}$-actions on Poisson manifolds, or, similarly, between complete Poisson maps into $\mathfrak{g}^*$ and Poisson manifolds carrying hamiltonian G-actions. This indicates that it may be useful to regard *arbitrary* (complete) Poisson maps as modules over Poisson manifolds; we will say more about this in Remarks 4.18 and 4.24.

We now define a tensor product operation on modules over a Poisson manifold. Let $J : S \to P$ be a right P-module, and let $J' : S' \to P$ be a left P-module. Just as, in algebra, we can think of the tensor product over $\mathcal{A}$ of a left module X and a right module Y as a quotient of their tensor product over the ground ring k, so in Poisson geometry we can define the tensor product of S and S' to be a "symplectic quotient" of $S \times S'$. Namely, the fibre product

$$S \times_{(J,J')} S' = \{(x, y) \in S \times S' \mid J(x) = J'(y)\} \tag{47}$$

is the inverse image of the diagonal under the Poisson map $(J, J') : S \times S' \to P \times \overline{P}$, hence, whenever it is smooth, it is a coisotropic submanifold. (Here $\overline{P}$ denotes P equipped with its Poisson structure multiplied by -1.) Let us assume then, that the fibre product is smooth; this is the case, for example, if either J or J' is a surjective submersion. Then we may define the **tensor product** $S * S'$ over P to be the quotient of this fibre product by its characteristic foliation. In general, even if the fibre product is smooth, $S * S'$ is still not a smooth manifold, but just a quotient of a manifold by a foliation. We will have to deal with this problem later, see Remark 4.40. But when the characteristic foliation is simple, $S * S'$ is a symplectic manifold. We may write $S *_P S'$ instead of $S * S'$ to identify the Poisson manifold over which we are taking the tensor product.

If one is given two left modules (one could do the same for right modules, of course), one can apply the tensor product construction by changing the "handedness" of one of them. Thus, if S and S' are left P-modules, then $\overline{S'}$ is a right module, and we can form the tensor product $\overline{S'} * S$. We call this the **classical intertwiner space** [95, 96] of S and S' and denote it by $\mathrm{Hom}(S, S')$. The name and notation come from the case of modules over an algebra, where the tensor product $\mathrm{Y}^* \otimes \mathrm{X}$ is naturally isomorphic to the space of module homomorphisms from Y to X when these modules are "finite dimensional". When the algebra is a group algebra, the modules are representations of the group, and the module homomorphisms are known as intertwining operators.

Example 4.3 (*Symplectic reduction*)

Let $J : S \to \mathfrak{g}^*$ be the momentum map for a hamiltonian action of a connected Lie group G on a symplectic manifold S. Let $S' = \mathcal{O}_\mu$ be the coadjoint orbit through $\mu \in \mathfrak{g}^*$, equipped with the symplectic structure induced by the Lie-Poisson structure on $\mathfrak{g}^*$, and let $\iota : \mathcal{O}_\mu \hookrightarrow \mathfrak{g}^*$ be the inclusion, which is a Poisson map. Then the classical intertwiner space $\mathrm{Hom}(S, \mathcal{O}_\mu)$ is equal to $J^{-1}(\mathcal{O}_\mu)/G \cong J^{-1}(\mu)/G_\mu$, i.e., the symplectic reduction of S at the momentum value μ.

A (P_1,P_2)-**bimodule** is a symplectic manifold S and a pair of maps $P_1 \stackrel{J_1}{\leftarrow} S \stackrel{J_2}{\rightarrow} P_2$ making S into a left P_1-module and a right P_2-module and satisfying the "commuting actions" condition:

$$\{J_1^* C^\infty(P_1), J_2^* C^\infty(P_2)\} = 0. \tag{48}$$

(Such geometric bimodules, without the completeness assumption, are called **dual pairs** in [88].) An **isomorphism** of bimodules is a symplectomorphism commuting with the Poisson maps.

Given bimodules $P_1 \stackrel{J_1}{\leftarrow} S \stackrel{J_2}{\rightarrow} P_2$ and $P_2 \stackrel{J_2'}{\leftarrow} S' \stackrel{J_3'}{\rightarrow} P_3$, we may form the tensor product $S *_{P_2} S'$, and it is easily seen that this tensor product, whenever it is smooth, becomes a (P_1, P_3)-bimodule [95, 53]. We think of this tensor product as the **composition** of S and S'.

Remark 4.4 (*Modules as bimodules and geometric Rieffel induction*) For any left P_2-module S', there is an associated bimodule $P_2 \leftarrow S' \rightarrow \mathrm{pt}$, where pt is just a point. Given a bimodule $P_1 \overset{J_1}{\leftarrow} S \overset{J_2}{\rightarrow} P_2$, we can form its tensor product with $P_2 \leftarrow S' \rightarrow \mathrm{pt}$ to get a (P_1, pt)-bimodule. In this way, the (P_1, P_2)-bimodule "acts" on P_2-modules to give P_1-modules. This is the geometric analogue of the functors (28) and (40), for unital and C^*-algebras, respectively.

Example 4.5 Following Example 4.3, suppose that the orbit space S/G is smooth, in which case it is a Poisson manifold in a natural way. Consider the bimodules $S/G \leftarrow S \overset{J}{\rightarrow} \overline{\mathfrak{g}^*}$ and $\overline{\mathfrak{g}^*} \overset{\iota}{\leftarrow} \overline{\mathcal{O}_\mu} \rightarrow \mathrm{pt}$. Their tensor product is the $(S/G, \mathrm{pt})$-bimodule $S/G \leftarrow S * \overline{\mathcal{O}_\mu} \rightarrow \mathrm{pt}$, where the map on the left is the inclusion of the symplectic reduced space $S * \overline{\mathcal{O}_\mu} = \mathrm{Hom}(S, \mathcal{O})$ as a symplectic leaf of S/G.

Following the analogy with algebras, it is natural to think of isomorphism classes of bimodules as generalized morphisms of Poisson manifolds. The extra technical difficulty in this geometric context is that tensor products do not always result in smooth spaces. So one needs a suitable notion of "regular bimodules", satisfying extra regularity conditions to guarantee that their tensor products are smooth and again "regular", see [19, 53], or an appropriate notion of bimodule modeled on "singular" spaces. We will come back to these topics in Section 4.7.

4.2 Symplectic groupoids

In order to regard geometric bimodules over Poisson manifolds as morphisms in a category, one needs to identify the bimodules which serve as identities, i.e., those satisfying

$$S * S' \cong S' \quad \text{and} \quad S'' * S \cong S''$$

for any other bimodules S' and S''. As we saw in Section 3.1, in the case of unital algebras, the identity bimodule of an object $\mathcal{A}$ in Alg is just $\mathcal{A}$ itself, regarded as an $(\mathcal{A}, \mathcal{A})$-bimodule in the usual way. This idea cannot work for Poisson manifolds, since they are generally not symplectic, and because we do not have commuting left and right actions of P on itself. Instead, it is the symplectic groupoids [89] which serve as such "identity bimodules" for Poisson manifolds, see [53]. If $P \stackrel{t}{\leftarrow} \mathcal{G} \stackrel{s}{\rightarrow} P$ is an identity bimodule for a Poisson manifold P, then there exists, in particular, a symplectomorphism $\mathcal{G} * \mathcal{G} \to \mathcal{G}$, and the composition

$$\mathcal{G} \times_{(s,t)} \mathcal{G} \to \mathcal{G} * \mathcal{G} \stackrel{\sim}{\to} \mathcal{G}$$

defines a map $m : \mathcal{G} \times_{(s,t)} \mathcal{G} \to \mathcal{G}$ which turns out to be a groupoid multiplication[1], compatible with the symplectic form on $\mathcal{G}$ in the sense that graph$(m) \subseteq \mathcal{G} \times \overline{\mathcal{G}} \times \overline{\mathcal{G}}$ is a *lagrangian* submanifold. If $p_i : \mathcal{G} \times_{(s,t)} \mathcal{G} \to \mathcal{G}$, $i = 1, 2$, are the natural projections, then the compatibility between m and ω is equivalent to the condition

$$m^*\omega = p_1^*\omega + p_2^*\omega. \tag{49}$$

A 2-form ω satisfying (49) is called **multiplicative** (note that if ω were a function, (49) would mean that $\omega(gh) = \omega(g) + \omega(h)$), and a groupoid equipped with a multiplicative symplectic form is called a **symplectic groupoid**.

If $(\mathcal{G}, \omega)$ is a symplectic groupoid over a manifold P, then the following important properties follow from the compatibility condition (49), see [26]:

i) The unit section $P \hookrightarrow \mathcal{G}$ is lagrangian;
ii) The inversion map $\mathcal{G} \to \mathcal{G}$ is an anti-symplectic involution;
iii) The fibres of the target and source maps, $t, s : \mathcal{G} \to P$, are the symplectic orthogonal of one another;
iv) At each point of $\mathcal{G}$, $\ker(Ts) = \{X_{t^*f} \,|\, f \in C^\infty(P)\}$ and $\ker(Tt) = \{X_{s^*f} \,|\, f \in C^\infty(P)\}$;
v) P carries a unique Poisson structure such that the target map t is a Poisson map (and the source map s is anti-Poisson).

A Poisson manifold (P, Π) is called **integrable** if there exists a symplectic groupoid $(\mathcal{G}, \omega)$ over P which induces Π in the sense of v), and we refer to $\mathcal{G}$ as an **integration** of P. As we will discuss later, not every Poisson manifold is integrable is this sense, see [30, 89]. But if P is integrable, then there exists a symplectic groupoid integrating it which

[1] For expositions on groupoids, we refer to [20, 63, 64]; we adopt the convention that, on a Lie groupoid $\mathcal{G}$ over P, with source s and target t, the multiplication is defined on $\{(g, h) \in \mathcal{G} \times \mathcal{G},\ s(g) = t(h)\}$, and we identify the Lie algebroid $A(\mathcal{G})$ with $\ker(Ts)|_P$, and Tt is the anchor map. The bracket on the Lie algebroid comes from identification with *right*-invariant vector fields, which is counter to a convention often used for Lie groups.

has simply-connected (i.e., connected with trivial fundamental group) source fibres [58], and this groupoid is unique up to isomorphism.

Remark 4.6 (*Integrability and complete symplectic realizations*)
If $(\mathcal{G}, \omega)$ is an integration of (P, Π), then the target map $t : \mathcal{G} \to P$ is a Poisson submersion which is always complete. On the other hand, as proven in [30], if a Poisson manifold P admits a complete symplectic realization $S \to P$ which is a submersion, then P must be integrable.

Remark 4.7 (*The Lie algebroid of a Poisson manifold*)
All the integrations of a Poisson manifold (P, Π) have (up to natural isomorphism) the same Lie algebroid. It is T^*P, with a Lie algebroid structure with anchor $\widetilde{\Pi} : T^*P \to TP$, and Lie bracket on $\Gamma(T^*P) = \Omega^1(P)$ defined by

$$[\alpha, \beta] := \mathcal{L}_{\widetilde{\Pi}(\alpha)}(\beta) - \mathcal{L}_{\widetilde{\Pi}(\beta)}(\alpha) - d\Pi(\alpha, \beta). \tag{50}$$

Note that (50) is uniquely characterized by $[df, dg] = d\{f, g\}$ and the Leibniz identity. Following Remark 2.18, we know that $L_\Pi = \mathrm{graph}(\widetilde{\Pi})$ also carries a Lie algebroid structure, induced by the Courant bracket. The natural projection $\mathrm{pr}_2 : TP \oplus T^*P \to T^*P$ restricts to a vector bundle isomorphism $L_\Pi \to T^*P$ which defines an isomorphism of Lie algebroids.

On the other hand, if $(\mathcal{G}, \omega)$ is a symplectic groupoid integrating (P, Π), then the bundle isomorphism

$$\ker(Ts)|_P \longrightarrow T^*P, \;\; \xi \mapsto i_\xi \omega|_{TP} \tag{51}$$

induces an isomorphism of Lie algebroids $A(\mathcal{G}) \xrightarrow{\sim} T^*P$, where $A(\mathcal{G})$ is the Lie algebroid of $\mathcal{G}$, so the symplectic groupoid $\mathcal{G}$ integrates T^*P in the sense of Lie algebroids. It follows from (51) that $\dim(\mathcal{G}) = 2\dim(P)$.

In the work of Cattaneo and Felder [22], symplectic groupoids arise as reduced phase spaces of Poisson sigma models. This means that one begins with the space of paths on T^*P, which has a natural symplectic structure, restricts to a certain submanifold of "admissible" paths, and forms the symplectic groupoid $\mathcal{G}(P)$ as a quotient of this submanifold by a foliation. This can also be described as an infinite-dimensional symplectic reduction. The resulting space is a groupoid but may not be a manifold. When it is a manifold, it is the source-simply-connected symplectic groupoid of P. When $\mathcal{G}(P)$ is not a manifold, as the leaf space of a foliation, it can be considered as a differentiable stack, and even as a symplectic stack. In the world of stacks [60], it is again a smooth groupoid; we will call it an **S-groupoid**. The first steps of this program have been carried out by Tseng and Zhu [83]. (See [93] for an exposition, as well as Remark 4.40 below.)

This construction of symplectic groupoids has been extended to general Lie algebroids, see [29, 79]. Crainic and Fernandes [29] describe explicitly the obstructions to the integrability of Lie algebroids and, in [30], identify these obstructions for the case of Poisson manifolds and symplectic groupoids. Integration by S-groupoids is done in [83].

The next three examples illustrate simple yet important classes of integrable Poisson manifolds and their symplectic groupoids.

Example 4.8 (*Symplectic manifolds*)

If (P, ω) is a symplectic manifold, then the pair groupoid $P \times P$ equipped with the symplectic form $\omega \times (-\omega)$ is a symplectic groupoid integrating P. In order to obtain a source-simply-connected integration, one should consider the fundamental groupoid $\pi(P)$, with symplectic structure given by the pull-back of the symplectic form on $P \times \overline{P}$ by the covering map $\pi(P) \to P \times \overline{P}$.

Example 4.9 (*Zero Poisson structures*)

If (P, Π) is a Poisson manifold with $\Pi = 0$, then $\mathcal{G}(P) = T^*P$. In this case, the source and target maps coincide with the projection $T^*P \to P$, and the multiplication on T^*P is given by fibrewise addition. There are, however, other symplectic groupoids integrating P, which may not have connected or simply-connected source fibres. For example, if T^*P admits a basis of closed 1-forms, we may divide the fibres of T^*P by the lattice generated by these forms to obtain a groupoid whose source and target fibres are tori. Or, if P is just a point, any discrete group is a symplectic groupoid for P. We refer to [19] for more details.

Example 4.10 (*Lie-Poisson structures*)

Let $P = \mathfrak{g}^*$ be the dual of a Lie algebra $\mathfrak{g}$, equipped with its Lie-Poisson structure, and let G be a Lie group with Lie algebra $\mathfrak{g}$. The transformation groupoid $G \ltimes \mathfrak{g}^*$ with respect to the coadjoint action, equipped with the symplectic form obtained from the identification $G \times \mathfrak{g}^* \cong T^*G$ by right translation, is a symplectic groupoid integrating $\mathfrak{g}^*$. This symplectic groupoid is source-simply-connected just when G is a (connected) simply-connected Lie group.

Remark 4.11 (*Lie's third theorem*)

Let $\mathfrak{g}$ be a Lie algebra. Example 4.10 shows that integrating $\mathfrak{g}$, in the usual sense of finding a Lie group G with Lie algebra $\mathfrak{g}$, yields an integration of the Lie-Poisson structure of $\mathfrak{g}^*$. On the other hand, one can use the integration of the Lie-Poisson structure of $\mathfrak{g}^*$ to construct a Lie group integrating $\mathfrak{g}$. Indeed, if $\mathcal{G}$ is a symplectic groupoid integrating $\mathfrak{g}^*$, then the map

$$\mathfrak{g} \to \mathcal{X}(\mathcal{G}), \quad v \mapsto X_{t^*v}$$

is a faithful representation of $\mathfrak{g}$ by vector fields on $\mathcal{G}$. Here $t : \mathcal{G} \to \mathfrak{g}^*$ is the target map, and we regard $v \in \mathfrak{g}$ as a linear function on $\mathfrak{g}^*$. We then use the flows of these vector fields to define a (local) Lie group integrating $\mathfrak{g}$. If we fix $x \in \mathcal{G}$, the "identity" of the local Lie group, so that $t(x) = 0$, then the Lie group sits in $\mathcal{G}$ as a lagrangian subgroupoid. So the two "integrations" are the same.

The idea of using a symplectic realization of $\mathfrak{g}^*$ to find a Lie group integrating $\mathfrak{g}$ goes back to Lie's original proof of "Lie's third theorem." A regular point of $\mathfrak{g}^*$ has a neighborhood U with coordinates $(q_1, \ldots, q_k, p_1, \ldots, p_k, e_1, \ldots, e_l)$ such that the Lie-Poisson structure can be written as

$$\sum_{i=1}^{k} \frac{\partial}{\partial q_i} \wedge \frac{\partial}{\partial p_i}$$

(see Section 2.4). The map $\mathfrak{g} \to \mathcal{X}(U)$, $v \mapsto X_v$ is a Lie algebra homomorphism, but not faithful in general. It suffices, though, to add l new coordinates $(f_1, \ldots, f_l)$ and consider the local symplectic realization $U \times \mathbb{R}^l \to U$, with symplectic Poisson structure

$$\Pi' = \sum_{i=1}^{k} \frac{\partial}{\partial q_i} \wedge \frac{\partial}{\partial p_i} + \sum_{i=1}^{l} \frac{\partial}{\partial e_i} \wedge \frac{\partial}{\partial f_i}.$$

The map $\mathfrak{g} \to \mathcal{X}(U \times \mathbb{R}^l)$, $v \mapsto X'_v := \widetilde{\Pi}'(v)$, is now a faithful Lie algebra homomorphism. Once again, we can use the flows of the hamiltonian vector fields of the coordinates on $\mathfrak{g}$ to construct a local Lie group.

More generally, if $\mathcal{G}$ is a Lie groupoid and A is its Lie algebroid, then $T^*\mathcal{G}$ is naturally a symplectic groupoid over A^*, see [26]. The induced Poisson structure on A^* is a generalization of a Lie-Poisson structure. Conversely, if A is an integrable Lie algebroid, then $\mathcal{G}(A)$, its source-simply-connected integration, can be constructed as a lagrangian subgroupoid of the symplectic groupoid $\mathcal{G}(A^*)$ integrating A^* [21].

The following is an example of a non-integrable Poisson structure.

Example 4.12 (*Nonintegrable Poisson structure*)

Let $P = S^2 \times \mathbb{R}$. Let Π_{S^2} be the natural symplectic structure on S^2. Then the product Poisson structure on P, $\Pi_{S^2} \times \{0\}$ is integrable. But if we multiply this Poisson structure by $(1+t^2)$, $t \in \mathbb{R}$ (or use any other nonconstant function which has a critical point), then the resulting Poisson structure $(1+t^2)(\Pi_{S^2} \times \{0\})$ is *not* integrable [30, 89]. In this case the symplectic S-groupoid $\mathcal{G}(P)$ is not a manifold.

We will have more to say about this example in Section 4.7.

Remark 4.13 (*Twisted presymplectic groupoids*)

Let $\mathcal{G}$ be a Lie groupoid over a manifold P. For each $k > 0$, let $\mathcal{G}_k$ be the manifold of composable sequences of k-arrows,

$$\mathcal{G}_k := \mathcal{G} \times_{(s,t)} \mathcal{G} \times_{(s,t)} \cdots \times_{(s,t)} \mathcal{G}, \qquad (k \text{ times})$$

and set $\mathcal{G}_0 = P$. The sequence of manifolds $\mathcal{G}_k$, together with the natural maps $\partial_i : \mathcal{G}_k \to \mathcal{G}_{k-1}$, $i = 0, \ldots, k$,

$$\partial_i(g_1, \ldots, g_k) = \begin{cases} (g_2, \ldots, g_k), & \text{if} \quad i = 0, \\ (g_1, \ldots, g_i g_{i+1}, \ldots, g_k), & \text{if} \quad 0 < i < k \\ (g_1, \ldots, g_{k-1}) & \text{if} \quad i = k. \end{cases}$$

defines a simplicial manifold $\mathcal{G}_\bullet$. The **bar-de Rham complex** of $\mathcal{G}$ is the total complex of the double complex $\Omega^\bullet(\mathcal{G}_\bullet)$, where the boundary maps are $d : \Omega^q(\mathcal{G}_k) \to \Omega^{q+1}(\mathcal{G}_k)$, the usual de Rham differential, and $\partial : \Omega^q(\mathcal{G}_k) \to \Omega^q(\mathcal{G}_{k+1})$, the alternating sum of the pull-back of the $k+1$ maps $\mathcal{G}_k \to \mathcal{G}_{k+1}$, as in group cohomology. For example, if $\omega \in \Omega^2(\mathcal{G})$, then

$$\partial\omega = p_1^*\omega - m^*\omega + p_2^*\omega.$$

(As before, m is the groupoid multiplication, and $p_i : \mathcal{G}_2 \to \mathcal{G}$, $i = 1, 2$, are the natural projections.) It follows that a 2-form ω is a 3-cocycle in the total complex if and only if it is multiplicative and closed; in particular, a symplectic groupoid can be defined as a Lie groupoid $\mathcal{G}$ together with a nondegenerate 2-form ω which is a 3-cocycle.

More generally, one can consider 3-cochains which are sums $\omega + \phi$, where $\omega \in \Omega^2(\mathcal{G})$ and $\phi \in \Omega^3(P)$. In this case, the coboundary condition is that $d\phi = 0$, ω is multiplicative, and

$$d\omega = s^*\phi - t^*\phi.$$

A groupoid $\mathcal{G}$ together with a 3-cocycle $\omega + \phi$ such that ω is nondegenerate is called a **ϕ-twisted symplectic groupoid** [80]. Just as symplectic groupoids are the global objects associated with Poisson manifolds, the twisted symplectic groupoids are the global objects associated with twisted Poisson manifolds [24].

Without non-degeneracy assumptions on ω, one has the following result concerning the infinitesimal version of 3-cocycles [15]: If $\mathcal{G}$ is source-simply connected and $\phi \in \Omega^3(P)$, $d\phi = 0$, then there is a one-to-one correspondence between 3-cocycles $\omega + \phi$ and bundle maps $\sigma : A \to T^*P$ satisfying the following two conditions:

$$\begin{aligned} \langle \sigma(\xi), \rho(\xi') \rangle &= -\langle \sigma(\xi'), \rho(\xi) \rangle; & (52) \\ \sigma([\xi, \xi']) &= \mathcal{L}_\xi(\sigma(\xi')) - \mathcal{L}_{\xi'}(\sigma(\xi)) \\ &\quad + d\langle \sigma(\xi), \rho(\xi') \rangle + i_{\rho(\xi)\wedge\rho(\xi')}(\phi), & (53) \end{aligned}$$

where A is the Lie algebroid of $\mathcal{G}$, $[\cdot,\cdot]$ is the bracket on $\Gamma(A)$, $\rho : A \to TP$ is the anchor, and $\xi, \xi' \in \Gamma(A)$. For one direction of this correspondence, given ω, the associated bundle map $\sigma_\omega : A \to T^*P$ is just $\sigma_\omega(\xi) = i_\xi \omega|_P$.

For a given $\sigma : A \to T^*P$ satisfying (52), (53), let us consider the bundle map

$$(\rho, \sigma) : A \to TP \oplus T^*P. \qquad (54)$$

A direct computation shows that if the rank of $L_\sigma := \mathrm{Image}(\rho, \sigma)$ equals $\dim(P)$, then $L_\sigma \subset TP \oplus T^*P$ is a ϕ-twisted Dirac structure on P. In this case, it is easy to check that (54) yields a (Lie algebroid) isomorphism $A \to L_\sigma$ if and only if

1) $\dim(\mathcal{G}) = 2\dim(P)$;
2) $\ker(\omega_x) \cap \ker(T_x s) \cap \ker(T_x t) = \{0\}$ for all $x \in P$.

A groupoid $\mathcal{G}$ over P satisfying 1) together with a 3-cocycle $\omega + \phi$ so that ω satisfies 2) is called a **ϕ-twisted presymplectic groupoid** [15, 97]. As indicated by the previous discussion, they are precisely the global objects integrating twisted Dirac structures. The 2-form ω is nondegenerate if and only if the associated Dirac structure is Poisson, recovering the known correspondence between (twisted) Poisson structures and (twisted) symplectic groupoids.

The following example describes presymplectic groupoids integrating Cartan-Dirac structures; it is analogous to Example 4.10.

Example 4.14 *(Cartan-Dirac structures and the AMM-groupoid)*
Let G be a Lie group with Lie algebra $\mathfrak{g}$, equipped with a nondegenerate bi-invariant quadratic form $(\cdot,\cdot)_\mathfrak{g}$. The **AMM groupoid** [8] is the action groupoid $\mathcal{G} = G \ltimes G$ with respect to the conjugation action,

together with the 2-form [3]

$$\omega_{(g,x)} = \frac{1}{2}\left((\mathrm{Ad}_x p_g^*\lambda, p_g^*\lambda)_{\mathfrak{g}} + (p_g^*\lambda, p_x^*(\lambda + \bar{\lambda}))_{\mathfrak{g}} \right),$$

where p_g and p_x denote the projections onto the first and second components of $G \times G$, and λ and $\overline{\lambda}$ are the left and right Maurer-Cartan forms. The AMM-groupoid is a ϕ^G-twisted presymplectic groupoid integrating L_G [15], the Cartan-Dirac structure on G defined in Example 2.13. If G is simply connected, then $(G \ltimes G, \omega)$ is isomorphic to $\mathcal{G}(L_G)$, the source-simply connected integration of L_G; in general, one must pull-back ω to $\widetilde{G} \ltimes G$, where $\widetilde{G}$ is the universal cover of G.

4.3 Morita equivalence for groups and groupoids

Since groupoids play such an important role in the Morita equivalence of Poisson manifolds, we will take some time to discuss Morita equivalence of groupoids in general. We begin with groups.

If we try to define Morita equivalence of groups as equivalence between their (complex linear) representation categories, then we are back to algebra, since representations of a group are the same as modules over its group algebra over $\mathbb{C}$. (This is straightforward for discrete groups, and more elaborate for topological groups.) Here, we just remark that nonisomorphic groups can have isomorphic group algebras (e.g. two finite abelian groups with the same number of elements), or more generally Morita equivalent group algebras (e.g. two finite groups with the same number of conjugacy classes, hence the same number of isomorphism classes of irreducible representations).

We obtain a more geometric notion of Morita equivalence for groups by considering actions on manifolds rather than on linear spaces. Thus, for Lie groups (including discrete groups) G and H, bimodules are (G,H)-"bispaces", i.e. manifolds where G acts on the left, H acts on the right, and the actions commute. The "tensor product" of such bimodules is defined by the orbit space

$${}_G\mathrm{X}_H * {}_H\mathrm{Y}_K := \mathrm{X} \times \mathrm{Y}/H,$$

where H acts on $\mathrm{X} \times \mathrm{Y}$ by $(x,y) \mapsto (xh, h^{-1}y)$. The result of this operation may no longer be smooth, even if X and Y are. Under suitable regularity assumptions, to be explained below, the tensor product is a smooth manifold, so we consider the category in which objects are groups and morphisms are isomorphism classes of "regular" bispaces, and we define **Morita equivalence** of groups as isomorphism in this category. Note that the identity morphisms are the groups themselves seen as bispaces with respect to left and right multiplication. Analogously to the case of algebras, we have an associated notion of **Picard group(oid)**.

Exercise

Show that a bispace ${}_G\mathrm{X}_H$ is invertible with respect to the tensor product operation if and only if the G and H-actions are free and transitive.

If ${}_G\mathrm{X}_H$ is invertible and we fix a point $x_0 \in \mathrm{X}$, by the result of the previous exercise, there exists for each $g \in G$ a unique $h \in H$ such that $gx_0h^{-1} = x_0$. The correspondence $g \mapsto h$ in fact establishes a group isomorphism $G \to H$. So, for groups, Morita equivalence induces the same equivalence relation as the usual notion of isomorphism. As we will see in Example 4.32 of Section 4.6, the situation for Picard groups resembles somewhat that for algebras, where outer automorphisms play a key role.

For a full discussion of Morita equivalence of Lie groupoids, we refer to the article of Moerdijk and Mrčun [64] in this volume. Here, we will briefly summarize the theory.

An action (from the left) of a Lie groupoid $\mathcal{G}$ over P on a manifold S consists of a map $J : S \to P$ and a map $\mathcal{G} \times_{(s,J)} S \to S$ (where s is the source map of $\mathcal{G}$) satisfying axioms analogous to those of a group action; J is sometimes called the **moment** of the action (see Example 4.16). The action is **principal** with respect to a map $p : S \to M$ if p is a surjective submersion and if $\mathcal{G}$ acts freely and transitively on each p-fibre; principal $\mathcal{G}$-bundles are also called **$\mathcal{G}$-torsors**.

Right actions and torsors are defined in the obvious analogous way. If groupoids $\mathcal{G}_1$ and $\mathcal{G}_2$ act on S from the left and right, respectively, and the actions commute, then we call S a **$(\mathcal{G}_1, \mathcal{G}_2)$-bibundle**. A bibundle is **left principal** when the left $\mathcal{G}_1$-action is principal with respect to the moment map for the right action of $\mathcal{G}_2$.

If S is a $(\mathcal{G}_1, \mathcal{G}_2)$-bibundle with moments $P_1 \overset{J_1}{\leftarrow} S \overset{J_2}{\rightarrow} P_2$, and if S' is a $(\mathcal{G}_2, \mathcal{G}_3)$-bibundle with moments $P_2 \overset{J_2'}{\leftarrow} S' \overset{J_3'}{\rightarrow} P_3$, then their "tensor product" is the orbit space

$$S * S' := (S \times_{(J_2, J_2')} S')/\mathcal{G}_2, \tag{55}$$

where $\mathcal{G}_2$ acts on $S \times_{(J_2, J_2')} S'$ diagonally. The assumption that S and S'

are left principal guarantees that $S * S'$ is a smooth manifold and that its natural $(\mathcal{G}_1, \mathcal{G}_3)$-bibundle structure is left principal.

Two $(\mathcal{G}_1, \mathcal{G}_2)$-bibundles are **isomorphic** if there is a diffeomorphism between them commuting with the groupoid actions and their moments. The "tensor product" (55) is associative up to natural isomorphism, so we may define a category LG in which the objects are Lie groupoids and morphisms are isomorphism classes of left principal bibundles. Just as in the case of algebras, we call two Lie groupoids **Morita equivalent** if they are isomorphic as objects in LG, and we define the associated notion of Picard group(oid) just as we do for algebras. We note that a $(\mathcal{G}_1, \mathcal{G}_2)$-bibundle S is "invertible" in LG if and only if it is **biprincipal**, i.e., principal with respect to both left and right actions; a biprincipal bibundle is also called a **Morita equivalence** or a **Morita bibundle**.

Example 4.15 (*Transitive Lie groupoids*)

Let $\mathcal{G}$ be a Lie groupoid over P. For a fixed $x \in P$, let $\mathcal{G}_x$ be the isotropy group of $\mathcal{G}$ at x, and let $E_x = s^{-1}(x)$. Then E_x is a $(\mathcal{G}, \mathcal{G}_x)$-bibundle. It is a Morita bibundle if and only if $\mathcal{G}$ is transitive, i.e., for any $x, y \in P$, there exists $g \in \mathcal{G}$ so that $s(g) = y$ and $t(g) = x$. In fact, a Lie groupoid is transitive if and only if it is Morita equivalent to a Lie group.

4.4 Modules over Poisson manifolds and symplectic groupoid actions

Example 4.1 shows that modules over $\mathfrak{g}^*$ are the same thing as hamiltonian G-manifolds, where G is the connected and simply connected Lie group with Lie algebra $\mathfrak{g}$. As we discuss in this section, this is a particular case of a much more general correspondence between modules over Poisson manifolds and symplectic groupoid actions.

Let $(\mathcal{G}, \omega)$ be a symplectic groupoid over P acting on a symplectic manifold (S, ω_S) with moment J. Let $a : \mathcal{G} \times_{(s,J)} S \to S$ denote the action. We call the action **symplectic** if it satisfies the property (analogous to the condition on multiplicative forms) that $\mathrm{graph}(a) \subset \mathcal{G}(P) \times S \times \overline{S}$ is lagrangian. Equivalently, a is symplectic if

$$a^* \omega_S = p_S^* \omega_S + p_{\mathcal{G}}^* \omega, \tag{56}$$

where $p_S : \mathcal{G} \times_{(s,J)} S \to S$ and $p_{\mathcal{G}} : \mathcal{G} \times_{(s,J)} S \to \mathcal{G}$ are the natural projections.

A key observation relating actions of symplectic groupoids to modules over Poisson manifolds is that if $J : S \to P$ is the moment map for a *symplectic* action of a symplectic groupoid $\mathcal{G}$ over P, then J is automatically a complete Poisson map [61], defining a module over P. On the other hand, a module $J : S \to P$ over an integrable Poisson manifold P automatically carries a *symplectic* action of the source-simply connected symplectic groupoid $\mathcal{G}(P)$. So there is a one-to-one correspondence between P-modules and symplectic actions of $\mathcal{G}(P)$ [26].

Example 4.16 (*Hamiltonian spaces*)

Let G be a simply-connected Lie group with Lie algebra $\mathfrak{g}$. Any complete symplectic realization $J : S \to \mathfrak{g}^*$ induces an action of the symplectic groupoid T^*G on S:

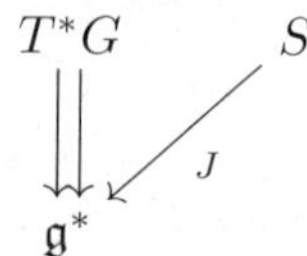

In this case, $T^*G = G \ltimes \mathfrak{g}^*$ is a transformation Lie groupoid, and, as such, its action is equivalent to an ordinary G-action on S for which J is G-equivariant. Moreover, the G-action corresponding to the symplectic T^*G-action induced by $J : S \to P$ is a hamiltonian G-action for which J is a momentum map. So we recover the result of Example 4.1 on the isomorphism (not only equivalence) between the categories of complete symplectic realizations of $\mathfrak{g}^*$ and hamiltonian G-manifolds. Notice that the momentum map for the group action is the moment map for the groupoid action; it is this example which motivates the term "moment" as applied to groupoid actions.

Remark 4.17 (*Infinitesimal actions*)

The relationship between complete symplectic realizations and symplectic groupoid actions has an infinitesimal counterpart. A symplectic realization (not necessarily complete) $J : S \to P$ induces a Lie algebra homomorphism

$$\Omega^1(P) \to \mathcal{X}(S), \quad \alpha \mapsto \widetilde{\Pi}_S(J^*\alpha), \tag{57}$$

where the bracket on 1-forms is the one of (50). This maps defines a Lie algebroid action of the Lie algebroid of P, T^*P, on S. The completeness of J allows this infinitesimal action to be integrated to an action of the source-simply-connected integration $\mathcal{G}(P)$ (see [62]), and this action turns out to be symplectic.

Remark 4.18 (*Symplectic groupoid actions on Poisson manifolds*)
As in Remark 4.2, the correspondence between P-modules and symplectic $\mathcal{G}(P)$-actions holds in more generality: a Poisson map $Q \to P$ from *any* Poisson manifold Q induces an infinitesimal T^*P-action on Q, by the same formula as in (57). When the Poisson map is complete (and P is integrable), it gives rise to an action of $\mathcal{G}(P)$ on Q, which preserves the symplectic leaves of Q; its restriction to each leaf is a symplectic action. The action is a Poisson action in the sense that its graph is lagrangian [90] in the appropriate product, see [14] for details.

Remark 4.19 (*Realizations of Dirac structures and presymplectic groupoid actions*)
The correspondence between modules over a Poisson manifold P and symplectic actions of $\mathcal{G}(P)$ extends to one between "modules" over Dirac manifolds and suitable actions of presymplectic groupoids [14, 15].

In order to introduce the notion of "realization" of a Dirac manifold, let us note that, if (P, Π) is a Poisson manifold, then the infinitesimal T^*P-action (57) induced by a Poisson map $J : Q \to P$ can be equivalently expressed in terms of L_Π by the Lie algebra homomorphism

$$\Gamma(L_\Pi) \to \mathcal{X}(Q), \quad (X, \alpha) \mapsto Y,$$

where Y is uniquely determined by the condition $(Y, J^*\alpha) \in L_{\Pi_Q}$. Since J is a Poisson map, it also follows that $X = TJ(Y)$.

If (P, L) and (Q, L_Q) are Dirac manifolds, and $J : Q \to P$ is a forward Dirac map, then (25) implies that for each $(X, \alpha) \in L$ over the point $J(y) \in P$, there exists $Y \in T_yQ$ such that $(Y, TJ^*(\alpha)) \in (L_Q)_y$ and $X = T_yJ(Y)$. However, unlike the situation of Poisson maps, Y is *not* uniquely determined by these conditions; this is the case if and only if

$$\ker(TJ) \cap \ker(L_Q) = \{0\}. \tag{58}$$

If (58) holds at all points of Q, then the induced map $\Gamma(L) \to \mathcal{X}(Q)$, $(X, \alpha) \mapsto Y$, defines an infinitesimal L-action on Q.

A **Dirac realization** [14] of a ϕ-twisted Dirac manifold (P, L) is a forward Dirac map $J : Q \to P$, where Q is a $J^*\phi$-twisted Dirac manifold and (58) is satisfied. If Q is a $J^*\phi$-twisted presymplectic manifold, then J is called a **presymplectic realization**. We call a Dirac realization **complete** if the induced infinitesimal action is complete (in the sense of Lie algebroid actions, see [62]). As in the case of Poisson maps, complete Dirac realizations $J : Q \to P$ are the same thing as global actions of the presymplectic groupoid $\mathcal{G}(L)$ on Q "compatible" with L_Q in a suitable way [14] (generalizing the conditions in (56) and Remark 4.18).

The next example illustrates the discussion in Remark 4.19 and the connection between Dirac geometry and group-valued momentum maps [3, 2].

Example 4.20 (*Modules over Cartan-Dirac structures and quasi-hamiltonian actions*)

As we saw in Example 4.1, symplectic realizations of (resp. Poisson maps into) the Lie-Poisson structure on $\mathfrak{g}^*$ are the same thing as hamiltonian $\mathfrak{g}$-actions on symplectic (resp. Poisson) manifolds; if the maps are complete, one gets a correspondence with global hamiltonian actions.

Analogously, let us consider a connected, simply-connected Lie group G equipped with L_G, the Cartan-Dirac structure associated with a nondegenerate bi-invariant quadratic form $(\cdot,\cdot)_{\mathfrak{g}}$. Then presymplectic realizations into G are exactly the same as quasi-hamiltonian $\mathfrak{g}$-manifolds, and complete realizations correspond to global quasi-hamiltonian G-actions (which can be seen as actions of the AMM-groupoid of Example 4.14, analogously to Example 4.16) [15]. More generally, (complete) Dirac realizations of (G, L_G) correspond to (global) hamiltonian quasi-Poison manifolds [14], in analogy with Remark 4.18.

In these examples, the realization maps are the group-valued momentum maps.

4.5 Morita equivalence of Poisson manifolds and symplectic groupoids

We now have all the ingredients which we need in order to define a geometric notion of Morita equivalence for Poisson manifolds which implies equivalence of their module categories.

A **Morita equivalence** between Poisson manifolds P_1 and P_2 is a (P_1, P_2)-bimodule $P_1 \stackrel{J_1}{\leftarrow} S \stackrel{J_2}{\rightarrow} P_2$ such that J_1 and J_2 are surjective submersions whose fibres are simply connected and symplectic orthogonals of each other. By Remark 4.6, Morita equivalence only applies to *integrable* Poisson structures. (The nonintegrable case can be handled with the use of symplectic S-groupoids. See Remark 4.40.) The bimodule $P_2 \stackrel{J_2}{\leftarrow} \overline{S} \stackrel{J_1}{\rightarrow} P_1$, where $\overline{S}$ has the opposite symplectic structure, is also a

Morita equivalence, and S and $\overline{S}$ satisfy

$$S *_{P_2} \overline{S} \cong \mathcal{G}(P_1), \quad \text{and} \quad \overline{S} *_{P_1} S \cong \mathcal{G}(P_2). \tag{59}$$

Since symplectic groupoids are "identity bimodules", (59) is analogous to the invertibility of algebraic bimodules (30).

Let us consider the category whose objects are complete symplectic realizations of an integrable Poisson manifold P, and morphisms are symplectic maps between symplectic realizations commuting with the realization maps. This category is analogous to the category of left modules over an algebra, and we call it the **category of modules over** P. If $P_1 \overset{J_1}{\leftarrow} S \overset{J_2}{\rightarrow} P_2$ is a Morita equivalence, then the regularity conditions on the maps J_1 and J_2 guarantee that if $S' \to P_2$ is a left P_2-module then the tensor product $S *_{P_2} S'$ is smooth and defines a left P_1-module [95]. So one can define a functor between categories of modules (i.e. complete symplectic realizations) just as one does for algebras, see (28) and (29), and prove that geometric Morita equivalence implies the equivalence of "representation" categories [53, 95]:

Theorem 4.21 *If P_1 and P_2 are Morita equivalent, then they have equivalent categories of complete symplectic realizations.*

Remark 4.22 (*The "category" of complete symplectic realizations*)
In the spirit of the symplectic "category" of [87], one can also define a larger "category" of complete symplectic realizations of P by considering the morphisms between two P-modules $J : S \to P$ and $J' : S' \to P$ to be lagrangian submanifolds in $S' \times_{(J',J)} \overline{S}$, see [94, 95], with composition given by composition of relations; the quotes in "category" are due to the fact that the composition of two such morphisms yields another morphism only under suitable transversality assumptions. Theorem 4.21 still holds in this more general setting [94]. Unlike in the case of algebras, though, the converse of Theorem 4.21 does not hold in general [95], see Remark 4.37. We will discuss ways to remedy this problem in Chapter 5 by introducing yet another category of representations of P (a "symplectic category").

Remark 4.23 (*Classical intertwiner spaces*)
As a consequence of (59), one can see that Morita equivalence, in addition to establishing an equivalence of module categories, preserves the classical intertwiner spaces.

Remark 4.24 (*More general modules*)
As indicated in Remarks 4.2 and 4.18, from the point of view of hamil-

tonian actions, it is natural to consider arbitrary complete Poisson maps (not necessarily symplectic realizations) as modules over Poisson manifolds. The "action" of (P_1, P_2)-bimodules on P_2-modules in Remark 4.4 naturally extends to an action on Poisson maps $Q \to P_2$; in fact, one can think of this more general tensor product as a leafwise version of the one in Section 4.1, and Theorem 4.21 still holds for these more general "representations". (This generalization is the analogue, in algebra, of considering homomorphisms of an algebra into direct sums of endomorphism algebras, rather than usual modules.)

The notion of Morita equivalence of Poisson manifolds is closely related to Morita equivalence of symplectic groupoids, which is a refinement of the notion of Morita equivalence for Lie groupoids, taking symplectic structures into account. If $\mathcal{G}_1$ and $\mathcal{G}_2$ are symplectic groupoids, then a $(\mathcal{G}_1, \mathcal{G}_2)$-bibundle is called **symplectic** if both actions are symplectic. The "tensor product" of two symplectic bibundles, as defined in Section 4.3, is canonically symplectic, so we may define a category SG in which the objects are symplectic groupoids and morphisms are isomorphism classes of left principal symplectic bibundles. (An isomorphism between symplectic bibundles is required to preserve the symplectic forms.) We call two symplectic groupoids $\mathcal{G}_1$ and $\mathcal{G}_2$ **Morita equivalent** [95] if they are isomorphic in SG, i.e. if there exists a biprincipal symplectic $(\mathcal{G}_1, \mathcal{G}_2)$-bibundle (see [53]). A **Morita equivalence** between symplectic groupoids is a symplectic bibundle which is biprincipal.

If P_1 and P_2 are Poisson manifolds, and if $P_1 \overset{J_1}{\leftarrow} S \overset{J_2}{\rightarrow} P_2$ is a (P_1, P_2)-bimodule, then we obtain a left symplectic action of the groupoid $\mathcal{G}(P_1)$ and right symplectic action of $\mathcal{G}(P_2)$,

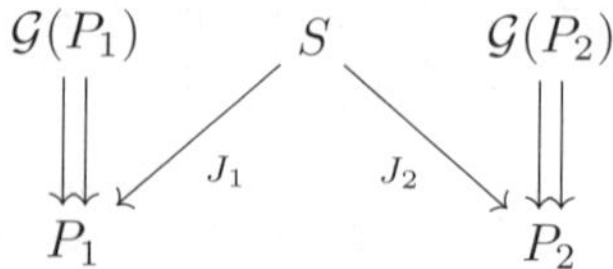

The property that $\{J_1^* C^\infty(P_1), J_2^* C^\infty(P_2)\} = 0$ implies that the actions of $\mathcal{G}(P_1)$ and $\mathcal{G}(P_2)$ commute, so that S is a symplectic $(\mathcal{G}(P_1), \mathcal{G}(P_2))$-bibundle. We say that a symplectic bimodule $P_1 \overset{J_1}{\leftarrow} S \overset{J_2}{\rightarrow} P_2$ is **regular** if the associated symplectic $(\mathcal{G}(P_1), \mathcal{G}(P_2))$-bibundle is left principal. The tensor product of symplectic bimodules defined in Section 4.1 coincides with their tensor product as symplectic bibundles. As a result, the tensor product of regular symplectic bimodules is smooth and regular.

Remark 4.25 (*Regular bimodules*)

Regular bimodules can be described with no reference to the symplectic groupoid actions: $P_1 \overset{J_1}{\leftarrow} S \overset{J_2}{\rightarrow} P_2$ is regular if and only if J_1 and J_2 are complete Poisson maps, J_1 is a submersion, J_2 is a surjective submersion with simply-connected fibres, and the J_1- and J_2-fibres are symplectic orthogonal of one another.

Exercise

Prove the equivalent formulation of regular bimodules in Remark 4.25. (Hint: this is a slight extension of [94, Thm. 3.2])

We define the category Poiss in which the objects are integrable Poisson manifolds and morphisms are isomorphism classes of regular symplectic bimodules.

If $P_1 \overset{J_1}{\leftarrow} S \overset{J_2}{\rightarrow} P_2$ is a Morita equivalence of Poisson manifolds, then the regularity assumptions on the maps J_1 and J_2 insure that S is biprincipal for the induced actions of $\mathcal{G}(P_1)$ and $\mathcal{G}(P_2)$, so that S is also a Morita equivalence for the symplectic groupoids $\mathcal{G}(P_1)$ and $\mathcal{G}(P_2)$. On the other hand, if $\mathcal{G}_1$ and $\mathcal{G}_2$ are source-simply-connected symplectic groupoids over P_1 and P_2, respectively, then a $(\mathcal{G}_1, \mathcal{G}_2)$-Morita equivalence is a (P_1, P_2)-Morita equivalence. So two integrable Poisson manifolds P_1 and P_2 are Morita equivalent if and only if their source-simply-connected integrations, $\mathcal{G}(P_1)$ and $\mathcal{G}(P_2)$, are Morita equivalent as symplectic groupoids.

Remark 4.26 (*Lie functor*)

It follows from the discussion above that there exists a natural equivalence between the category of *source-simply-connected* symplectic groupoids with morphisms being Morita equivalences (resp. left principal symplectic bibundles), and the category of integrable Poisson manifolds with morphisms being Morita equivalences (resp. regular bimodules). These equivalences are similar to the one between the categories of Lie algebras and simply-connected Lie groups, with their usual morphisms.

Example 4.27 (*Symplectic manifolds*)

Let P be a connected symplectic manifold. The universal cover of P with base point x, denoted $\widetilde{P}$, is a Morita equivalence between the symplectic groupoid $\mathcal{G}(P)$, which in this case is the fundamental groupoid of P, and $\pi_1(P,x)$:

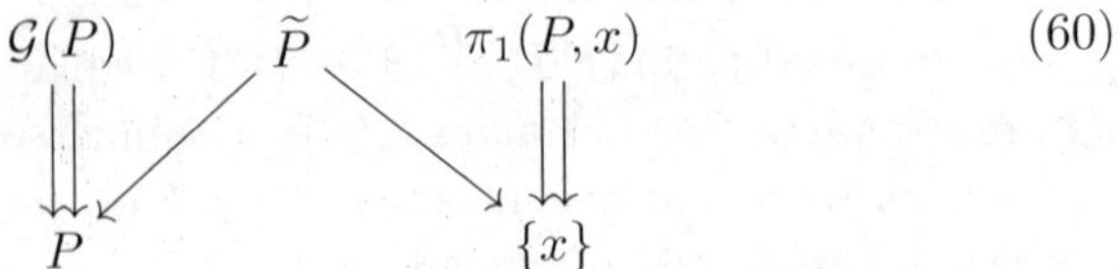
(60)

Note that $\pi_1(P,x)$ is a symplectic groupoid for the zero-dimensional Poisson manifold $\{x\}$, though generally not the source-simply-connected one.

In analogy with Example 4.16 on hamiltonian actions, there is an equivalence of categories between complete symplectic realizations of P and symplectic actions of $\pi_1(P,x)$. This suggests the slogan that "a (connected) symplectic manifold P with fundamental group $\pi_1(P)$ is the dual of the Lie algebra of $\pi_1(P)$".

It follows from the Morita equivalence (60) and the discussion about Morita equivalence of groups in Section 3.1 that connected symplectic manifolds P_1 and P_2 are Morita equivalent if and only if $\pi_1(P_1) \cong \pi_1(P_2)$.

Example 4.28 (*Symplectic fibrations*)

It follows from the previous example that every simply-connected symplectic manifolds is Morita equivalent to a point. Similarly, if (Q, Π) is a Poisson manifold with $\Pi = 0$, then Q is Morita equivalent to any product $Q \times S$ where S is a simply-connected symplectic manifold. In fact, $Q \times S \overset{\mathrm{pr}_1}{\leftarrow} T^*Q \times S \overset{\mathrm{pr}_2}{\rightarrow} Q$ is a Morita bimodule, where pr_1 and pr_2 are the natural projections.

More generally, let us assume that P is a Poisson manifold whose symplectic foliation is a fibration $P \to Q$ with simply-connected fibres. In general, there are obstructions to P being Morita equivalent to Q [94]: P is Morita equivalent to Q if and only if there exists a closed 2-form on P which restricts to the symplectic form on each fibre. We will have more to say about "fibrating" Poisson manifolds and their Morita invariants in Section 4.7.

Example 4.29 (*Lie-Poisson structures*)

Let us consider $\mathfrak{g}_1^*$ and $\mathfrak{g}_2^*$, the duals of the Lie algebras $\mathfrak{g}_1$ and $\mathfrak{g}_2$, equipped with their Lie-Poisson structures. Then $\mathfrak{g}_1^*$ and $\mathfrak{g}_2^*$ are Morita equivalent if and and only they are isomorphic. Indeed, suppose that $\mathfrak{g}_1^* \overset{J_1}{\leftarrow} S \overset{J_2}{\rightarrow} \mathfrak{g}_2^*$ is a Morita bimodule, and let $\mathrm{X} = J_2^{-1}(0)$. A dimension count shows that there exists $\mu \in \mathfrak{g}_1^*$ such that $\mathrm{X} = J_1^{-1}(\mu)$. Since S is a biprincipal bibundle for the symplectic groupoids $\mathcal{G}(\mathfrak{g}_i^*) = T^*G_i$,

$i = 1, 2$, it follows that X is a (G_1, G_2)-Morita bibundle. Therefore G_1 and G_2 are isomorphic, and so are $\mathfrak{g}_1^*$ and $\mathfrak{g}_2^*$.

This example also follows from the Morita invariants discussed in Section 4.7.

Example 4.30 (*Topologically stable structures on surfaces*)

Let Σ be a compact, connected, oriented surface equipped with a Poisson structure Π which has at most linear degeneracies and whose zero set consists of n smooth, disjoint, closed curves, for $n \geq 0$. These are called **topologically stable structures** (TSS) [71].

Any two modular vector fields for Π [92] coincide at points where Π vanishes, so the curves in the zero set carry a natural orientation. We denote the zero set of Π, regarded as an oriented 1-manifold, by $Z(\Sigma, \Pi)$. Two TSS (Σ, Π) and (Σ, Π') are **topologically equivalent** if there is an orientation-preserving diffeomorphism $\psi : \Sigma \to \Sigma'$ such that $\psi(Z(\Sigma, \Pi)) = Z(\Sigma', \Pi')$. We denote the equivalence class of $Z(\Sigma, \Pi)$ by $[Z(\Sigma, \Pi)]$. This class can be represented by an oriented labeled graph $\mathfrak{G}(\Sigma, \Pi)$: each vertex corresponds to a 2-dimensional leaf of the structure, two vertices being connected by an edge for each boundary zero curve they share; each edge is oriented to point toward the vertex for which Π is positive with respect to the orientation of Σ. We then label each vertex by the genus of the corresponding leaf.

It turns out that the topology of the zero set plus the modular periods (periods of a modular vector field around the zero curves) completely determine the Morita equivalence class of TSS [16, 19]. In fact, let us define a more elaborate graph $\mathfrak{G}_T(\Sigma, \Pi)$, obtained from $\mathfrak{G}(\Sigma, \Pi)$ by labeling each of its edges by the modular period around the corresponding zero curve. Then two TSS (Σ, Π) and (Σ', Π') are Morita equivalent if and only if there is an isomorphism of labeled graphs $\mathfrak{G}_T(\Sigma, \Pi) \cong \mathfrak{G}(\Sigma', \Pi')$. (It follows from the results in [29] that TSS are always integrable.)

The classification of TSS up to Morita equivalence was preceded by (and builds on) their classification up to orientation-preserving Poisson diffeomorphisms by Radko [71], who shows that the topological class of the zero set and the modular periods, together with a certain volume invariant (generalizing the Liouville volume when the TSS is symplectic), form a complete set of invariants.

Remark 4.31 (*Morita equivalence of presymplectic groupoids and "momentum map theories"*)

As we noted in Examples 4.16 and 4.20, hamiltonian spaces can be seen as modules over Lie-Poisson structures on duals of Lie algebras, whereas quasi-hamiltonian (or hamiltonian quasi-Poisson) manifolds are modules over Cartan-Dirac structures on Lie groups. Thus, the category of modules over an arbitrary (integrable) Poisson or Dirac manifold can be regarded as the category of "hamiltonian spaces" for some generalized "momentum map theory". Since Morita equivalence establishes an equivalence of categories of modules, it provides a precise notion of equivalence for "momentum map theories" and automatically implies the existence of other invariants (such as classical intertwiner spaces–see Remark 4.23).

An extended notion of Morita equivalence for ϕ-twisted presymplectic groupoids (or, infinitesimally, ϕ-twisted Dirac structures) was developed by Xu in [97]. In Xu's work, it is shown that various known correspondences of "momentum map theories" can be described by appropriate Morita equivalences. Examples include the equivalence between ordinary momentum maps and momentum maps for actions of Poisson-Lie groups (taking values in the dual group) [1, 42] and the one between quasi-hamiltonian spaces for groups and ordinary hamiltonian spaces for their loop groups [3]. An interesting feature of Morita equivalence for presymplectic groupoids is that the bimodules are not simply a pair of modules structures which commute.

Besides relating "momentum map theories", Morita equivalence of groupoids plays a central role in certain approaches to geometric quantization of these generalized hamiltonian spaces, where the usual line bundles are replaced by gerbes [8, 54].

4.6 Picard groups

Just as for algebras, there are Picard groupoids associated with the categories Poiss and SG. In particular, the isomorphism classes of Morita self-equivalences of a Poisson manifold P (resp. symplectic groupoid $\mathcal{G}$) form a group $\mathrm{Pic}(P)$ (resp. $\mathrm{Pic}(\mathcal{G})$), called the **Picard group**. It follows from the discussion in the previous section that $\mathrm{Pic}(P) = \mathrm{Pic}(\mathcal{G}(P))$.

We now discuss some examples of "geometric" Picard groups; see [19] for details.

Example 4.32 (*Picard groups of groups*)

As we saw in Section 4.3, geometric Morita equivalences between groups are closely related to group isomorphisms. A closer analysis shows that the Picard group of a group G is naturally isomorphic to its group $\mathrm{OutAut}(G) := \mathrm{Aut}(G)/\mathrm{InnAut}(G)$ of outer automorphisms.

It follows from Example 4.15 and the invariance of Picard groups under Morita equivalence that, if $\mathcal{G}$ is a transitive groupoid over P, then $\mathrm{Pic}(\mathcal{G}) \cong \mathrm{OutAut}(\mathcal{G}_x)$, where $\mathcal{G}_x$ is the isotropy group at a point $x \in P$. This isomorphism is natural, so the outer automorphism groups attached to different points are all naturally isomorphic to one another.

Example 4.33 (*Picard groups of symplectic manifolds*)

Since, according to Example 4.27, the fundamental groupoid of a connected symplectic manifold P is Morita equivalent to any of its fundamental groups $\pi_1(P, x)$, it follows from Example 4.32 that, for such a manifold, $\mathrm{Pic}(P)$ is naturally isomorphic to $\mathrm{OutAut}(\pi_1(P, x))$ for any x in P.

The Picard group of a Poisson manifold or symplectic manifold is also related to a group of outer automorphisms of the manifold itself. For a Poisson manifold P, let $\mathrm{Aut}(P)$ denote its group of Poisson diffeomorphisms. There is a natural map

$$j : \mathrm{Aut}(P) \to \mathrm{Pic}(P), \tag{61}$$

analogous to (34), which assigns to each $\psi \in \mathrm{Aut}(P)$ the isomorphism class of the bimodule $P \xleftarrow{t} \mathcal{G}(P) \xrightarrow{\psi^{-1}\circ s} P$. Any lagrangian bisection of $\mathcal{G}(P)$ (which is the analogue of a group element) naturally induces a Poisson diffeomorphism of P that we call an **inner automorphism**. It turns out that $\ker(j) = \mathrm{InnAut}(P)$, the group of inner automorphisms of P, just as in the algebraic setting discussed in Section 3.1.

The situation for symplectic groupoids is completely analogous [19].

Exercise

Let P be the $2n$ dimensional torus $\mathbb{R}^{2n}/(2\pi\mathbb{Z})^{2n}$ with a symplectic structure of the form $\frac{1}{2}\omega_{ij} d\theta^i \wedge d\theta^j$, where ω is a nondegenerate antisymmetric matrix of real constants. Show that the Picard group of P is independent of the choice of ω, while the subgroup of $\mathrm{Pic}(P)$ arising from outer automorphisms of (P, ω) does depend on ω.

Exercise

Compare $\mathrm{OutAut}(P)$ with $\mathrm{Pic}(P)$ when P is the disjoint union of several 2-dimensional spheres, possibly with different symplectic areas. Hint: use the theorems of Smale [81] and Moser [66] to show that every symplectomorphism of S^2 is inner.

There are geometric versions of the maps (33) and (35). Let $P_1 \xleftarrow{J_1} S \xrightarrow{J_2} P_2$ be a Morita equivalence. If $\mathcal{O} \subseteq P_2$ is a symplectic leaf, then $J_1(J_2^{-1}(\mathcal{O}))$ is a symplectic leaf of P_1, and this is a bijective correspondence between symplectic leaves. So, for a Poisson manifold P, we have a map

$$\mathrm{Pic}(P) \to \mathrm{Aut}(\mathrm{Leaf}(P)), \tag{62}$$

where $\mathrm{Leaf}(P)$ is the leaf space of P, analogous to the map (35). We define the **static Picard group** $\mathrm{SPic}(P)$ of P as the kernel of (62), i.e., the self-Morita equivalences inducing the identity map on the leaf space. Note that functions on the leaf space constitute the center of the Poisson algebra of functions on P, hence the terminology analogous to that for algebras.

Example 4.34 (*Zero Poisson structures*)

As we saw in Example 4.9, in this case $\mathcal{G}(P) = T^*P$, and $\mathrm{Pic}(P) = \mathrm{Pic}(T^*P)$. Since $\mathrm{Leaf}(P) = P$, (62) implies that each self-Morita bimodule S induces a diffeomorphism ψ of P. So composing S with ψ^{-1} defines an element of the static Picard group $\mathrm{SPic}(P)$. A direct computation shows that the map (62) is split by the map $\mathrm{Aut}(P) \to \mathrm{Pic}(P)$ (61), hence

$$\mathrm{Pic}(P) = \mathrm{Diff}(P) \ltimes \mathrm{SPic}(P),$$

in complete analogy with (36). Bimodules in $\mathrm{SPic}(P)$ are of the form

$$\begin{array}{c} S \\ p \Big\downarrow\Big\downarrow p \\ P \end{array} \tag{63}$$

so each fibre $p^{-1}(x)$ is lagrangian and simply-connected; moreover, the fact that p is a *complete* Poisson map implies that the p-fibres are complete with respect to their natural affine structure.

Since $P \leftarrow S \to P$ is a Morita bimodule, the p-fibres are isomorphic to the fibres of the symplectic groupoid target map $T^*P \to P$, so they are contractible. As a result, there exists a cross section $P \to S$, which implies that there is a diffeomorphism $S \cong T^*P$ preserving the fibres [19, Sec. 3]. Hence, in order to characterize a bimodule (63), the only remaining freedom is on the choice of symplectic structure on T^*P. It turns out that the most general possible symplectic structure on T^*P with respect to which the fibres of $T^*P \to P$ are lagrangian and complete is of the form:

$$\omega + p^*B,$$

where ω is the canonical symplectic form on T^*P and B is a closed 2-form on P (a "magnetic" term). One can show that two such bimodules are isomorphic if and only if B is exact. Hence

$$\mathrm{SPic}(P) \cong H^2(P, \mathbb{R}), \tag{64}$$

and

$$\mathrm{Pic}(P) \cong \mathrm{Diff}(P) \ltimes H^2(P, \mathbb{R}), \tag{65}$$

where the semi-direct product is with respect the natural action of $\mathrm{Diff}(P)$ on $H^2(P, \mathbb{R})$ by pull back. The reader can find the details in [19, Sec. 6.2].

Remark 4.35 (*An intriguing resemblance*)
Recall from Example 3.5 that the Picard group of the algebra $C^\infty(P)$ (which can be seen as a trivial quantization of (P, Π), if $\Pi = 0$) is $\mathrm{Diff}(P) \ltimes H^2(P, \mathbb{Z})$. Is there a theorem relating classical and quantum Picard groups which would explain the similarity between this fact and (65)?

4.7 Fibrating Poisson manifolds and Morita invariants

In this section, we will discuss "rigidity" aspects of geometric Morita equivalence. As we saw in Theorem 4.21, Morita equivalence preserves categories of "geometric representations". We point out a few other invariants, some of which have already appeared in previous sections.

1. As shown in Example 4.27, the Morita equivalence class of a symplectic manifold is completely determined by the isomorphism class of its fundamental group;
2. As remarked in Section 4.6, Morita equivalence induces a one-to-one correspondence of symplectic leaves, which is a diffeomorphism whenever the leaf spaces are smooth; moreover, corresponding symplectic leaves are themselves Morita equivalent [16, 30] and have isomorphic transverse Poisson structures [88];
3. Morita equivalence preserves first Poisson cohomology groups [28, 41], and modular classes [28, 92];

4. The monodromy groups and isotropy Lie algebras are Morita invariant [30].

As remarked in [30], all the invariants listed above turn out to be preserved by a notion of equivalence which is much weaker than Morita equivalence, called weak Morita equivalence, which does not require the integrability of Poisson manifolds. We do not know any Morita invariant which is not a weak Morita invariant.

By 1. above, the only Morita invariant of a connected symplectic manifold is its fundamental group. For a disjoint union of symplectic components, it is the unordered list of fundamental groups which counts; in particular, if all the components are simply connected, the number of components is a complete invariant. In this section, we will see that the Morita invariant structure is much richer for a Poisson manifold which is a smooth family of (diffeomorphic) symplectic manifolds.

We will say that a Poisson manifold P is **fibrating** if its symplectic leaves are the fibres of a smooth locally trivial fibration from P to $\mathrm{Leaf}(P)$. Here, locally triviality is meant in the differentiable rather than symplectic sense; in fact, it is the variation in symplectic structure from fibre to fibre which will concern us.

When P is fibrating, the fibrewise homology groups $H_2(\mathrm{Fib}, \mathbb{Z})$ form a locally trivial bundle of abelian groups over $\mathrm{Leaf}(P)$. Pairing with the fibrewise symplectic structure gives a map $H_2(\mathrm{Fib}, \mathbb{Z}) \to \mathbb{R}$, which encodes the variation of the symplectic cohomology class from fibre to fibre. The derivative of this map with respect to the base point in $\mathrm{Leaf}(P)$ gives rise to a map $\nu : H_2(\mathrm{Fib}, \mathbb{Z}) \to T^*\mathrm{Leaf}(P)$.

The map ν vanishes on torsion elements of $H_2(\mathrm{Fib}, \mathbb{Z})$, so its image is a family of embedded abelian groups in the fibres of $T^*\mathrm{Leaf}(P)$, called the **variation lattice** of P. Dazord [33] proves that, if P is integrable and has simply connected fibres, the variation lattice must be topologically closed with constant rank, having local bases of closed 1-forms. Failure of the variation lattice to have these properties provides an obstruction to integrability which was extended to general Poisson manifolds in [30].

A nice application of the variation lattice is to the study of the Picard groups of the duals of Lie algebras of compact groups [19], in which the lattice imparts a flat affine structure to the regular part of the symplectic leaf space.

Example 4.36 (*Nonintegrable Poisson structures revisited*)

Let us again consider $P = \mathbb{R} \times S^2$ from Example 4.12, with Poisson structure $(1/f(t))\Pi_{S^2} \times 0$, $f(t) > 0$. The area of the symplectic leaf over

$t \in \mathbb{R}$ is $4\pi f(t)$. The variation lattice is spanned by $4\pi f'(t)dt$, so it has constant rank if and only if $f'(t) \equiv 0$ or $f'(t)$ is not zero for all t.

If P_1 and P_2 are Morita equivalent fibrating Poisson manifolds with simply connected leaves, then the induced diffeomorphism $\mathrm{Leaf}(P_1) \to \mathrm{Leaf}(P_2)$ preserves the variation lattice; this can be seen as a special case of 4. above. So, although Morita equivalence does not determine the fibrewise symplectic structures, it is sensitive to how symplectic structures vary from fibre to fibre. This sensitivity leads to the following example [94], of Poisson manifolds which are representation equivalent but not Morita equivalent (see Remark 4.22).

Example 4.37 (*Representation equivalence vs. Morita equivalence*)
Consider $(0,1) \times S^2$ with Poisson structures Π_1 and Π_2 determined by the fibrewise symplectic structures $(1/t)\Pi_{S^2}$ and $(1/2t)\Pi_{S^2}$, respectively. Their variation lattices are spanned by $4\pi dt$ and $8\pi dt$, respectively. Since there is no diffeomorphism of $(0,1)$ taking dt to $2dt$, these structures cannot be Morita equivalent. Note however that these structures are representation equivalent: representations of Π_1 and Π_2 can be interchanged by dividing or multiplying the symplectic form on the realizations by 2.

Remark 4.38 (*A complete invariant?*)
Xu [94] shows that the leaf space with its variation lattice completely determines the Morita equivalence class of a fibrating Poisson manifold for which the symplectic leaves are simply connected and form a differentiably *globally* trivial fibration. It does not seem to be known whether this result persists without the global triviality assumption. The attempt to attack this problem by "gluing" together applications of the known case to local trivializations seems to lead to the problem of computing the static Picard group of a fibrating Poisson manifold.

To extend the discussion above to the case where the leaves are not simply connected, it seems that the variation lattice should be replaced by its "spherical" part, obtained by replacing $H_2(\mathrm{Fib}, \mathbb{Z})$ by the subgroup consisting of the spherical classes, i.e. the image of the Hurewicz homomorphism from the bundle $\pi_2(\mathrm{Fib})$ of homotopy groups. This spherical variation lattice is very closely related to the monodromy groups in [30]. Details in this case should be interesting to work out, particularly when the symplectic leaf fibration is not globally trivial.

Remark 4.39 (*Noncompact fibres*)
If the leaves of a fibrating Poisson manifold are compact, Moser's

theorem [66] implies that the variation lattice actually measures how the isomorphism class of the symplectic structure varies from leaf to leaf. If the leaves are noncompact, e.g. if they are discs in $\mathbb{R}^2$, then their area can vary without this being detected by any Morita invariant. Is there another notion of Morita invariance which would detect the variation from fibre to fibre of symplectic volume or other invariants, such as capacities?

Remark 4.40 (*Morita equivalence for nonintegrable Poisson manifolds*).

For a fibrating Poisson manifold which is nonintegrable, the variation lattice still exists, so one might hope that it is still a Morita invariant when the leaves are simply connected. But there is no Morita equivalence between such a manifold and itself, much less another one. To remedy this problem, we should extend the notion of Morita equivalence to admit as bimodules smooth stacks which are not manifolds, as we did for self-equivalences in Section 4.2. If we do this, then the variation lattice is indeed Morita invariant. In particular, this shows that integrability is an invariant property under this broadened notion of Morita invariance. Moreover, it turns out that any "S"-Morita equivalence between integrable Poisson manifolds is given by a manifold, so that the integrable part of the Picard groupoid remains unchanged. It would be interesting to see how S-Morita equivalence is related to weak Morita equivalence.

4.8 Gauge equivalence of Poisson structures

Let P be a manifold, and let $\phi \in \Omega^3(P)$ be closed. There is a natural way in which *closed* 2-forms on P act on ϕ-twisted Dirac structures: if $B \in \Omega^2(P)$ is closed and L is a ϕ-twisted Dirac structure on P, then we set

$$\tau_B(L) := \{(X, \alpha + \widetilde{B}(X)) \,|\, (X, \alpha) \in L\},$$

which is again a ϕ-twisted Dirac structure. We call this operation on Dirac structures a **gauge transformation** associated with a 2-form [80]. (More generally, for an arbitrary B, $\tau_B(L)$ is a $(\phi - dB)$-twisted Dirac structure.) Geometrically, a gauge transformation changes a Dirac structure L by adding the pull-back of a closed 2-form to its leafwise presymplectic form.

Remark 4.41 (*Gauge transformations and B-fields*)
In a completely similar way, complex closed 2-forms act on complex Dirac structures. If $B \in \Omega^2(P)$ is a *real* 2-form, and L is a generalized complex structure on P (see Remark 2.11), then one can show that $\tau_B(L)$ is again a generalized complex structure, and this operation is called a **B-field transform** [43, 46].

If Π is a Poisson structure on P, then changing it by a gauge transformation will generally result in a Dirac structure which is no longer Poisson. In fact, if $B \in \Omega^2(P)$ is closed, then $\tau_B(L_\Pi)$ is a Poisson structure if and only if the bundle map

$$\mathrm{Id} + \widetilde{B}\widetilde{\Pi} : T^*P \to T^*P \tag{66}$$

is invertible. In this case, the resulting Poisson structure is the one associated with the bundle map

$$\widetilde{\Pi}(\mathrm{Id} + \widetilde{B}\widetilde{\Pi})^{-1} : T^*P \to TP,$$

and we denote it by $\tau_B(\Pi)$.

Let (P, Π) be a fibrating Poisson manifold, as in Section 4.7. Since a gauge transformation adds the pull-back of a closed 2-form on P to the symplectic form on each fibre, the cohomology classes of fibrewise symplectic forms may change in this operation; however, the way they vary from fibre to fibre does not. This suggests that gauge transformations preserve the Morita equivalence class of (P, Π). In fact, this holds in complete generality [16]:

Theorem 4.42 *Gauge equivalence of integrable Poisson structures implies Morita equivalence.*

Since gauge transformations do not change the foliation of a Poisson structure, there is no hope that the converse of Theorem 4.42 holds, since even Poisson diffeomorphic structures may have different foliations.

We call two Poisson manifolds (P_1, Π_1) and (P_2, Π_2) **gauge equivalent up to Poisson diffeomorphism** if there exists a Poisson diffeomorphism $\psi : (P_1, \Pi_1) \to (P_2, \tau_B(\Pi_2))$ for some closed 2-form $B \in \Omega^2(P_2)$. It clearly follows from Theorem 4.42 that if two integrable Poisson manifolds are gauge equivalent up to a Poisson diffeomorphism, then they are Morita equivalent. The following properties are clear:

1. Two symplectic manifolds are gauge equivalent up to Poisson diffeomorphism if and only if they are symplectomorphic;

2. If two Poisson manifolds are gauge equivalent up to Poisson diffeomorphism, then they have isomorphic foliations (though generally different leafwise symplectic structures);
3. The Lie algebroids associated with gauge equivalent Dirac structures are isomorphic [80]; as a result, two Poisson manifolds which are gauge equivalent up to Poisson diffeomorphism have isomorphic Poisson cohomology groups.

A direct comparison between the properties above and the Morita invariants listed in Section 4.7 suggests that Morita equivalence should still be a weaker notion of equivalence. Indeed, two nonisomorphic symplectic manifolds with the same fundamental group are Morita equivalent, but not gauge equivalent up to Poisson diffeomorphism. In [16, Ex. 5.2], one can also find examples of Morita equivalent Poisson structures on the *same* manifold which are not gauge equivalent up to Poisson diffeomorphism by finding nonequivalent symplectic fibrations with diffeomorphic total space and base (and using Example 4.28). Nevertheless, there are interesting classes of Poisson structures for which both notions of equivalence coincide, such as the topologically stable structures of Example 4.30 [16, 19].

Remark 4.43 (*Gauge transformations and Morita equivalence of quantum algebras*)

As mentioned in Remark 3.14, gauge transformations associated with *integral* 2-forms define an action of $H^2(P,\mathbb{Z})$ on formal Poisson structures on P which can be "quantized" (via Kontsevich's quantization [49]) to Morita equivalent deformation quantization algebras.

On the other hand, gauge transformations of translation-invariant Poisson structures on tori are particular cases of the linear fractional transformations (45), which quantize, according to Theorem 3.15, to strongly Morita equivalent quantum tori. As we already mentioned in Remark 4.35, it would be very interesting to have a unified picture relating Morita equivalence of quantum algebras to geometric Morita equivalence.

5

Geometric representation equivalence

In Chapter 4, we considered the category of P-modules (i.e. complete symplectic realizations) over a Poisson manifold P, the geometric analogue of the category of left modules over an algebra. We observed in Remark 4.37 that, unlike the category of representations of an algebra, this category does not determine the Morita equivalence class of P. In this chapter, we will discuss refinements of the notion of category of representations of a Poisson manifold in order to remedy this defect.

The contents of Sections 5.2, 5.3 are preliminary ideas, and a fuller treatment is in progress.

5.1 Symplectic torsors

The first refinement we discuss is motivated by the theory of differentiable stacks [7, 60, 70].

Given a Lie groupoid $\mathcal{G}$, let $B\mathcal{G}$ denote the category of (left) $\mathcal{G}$-torsors. If two Lie groupoids $\mathcal{G}_1$ and $\mathcal{G}_2$ are Morita equivalent, then the natural functor $B\mathcal{G}_1 \to B\mathcal{G}_2$ induced by any Morita bibundle establishes an equivalence of these categories.

However, to recover the Morita equivalence class of $\mathcal{G}$ from $B\mathcal{G}$, one needs to consider another piece of information: the natural "projection" functor $B\mathcal{G} \to \mathcal{C}$, where $\mathcal{C}$ denotes the category of smooth manifolds, which assigns to a $\mathcal{G}$-torsor $S \to M$ the manifold $M = S/\mathcal{G}$. The category $B\mathcal{G}$ *together* with this projection functor is an example of a differentiable stack. Taking this extra structure into account, one defines $B\mathcal{G}_1$ and $B\mathcal{G}_2$ to be **isomorphic** if there is an equivalence of categories $B\mathcal{G}_1 \to B\mathcal{G}_2$ commuting with the respective "projections" into $\mathcal{C}$.

It is clear that a functor induced by a Morita bibundle establishes an isomorphism of stacks of torsors. It turns out that the converse is also true: if $B\mathcal{G}_1$ and $B\mathcal{G}_2$ are isomorphic in this refined sense, then the Lie groupoids $\mathcal{G}_1$ and $\mathcal{G}_2$ are Morita equivalent. As we will see, much of this discussion can be adapted to the context of Poisson manifolds and symplectic groupoids.

Let P be an integrable Poisson manifold. A **symplectic P-torsor** is a complete symplectic realization $J : S \to P$ with the additional property that the induced left action of the symplectic groupoid $\mathcal{G}(P)$ on S is principal. Note that, in this case, the manifold $M = S/\mathcal{G}(P)$ has a natural Poisson structure. (As with the regular bimodules in Remark 4.25, we can also describe symplectic torsors without reference to groupoids: $J : S \to P$ should be a surjective submersion, and the symplectic orthogonal leaves to the J-fibres should be simply-connected and form a simple foliation.)

Instead of considering the category of all complete symplectic realizations over a Poisson manifold P, let us consider the category BP of symplectic P-torsors, as we did for Lie groupoids. If we restrict the morphisms in BP to symplectomorphisms, then there is a well-defined "projection" functor $BP \to \mathcal{C}_{Pois}$, where $\mathcal{C}_{Pois}$ denotes the category of Poisson manifolds, with *ordinary* (invertible) Poisson maps as morphisms. As in the case of Lie groupoids, we refine the notion of isomorphism of categories to include the "projection" functors: BP_1 and BP_2 are **isomorphic** if there is an equivalence of categories $BP_1 \to BP_2$ commuting with the projections $BP_i \to \mathcal{C}_{Pois}$, $i = 1, 2$. In this setting, it is also clear that a Morita equivalence of P_1 and P_2 induces an isomorphism between BP_1 and BP_2. The following is a natural question: If BP_1 and BP_2 are isomorphic, must P_1 and P_2 be Morita equivalent Poisson manifolds?

In Remark 4.37, we saw that the Poisson manifolds $P_1 = ((0,1) \times S^2, (1/t)\Pi_{S^2})$ and $P_2 = ((0,1) \times S^2, (1/2t)\Pi_{S^2})$ are not Morita equivalent, but there is an equivalence of categories $BP_1 \to BP_2$. However, this equivalence does *not* commute with the "projection" functors, so it is not an isomorphism in the refined sense. Thus there is some hope that the answer to the question above is "yes," though we do not yet have a complete proof.

5.2 Symplectic categories

The next approach to find a "category of representations" that determines the Morita equivalence class of a Poisson manifold is based on the notion of "symplectic category". One can think of it as the classical limit of the usual notion of abelian category, in the sense that the vector spaces (or modules) of morphisms in the theory of abelian categories are replaced by symplectic manifolds. Notice that we are referring to *a* "symplectic category", rather than *the* symplectic "category" of [87]. From now on, we will drop the quotation marks when referring to the new notion.

In a **symplectic category**, one has a class of objects, and, for any two objects A and B, a symplectic manifold, denoted by $\mathrm{Hom}(A, B)$, which plays the role of the space of morphisms from B to A. Given three objects A, B and C, the "composition operation" $\mathrm{Hom}(A, C) \leftarrow \mathrm{Hom}(A, B) \times \mathrm{Hom}(B, C)$ is a lagrangian submanifold

$$L_{ABC} \subset \mathrm{Hom}(A, C) \times \overline{\mathrm{Hom}(A, B)} \times \overline{\mathrm{Hom}(B, C)}.$$

This may *not* be the graph of a map, but just a canonical relation, so we will refer to it as the **composition relation**. So, unlike in ordinary categories, $\mathrm{Hom}(A, B)$ should not be thought of as a set of points. Instead, certain lagrangian submanifolds of $\mathrm{Hom}(A, B)$ will play the role of "invertible elements", so that we can talk about "isomorphic" objects (this can also be done intrinsically via lagrangian calculus). In other words, the guiding principle is to think of *a* symplectic category as a category in *the* symplectic "category."

A **functor** between symplectic categories should consist of a map F between objects together with symplectic maps (or, more generally, canonical relations) $\mathrm{Hom}(A, B) \to \mathrm{Hom}(F(A), F(B))$, so that the induced map from

$$\mathrm{Hom}(A, C) \times \mathrm{Hom}(A, B) \times \mathrm{Hom}(B, C)$$

to

$$\mathrm{Hom}(F(A), F(C)) \times \mathrm{Hom}(F(A), F(B)) \times \mathrm{Hom}(F(B), F(C))$$

maps the composition relation L_{ABC} to $L_{F(A)F(B)F(C)}$. It is also natural to require that if $\mathrm{Hom}(A, B)$ contains "invertible elements", then so does $\mathrm{Hom}(F(A), F(B))$.

If $\mathcal{S}$ and $\mathcal{S}'$ are symplectic categories, then a functor $\mathcal{S} \to \mathcal{S}'$ is an **equivalence of symplectic categories** if for any object A' in $\mathcal{S}'$, there exists an object A such that $F(A)$ and A are "isomorphic" (in the sense that $\mathrm{Hom}(F(A), A')$ contains an "invertible element"), and the maps $\mathrm{Hom}(A, B) \to \mathrm{Hom}(F(A), F(B))$ are symplectomorphisms.

We have not answered some questions about symplectic categories which arise naturally. Is $\mathrm{Hom}(A, A)$ always a symplectic groupoid? If not, what are sufficient conditions? Is there always a "base" functor from a given symplectic category to the category of Poisson manifolds and Morita morphisms? Nevertheless, we can still discuss interesting examples, such as the one which follows.

5.3 Symplectic categories of representations

In the theory of abelian categories, a model example is the category of modules over a ring (for instance, the group ring of a group, in which case we have a category of representations). The morphisms are module homomorphisms (or intertwining operators in the case of representations). The symplectic analogue of this example is the "symplectic category" of representations of a Poisson manifold, in which objects are symplectic realizations and spaces of morphisms are the classical intertwiner spaces.

To avoid smoothness issues, we will be more restrictive and define the **representation category** of a Poisson manifold P to be the symplectic category in which the objects are symplectic P-torsors $S \to M$ which are (P, M)-Morita equivalences, and the morphism spaces are classical intertwiner spaces, $\mathrm{Hom}(S_1, S_2) := \overline{S_2} *_P S_1$.

Composition relations are given by

$$\begin{aligned} L_{S_1 S_2 S_3} &:= \{([(z,x)], [(y,x)], [(z,y)])\} \\ &\qquad \subset \overline{S_3} *_P S_1 \times S_2 *_P \overline{S_1} \times S_3 *_P \overline{S_2} \end{aligned} \tag{67}$$

where $[(a,b)] \in \overline{S'} *_P S$ denotes the image of $(a,b) \in \overline{S'} \times_P S$ under the natural projection.

Exercise

Check that the composition relation (67) is a lagrangian submanifold. (Hint: first prove it when P is a point, then use coisotropic reduction for the general case.)

Note that $\mathrm{Hom}(S, S) = \overline{S} *_P S$ is symplectomorphic to the symplectic groupoid $\mathcal{G}(M)$, where $M = S/\mathcal{G}(P)$.

Exercise

Show that the composition relation in $\mathrm{Hom}(S,S) = \mathcal{G}(M)$, where $M = S/\mathcal{G}(P)$, is just the graph of the groupoid multiplication

$$\mathcal{G}(M) \leftarrow \mathcal{G}(M) \times_{(s,t)} \mathcal{G}(M)$$

in $\mathcal{G}(M) \times \overline{\mathcal{G}(M)} \times \mathcal{G}(M)$.

Finally, we define "invertible elements" in $\mathrm{Hom}(S_1,S_2) = \overline{S_2} *_P S_1$ to be those lagrangian submanifolds which are the reductions of graphs of isomorphisms of symplectic realizations $S_1 \to S_2$ via the coisotropic submanifold $\overline{S_2} \times_P S_1$ of $\overline{S_2} \times S_1$. In particular, two symplectic realizations are "isomorphic" in the representation category of P if and only if they are isomorphic in the usual sense.

Proposition 5.1 *Two Poisson manifolds are Morita equivalent if and only if they have equivalent representation categories.*

Proof Suppose that P_1 and P_2 are Morita equivalent, and let X be a (P_1,P_2)-Morita bimodule. Let $\mathcal{S}(P_i)$ denote the representation category of P_i, $i = 1,2$. Then X induces an equivalence of symplectic categories $\mathcal{S}(P_2) \to \mathcal{S}(P_1)$: at the level of objects, a Morita bimodule $P_2 \leftarrow S \to M$ is mapped to the Morita bimodule $P_1 \leftarrow \mathrm{X} *_{P_2} S \to M$; if $P_1 \leftarrow S' \to M'$ is an object in $\mathcal{S}(P_1)$, then $\mathrm{X} *_{P_1} \overline{S'}$ is an object in $\mathcal{S}(P_1)$ such that S' and $\mathrm{X} *_{P_2} \overline{\mathrm{X}} *_{P_1} S'$ are isomorphic; at the level of morphisms, because $\overline{S_2} *_{P_2} S_1 \cong \overline{S_2} *_{P_2} \overline{\mathrm{X}} *_{P_1} \mathrm{X} *_{P_2} S_1$, we have a natural symplectomorphism

$$\mathrm{Hom}(S_1,S_2) \cong \mathrm{Hom}(\mathrm{X} *_{P_1} S_1, \mathrm{X} *_{P_1} S_2).$$

Conversely, suppose that $F : \mathcal{S}(P_2) \to \mathcal{S}(P_1)$ is an equivalence of symplectic categories, and let $P_2 \leftarrow S \to M$ be an object in $\mathcal{S}(P_2)$. Then there is a symplectomorphism from $\mathrm{Hom}(S,S) = \mathcal{G}(M)$ to $\mathrm{Hom}(F(S),F(S)) = \mathcal{G}(M')$, where $M' = F(S)/\mathcal{G}(P_1)$. Since this symplectomorphism preserves the composition relation, it is a symplectic groupoid isomorphism. In particular, M and M' are isomorphic as Poisson manifolds, which implies that $F(S)$ is a (P_1,M)-Morita bimodule. If we take $S = \mathcal{G}(P_2)$, then $M = P_2$ and $F(S)$ is a (P_1,P_2)-Morita bimodule. □

The equivalence in Remark 4.37 does not preserve intertwiner spaces (their symplectic structures are related by a factor of 2), so it does not contradict the result above.

Bibliography

1. Alekseev, A.: *On Poisson actions of compact Lie groups on symplectic manifolds.* J. Diff. Geom. **45** (1997), 241–256.
2. Alekseev, A., Kosmann-Schwarzbach, Y., Meinrenken, E.: *Quasi-Poisson manifolds.* Canadian J. Math. **54** (2002), 3–29.
3. Alekseev, A., Malkin, A., Meinrenken, E.: *Lie group valued moment maps.* J. Diff. Geom. **48** (1998), 445–495.
4. Bass, H.: *Algebraic K-theory.* W. A. Benjamin, Inc., New York-Amsterdam, 1968.
5. Bayen, F., Flato, M., Frønsdal, C., Lichnerowicz, A., Sternheimer, D.: *Deformation Theory and Quantization.* Ann. Phys. **111** (1978), 61–151.
6. Beer, W.: *On Morita equivalence of nuclear C^*-algebras.* J. Pure Appl. Algebra **26** (1982), 249–267.
7. Behrend. K., Xu, P.: *Differentiable stacks and gerbes.* In preparation.
8. Behrend. K., Xu, P., Zhang, B.: *Equivariant gerbes over compact simple Lie groups.* C.R. Acad. Sci. Paris **336** (2003), 251–256.
9. Bénabou, J.: *Introduction to bicategories.* In: *Reports of the Midwest Category Seminar*, 1–77. Springer, Berlin, 1967.
10. Berezin, F.A.: *Some remarks about the associated envelope of a Lie algebra,* Funct. Anal. Appl. **1** (1967), 91–102.
11. Blecher, D.: *On Morita's fundamental theorem for C^*-algebras.* Math. Scand. **88** (2001), 137–153.
12. Brown, L.G., Green, P., Rieffel, M.: *Stable isomorphism and strong Morita equivalence of C^*-algebras.* Pacific J. Math **71** (1977), 349–363.
13. Bursztyn, H.: *Semiclassical geometry of quantum line bundles and

Morita equivalence of star products. Int. Math. Res. Not. **16** (2002), 821–846.

14. BURSZTYN, H., CRAINIC, M.: *Dirac structures, moment maps and quasi-Poisson manifolds.* Progress in Mathematics, Festschrift in Honor of Alan Weinstein, Birkhäuser, to appear. Math.DG/0310445.
15. BURSZTYN, H., CRAINIC, M., WEINSTEIN, A., ZHU, C.: *Integration of twisted Dirac brackets.* Duke Math. J., to appear. Math.DG/0303180.
16. BURSZTYN, H., RADKO, O.: *Gauge equivalence of Dirac structures and symplectic groupoids.* Ann. Inst. Fourier (Grenoble) **53** (2003), 309–337.
17. BURSZTYN, H., WALDMANN, S.: *The characteristic classes of Morita equivalent star products on symplectic manifolds.* Comm. Math. Phys. **228** (2002), 103–121.
18. BURSZTYN, H., WALDMANN, S.: *Completely positive inner products and strong Morita equivalence*, to appear in Pacific J. Math. Math.QA/0309402.
19. BURSZTYN, H., WEINSTEIN, A.: *Picard groups in Poisson geometry.* Moscow Math. J. **4** (2004), 39–66.
20. CANNAS DA SILVA, A., WEINSTEIN, A.: *Geometric models for noncommutative algebras.* American Mathematical Society, Providence, RI, 1999.
21. CATTANEO, A.: *On the integration of Poisson manifolds, Lie algebroids, and coisotropic submanifolds.* Lett. Math. Phys., to appear. Math.SG/0308180.
22. CATTANEO, A., FELDER, G.: *Poisson sigma models and symplectic groupoids.* In: N.P.Landsman, M. Pflaum, and M. Schlichenmaier, eds., *Quantization of singular symplectic quotients*, Progr. Math. 198, 61–93, Birkhäuser, Basel, 2001.
23. CATTANEO, A., INDELICATO, D.: *Formality and Star Products*, in this volume.
24. CATTANEO, A., XU, P.: *Integration of twisted Poisson structures.* J. Geom. Phys., to appear. Math.SG/0302268.
25. CONN, J.: *Normal forms for smooth Poisson structures.* Ann. of Math. **121** (1995), 565–593.
26. COSTE, A., DAZORD, P., WEINSTEIN, A.: *Groupoïdes symplectiques.* In: *Publications du Département de Mathématiques. Nouvelle Série. A, Vol. 2*, i–ii, 1–62. Univ. Claude-Bernard, Lyon, 1987. Scanned version available at `math.berkeley.edu/~alanw/cdw.pdf`.

27. Courant, T.: *Dirac manifolds.* Trans. Amer. Math. Soc. **319** (1990), 631–661.
28. Crainic, M.: *Differentiable and algebroid cohomology, van Est isomorphisms, and characteristic classes.* Comment. Math. Helv. **78** (2003), 681–721.
29. Crainic, M., Fernandes, R.L.: *Integrability of Lie brackets.* Ann. of Math. **157** (2003), 575–620.
30. Crainic, M., Fernandes, R.L.: *Integrability of Poisson brackets.* J. Diff. Geom., to appear. Math.DG/0210152.
31. Crainic, M., Fernandes, R.L.: *Linearization of Poisson structures.* Talks given at "PQR2003", Brussels, and "AlanFest", Vienna, 2003.
32. Davis, B.L., Wade, A. : *Vorobjev linearization of Poisson manifolds*, in preparation.
33. Dazord, P.: *Groupoïdes symplectiques et troisième théorème de Lie "non linéaire".* Lect. Notes in Math. **1416**, 39–74, Springer-Verlag, Berlin, 1990.
34. Dufour, J.-P, Wade, A.: *On the local structure of Dirac manifolds.* Math.SG/0405257.
35. Dufour, J.-P., Zung, N.T.: *Poisson Structures and their Normal Forms*, book in preparation.
36. Elliot, G. A., Li, H.: *Morita equivalence of smooth noncommutative tori*, Math.OA/0311502.
37. Fedosov, B.: *A simple geometrical construction of deformation quantization, J. Diff. Geom.* **40** (1994), 213–238.
38. Fernandes, R.L., Monnier, P. : *Linearization of Poisson brackets.* To appear in Lett. Math. Phys. Math.SG/0401273.
39. Gerstenhaber, M., Schack, S.D., : *Algebraic cohomology and deformation theory.* In: M. Hazewinkel and M. Gerstenhaber (eds.): Deformation Theory of Algebras and Structures and Applications, Kluwer Acad. Publ., Dordrecht, 1988, pp. 13–264.
40. Ginzburg, V. L., Golubev, A.: *Holonomy on Poisson manifolds and the modular class.* Israel J. Math **122** (2001), 221–242.
41. Ginzburg, V. L., Lu, J.-H.: *Poisson cohomology of Morita-equivalent Poisson manifolds.* Internat. Math. Res. Notices **10** (1992), 199–205.
42. Ginzburg, V., Weinstein, A.: *Lie-Poisson structure on some Poisson Lie groups.* J. Amer. Math. Soc **5** (1992), 445–453.
43. Gualtieri, M.: *Generalized complex geometry.* Math.DG/0401221.
44. Guruprasad, K., Huebschmann, J., Jeffrey, L., Weinstein,

A.: *Group systems, groupoids, and moduli spaces of parabolic bundles* Duke Math. J. **89** (1997), 377–412.

45. Gutt, S.: *An explicit *-product on the cotangent bundle of a Lie group*, Lett. Math. Phys **7** (1983), 249–258.
46. Hitchin, N. : *Generalized Calabi-Yau manifolds.* Q. J. Math. **54** (2003), 281–308.
47. Jurco, B., Schupp, P., Wess, J.: *Noncommutative line bundle and Morita equivalence.* Lett. Math. Phys. **61** (2002), 171–186.
48. Klimčík, C., Strobl,T.: *WZW-Poisson manifolds.* J. Geom. Phys. **4** (2002), 341–344.
49. Kontsevich, M.: *Deformation quantization of Poisson manifolds.* Lett. Math. Phys. **66** (2003), 157–216.
50. Kosmann-Schwarzbach, Y.: *Quasi, twisted, and all that... in Poisson geometry and Lie algebroid theory* . Progress in Mathematics, Festschrift in Honor of Alan Weinstein, Birkhäuser, to appear. Math.SG/0310359.
51. Lance, E. C.: *Hilbert C^*-modules. A toolkit for operator algebraists.* London Mathematical Society Lecture Note Series, 210. Cambridge University Press, Cambridge, 1995.
52. Landsman, N. P.: *Mathematical Topics between Classical and Quantum Mechanics. Springer Monographs in Mathematics.* Springer-Verlag, Berlin, Heidelberg, New York, 1998.
53. Landsman, N. P.: *Quantized reduction as a tensor product* In: *Quantization of Singular Symplectic Quotients*, N.P. Landsman, M. Pflaum, M. Schlichenmaier, eds., 137–180, Birkhäuser, Basel, 2001.
54. Laurent-Gengoux, C., Xu, P.: *Quantization of quasi-presymplectic groupoids and their hamiltonian spaces.* Progress in Mathematics, Festschrift in Honor of Alan Weinstein, Birkhäuser, to appear. Math.SG/0311154.
55. Li, H.: *Strong Morita equivalence of higher-dimensional noncommutative tori.* J. Reine Angew. Math., to appear. Math.OA/0303123.
56. Lie, S.: *Theorie der Transformationsgruppen*, (Zweiter Abschnitt, unter Mitwirkung von Prof. Dr. Friedrich Engel), Teubner, Leipzig, 1890.
57. Liu, Z.-J., Weinstein, A., Xu , P.: *Manin triples for Lie bialgebroids.* J. Diff. Geom. **45** (1997), 547–574.
58. Mackenzie, K., Xu , P.: *Integration of Lie bialgebroids.* Topology **39** (2000), 445–467.
59. Marsden, J., Ratiu, T.: *Introduction to Mechanics and Symmetry*, Text in Applied Mathematics, Vol. 17, Springer-Verlag, 1994.

60. METZLER, D.: *Topological and smooth stacks.* Math.DG/0306176.
61. MIKAMI, K., WEINSTEIN, A.: *Moments and reduction for symplectic groupoid actions.* Publ. RIMS, Kyoto Univ. **24** (1988), 121–140.
62. MOERDIJK, I., MRČUN., J.: *On integrability of infinitesimal actions.* Amer. J. Math. **124** (2002), 567–593.
63. MOERDIJK, I., MRČUN., J.: *Introduction to Foliations and Lie Groupoids*, Cambridge U. Press, Cambridge, 2003.
64. MOERDIJK, I., MRČUN., J.: *Lie groupoids, sheaves, and cohomology,* in this volume.
65. MORITA, K.: *Duality for modules and its applications to the theory of rings with minimum condition.* Sci. Rep. Tokyo Kyoiku Daigaku Sect. A **6** (1958), 83–142.
66. MOSER, J.: *On the volume elements on a manifold*, Trans. Amer. Math. Soc. **120** (1965), 286–294.
67. MRČUN, J.: *On isomorphisms of algebras of smooth functions.* Preprint math.DG/0309179.
68. NEST, R., TSYGAN, B.: *Algebraic index theorem.* Comm. Math. Phys. **172** (1995), 223–262.
69. PARK,J.-S., *Topological open p-branes*, Symplectic geometry and Mirror symmetry (Seoul 2000), 311–384, World Sci. Publishing, River Edge, NJ, 2001.
70. PRONK, D.: *Etendues and stacks as bicategories of fractions.* Compositio Math. **102** (1996), 243–303.
71. RADKO, O.: *A classification of topologically stable Poisson structures on a compact oriented surface.* J. Symp. Geometry. **1** (2002), 523–542.
72. RAEBURN, I., WILLIAMS, D.: *Morita equivalence and continuous-trace C^*-algebras.* Amer. Math. Soc., Providence, RI, 1998.
73. RIEFFEL, M. A.: *Induced representations of C^*-algebras.* Adv. Math. **13** (1974), 176–257.
74. RIEFFEL, M. A.: *Morita equivalence for C^*-algebras and W^*-algebras.* J. Pure Appl. Algebra **5** (1974), 51–96.
75. RIEFFEL, M. A.: *C^*-algebras associated with irrational rotations.* Pacific J. Math. **93** (1981), 415–429.
76. RIEFFEL, M. A.: *Deformation Quantization of Heisenberg Manifolds.* Commun. Math. Phys. **122** (1989), 531–562.
77. RIEFFEL, M. A., SCHWARZ, A.: *Morita equivalence of multidimensional noncommutative tori.* Internat. J. Math. **10** (1999), 289–299.
78. SCHWARZ, A.: *Morita equivalence and duality.* Nuclear Phys. B **534**

(1998), 720–738.

79. ŠEVERA, P.: *Some title containing the words "homotopy" and "symplectic", e.g. this one.* Math.SG/0105080.
80. ŠEVERA, P., WEINSTEIN, A.: *Poisson geometry with a 3-form background.* Prog. Theo. Phys. Suppl. **144** (2001), 145–154.
81. SMALE, S.: *Diffeomorphisms of the 2-sphere*, Proc. Amer. Math. Soc. **10** (1959), 621–626.
82. TANG, X., WEINSTEIN, A.: *Quantization and Morita equivalence for constant Dirac structures on tori.* To appear in Ann. Inst. Fourier (Grenoble). Math.QA/0305413.
83. TSENG, H.-H., ZHU, C.: *Integrating Lie algebroids via stacks.* Math.QA/0405003.
84. UCHINO, K.: *Lagrangian calculus on Dirac manifolds.* Preprint.
85. VAISMAN, I.: *Lectures on the geometry of Poisson manifolds.* Birkhäuser, Basel, 1994.
86. VOROBJEV, Y.: *Coupling tensors and Poisson geometry near a single symplectic leaf*, in *Lie algebroids and related topics in differential geometry (Warsaw, 2000)*, Banach Center Publ., **54** , Polish Acad. Sci., Warsaw, 2001, 249–274.
87. WEINSTEIN, A.: *The symplectic "category"*. In: Differential geometric methods in mathematical physics (Clausthal, 1980), Springer, Berlin, 1982, 45–51.
88. WEINSTEIN, A.: *The local structure of Poisson manifolds.* J. Diff. Geom. **18** (1983), 523–557.
89. WEINSTEIN, A.: *Symplectic groupoids and Poisson manifolds.* Bull. Amer. Math. Soc. (N.S.) **16** (1987), 101–104.
90. WEINSTEIN, A.: *Coisotropic calculus and Poisson groupoids.* J. Math. Soc. Japan **40** (1988), 705–727.
91. WEINSTEIN, A.: *Lagrangian mechanics and groupoids*, Mechanics Day, W.F. Shadwick, P.S. Krishnaprasad, and T.S. Ratiu, eds., Fields Institute Communications, v. 7., Amer. Math. Soc., Providence, 1995, pp. 207–231.
92. WEINSTEIN, A.: *The modular automorphism group of a Poisson manifold.* J. Geom. Phys. **23** (1997), 379–394.
93. WEINSTEIN, A.: *Integrating the nonintegrable.* Preprint, to appear in proceedings of "Feuilletages – Quantification géométrique," Maison des Sciences de l'Homme, Paris, 2004.
94. XU, P.: *Morita equivalence of Poisson manifolds.* Comm. Math. Phys. **142** (1991), 493–509.

95. Xu, P.: *Morita equivalent symplectic groupoids.* In: *Symplectic geometry, groupoids, and integrable systems (Berkeley, CA, 1989)*, 291–311. Springer, New York, 1991.
96. Xu, P.: *Classical intertwiner spaces and quantization.* Comm. Math. Phys. **164** (1994), 473–488.
97. Xu, P.: *Momentum maps and Morita equivalence.* Math.SG/0307319

PART TWO

Formality and Star Products

Alberto S. Cattaneo
asc@math.unizh.ch

Lecture Notes taken by D. Indelicato
inde@math.unizh.ch

Institut für Mathematik
Universität Zürich–Irchel
Winterthurerstrasse 190
CH-8057 Zürich
Switzerland

We thank the hospitality and financial support at Euroschool and Euroconference PQR2003. A. S. C. acknowledges partial support of SNF Grant No. 20-100029/1.
We are very grateful to Carlo A. Rossi and Jim Stasheff for carefully reading the manuscript and making several suggestions to improve the text. We also want to thank Benoit Dherin and Luca Stefanini for the stimulating discussions we have had with them.

1
Introduction

This work is based on the course given during the international Euroschool on *Poisson Geometry, Deformation Quantisation and Group Representations* held in Brussels in 2003.

The main goal is to describe Kontsevich's proof of the formality of the (differential graded) Lie algebra of multidifferential operators on $\mathbb{R}^d$ and its relationship to the existence and classification of star products on a given Poisson manifold. We start with a survey of the physical background which gave origin to such a problem and a historical review of the subsequent steps which led to the final solution.

1.1 Physical motivation

In this Section we give a brief overview of physical motivations that led to the genesis of the deformation quantization problem, referring to the next Sections and to the literature cited throughout the paper for a precise definition of the mathematical structures we introduce.

In the hamiltonian formalism of classical mechanics, a physical system is described by an even-dimensional manifold M — the phase space — endowed with a symplectic (or more generally a Poisson) structure together with a smooth function H — the hamiltonian function — on it. A physical state of the system is represented by a point in M while the physical observables (energy, momentum and so on) correspond to (real) smooth functions on M. The time evolution of an observable O is governed by an equation of the form

$$\frac{dO}{dt} = \{H\, , O\}$$

where $\{\ ,\ \}$ is the Poisson bracket on $C^\infty(M)$. This bracket is completely determined by its action on the coordinate functions

$$\{ p_i \,,\, q_j \} = \delta_{ij}$$

(together with $\{ p_i \,,\, p_j \} = \{ q_i \,,\, q_j \} = 0$) where $(q_1, \ldots, q_n, p_1, \ldots, p_n)$ are local coordinates on the $2n$-dimensional manifold M.

On the other hand, a quantum system is described by a complex Hilbert space $\mathcal{H}$ together with an operator $\widehat{H}$. A physical state of the system is represented by a vector[1] in $\mathcal{H}$ while the physical observables are now self-adjoint operators in $\mathcal{L}(\mathcal{H})$. The time evolution of such an operator in the Heisenberg picture is given by

$$\frac{d\widehat{O}}{dt} = \frac{i}{\hbar}\Big[\, \widehat{H} \,,\, \widehat{O} \,\Big]$$

where $[\ ,\ ]$ is the usual commutator which endows $\mathcal{L}(\mathcal{H})$ with a Lie algebra structure. The correspondence with classical mechanics is completed by the introduction of the position $\widehat{q}_i$ and momentum $\widehat{p}_j$ operators, which satisfy the canonical commutation relations:

$$[\, \widehat{p}_i \,,\, \widehat{q}_j \,] = \frac{i}{\hbar}\delta_{ij}.$$

This correspondence is by no means a mere analogy, since quantum mechanic was born to replace the hamiltonian formalism in such a way that the classical picture could still be recovered as a "particular case". This is a general principle in the development of a new physical theory: whenever experimental phenomena contradict an accepted theory, a new one is sought which can account for the new data, but still reduces to the previous formalism when the new parameters introduced go to zero. In this sense, classical mechanics can be regained from the quantum theory in the limit where $\hbar$ goes to zero.

The following question naturally arises: is there a precise mathematical formulation of this quantization procedure in the form of a well-defined map between classical objects and their quantum counterpart?

Starting from the canonical quantization method for $\mathbb{R}^{2n}$, in which the central role is played by the canonical commutation relation, a first approach was given by geometric quantization: the basic idea underlying this theory was to set a relation between the phase space $\mathbb{R}^{2n}$ and the

[1] Actually, due to the linearity of the dynamical equations, there is a non-physical multiplicity which can be avoided rephrasing the quantum formalism on a projective Hilbert space, thus identifying a physical state with a ray in $\mathcal{H}$

corresponding Hilbert space $\mathcal{L}(\mathbb{R}^n)$ on which the Schrödinger equation is defined. The first works on geometric quantization are due to Souriau [43], Kostant [32] and Segal [41], although many of their ideas were based on previous works by Kirillov [31]. We will not discuss further this approach, referring the reader to the cited works.

On the other hand, one can focus attention on the observables instead of the physical states, looking for a procedure to get the non-commutative structure of the algebra of operators from the commutative one on $C^\infty(\mathbb{R}^{2n})$. However, one of the first result achieved was the "no go" theorem by Groenwold [25] which states the impossibility of quantizing the Poisson algebra $C^\infty(\mathbb{R}^{2n})$ in such a way that the Poisson bracket of any two functions is sent onto the Lie bracket of the two corresponding operators. Nevertheless, instead of mapping functions to operators, one can "deform" the pointwise product on functions into a non-commutative one, realizing, in an autonomous manner, quantum mechanics directly on $C^\infty(\mathbb{R}^{2n})$: this is the content of the deformation quantization program promoted by Flato in collaboration with Bayen, Fronsdal, Lichnerowicz and Sternheimer,

1.2 Historical review of deformation quantization

The origins of the deformation quantization approach can be traced back to works of Weyl's [46], who gave an explicit formula for the operator $\Omega(f)$ on $\mathcal{L}(\mathbb{R}^n)$ associated to a smooth function f on the phase space $\mathbb{R}^{2n}$:

$$\Omega(f) := \int_{\mathbb{R}^{2n}} \check{f}(\xi,\eta)\, e^{\frac{i}{\hbar}(P\cdot\xi+Q\cdot\eta)} d^n\xi\, d^n\eta,$$

where $\check{f}$ is the inverse Fourier transform of f, P_i and Q_j are operators satisfying the canonical commutation relations and the integral is taken in the weak sense. The arising problem of finding an inverse formula was solved shortly afterwards by Wigner [47], who gave a way to recover the classical observable from the quantum one taking the symbol of the operator. It was then Moyal [37] who interpreted the symbol of the commutator of two operators corresponding to the functions f and g as what is now called a Moyal bracket $\mathcal{M}$:

$$\mathcal{M}(f,g) = \frac{\sinh(\epsilon\, P)}{\epsilon}(f,g) = \sum_{k=0}^{\infty} \frac{\epsilon^{2k}}{(2k+1)!}\, P^{2k+1}(f,g),$$

where $\epsilon = \frac{i\hbar}{2}$ and P^k is the k-th power of the Poisson bracket on $C^\infty(R^{2n})$. A similar formula for the symbol of a product $\Omega(f)\Omega(g)$ had already been found by Groenewold [25] and can now be interpreted as the first appearance of the Moyal star product $\star$, in terms of which the above bracket can be rewritten as

$$\mathcal{M}(f,g) = \frac{1}{2\epsilon}(f \star g - g \star f).$$

However, it was not until Flato gave birth to his program for deformation quantization that this star product was recognized as a non commutative deformation of the (commutative) pointwise product on the algebra of functions. This led to the first paper [20] in which the problem was posed of giving a general recipe to deform the product in $C^\infty(M)$ in such a way that $\frac{1}{2\epsilon}(f \star g - g \star f)$ would still be a deformation of the given Poisson structure on M. Shortly afterward Vey [45] extended the first approach, which considered only 1-differentiable deformation, to more general differentiable deformations, rediscovering in an independent way the Moyal bracket. This opened the way to subsequent works ([21] and [5]) in which quantum mechanics was formulated as a deformation (in the sense of Gerstenhaber theory) of classical mechanics and the first significant applications were found.

The first proof of the existence of star products on a generic symplectic manifold was given by DeWilde and Lecomte [16] and relies on the fact that locally any symplectic manifold of dimension $2n$ can be identified with $\mathbb{R}^{2n}$ via a Darboux chart. A star product can thus be defined locally by the Moyal formula and these local expressions can be glued together by using cohomological arguments.

A few years later and independently of this previous result, Fedosov [19] gave an explicit algorithm to construct star products on a given symplectic manifold: starting from a symplectic connection on M, he defined a flat connection D on the Weyl bundle associated to the manifold, to which the local Moyal expression for $\star$ is extended; the algebra of (formal) functions on M can then be identified with the subalgebra of horizontal sections w.r.t. D. We refer the reader to Fedosov's book for the details. This provided a new proof of existence which could be extended to regular Poisson manifolds and opened the way to further developments.

Once the problem of existence was settled, it was natural to focus on the classification of equivalent star products, where the equivalence of two star products has to be understood in the sense that they give rise to the same algebra up to the action of formal automorphisms which are deformations of the identity. Several authors came to the same classification result using very different approaches, confirming what was already in the seminal paper [5] by Flato et al. namely that the obstruction to equivalence lies in the second de Rham cohomology of the manifold M. For a comprehensive enumeration of the different proofs we address the reader to [17].

The ultimate generalization to the case of a generic Poisson manifold relies on the formality theorem Kontsevich announced in [33] and subsequently proved in [34]. In this last work he derived an explicit formula for a star product on $\mathbb{R}^d$, which can be used to define it locally on any M. Finally, Cattaneo, Felder and Tomassini [13] gave a globalization procedure to realize explicitly what Kontsevich proposed, thus completing the program outlined some thirty years before by Flato.

For a complete overview of the process which led from the origins of quantum mechanics to this last result and over, we refer to the extensive review given by Dito and Sternheimer in [17].

As a concluding remark, we would like to mention that the Kontsevich formula can also be expressed as the perturbative expression of the functional integral of a topological field theory — the so-called Poisson sigma model ([29], [39]) — as Cattaneo and Felder showed in [11]. The diagrams Kontsevich introduced for his construction of the local expression of the star product arise naturally in this context as Feynman diagrams corresponding to the perturbative evaluation of a certain observable.

1.3 Plan of the work

In the first Section we introduce the basic definition and properties of the star product in the most general setting and give the explicit expression of the Moyal product on $\mathbb{R}^{2d}$ as an example. The equivalence relation on star products is also discussed, leading to the formulation of the classification problem.

In the subsequent Section we establish the relation between the existence of a star product on a given manifold M and the formality of the (differential graded) Lie algebra $\mathcal{D}$ of multidifferential operators on M. We introduce the main tools used in Kontsevich's construction and present the fundamental result of Hochschild, Kostant and Rosenberg on which the formality approach is based.

A brief digression follows, in which the formality condition is examined from a dual point of view. The equation that the formality map from the (differential graded) Lie algebra $\mathcal{V}$ of multivector fields to $\mathcal{D}$ must fulfill is rephrased in terms of an infinite family of equations on the Taylor coefficients of the dual map.

In the third Section Kontsevich's construction is worked out explicitly and the formality theorem for $\mathbb{R}^d$ is proved following the outline given in [34]. Finally, the result is generalized to any Poisson manifold M with the help of the globalization procedure contained in [13].

2
The star product

In this Section we will briefly give the definition and main properties of the star product.

Morally speaking, a star product is a formal non-commutative deformation of the usual pointwise product of functions on a given manifold. To give a more general definition, one can start with a commutative associative algebra A with unity over a base ring $\mathbb{K}$ and deform it to an algebra $\mathsf{A}[[\epsilon]]$ over the ring of formal power series $\mathbb{K}[[\epsilon]]$. Its elements are of the form

$$C = \sum_{i=0}^{\infty} c_i \, \epsilon^i \qquad\qquad c_i \in \mathsf{A}$$

and the product is given by the Cauchy formula, multiplying the coefficients according to the original product on A

$$\Big(\sum_{i=0}^{\infty} a_i \, \epsilon^i\Big) \bullet_\epsilon \Big(\sum_{j=0}^{\infty} b_j \, \epsilon^j\Big) = \sum_{k=0}^{\infty} \Big(\sum_{l=0}^{k} a_{k-l} \cdot b_l\Big) \, \epsilon^k$$

The star product is then a $\mathbb{K}[[\epsilon]]$-linear associative product $\star$ on $\mathsf{A}[[\epsilon]]$ which deforms this trivial extension $\bullet_\epsilon \colon \mathsf{A}[[\epsilon]] \otimes_{\mathbb{K}[[\epsilon]]} \mathsf{A}[[\epsilon]] \to \mathsf{A}[[\epsilon]]$ in the sense that for any two $v, w \in \mathsf{A}[[\epsilon]]$

$$v \star w = v \bullet_\epsilon w \quad \mod \epsilon.$$

In the following we will restrict our attention to the case in which A is the Poisson algebra $C^\infty(M)$ of smooth functions on M endowed with the usual pointwise product

$$f \cdot g(x) := f(x) \, g(x) \qquad\qquad \forall x \in M$$

and $\mathbb{K}$ is $\mathbb{R}$.

With these premises we can give the following

Definition 2.1 A star product on M is an $\mathbb{R}[[\epsilon]]$-bilinear map

$$\begin{array}{ccc} C^\infty(M)[[\epsilon]] \times C^\infty(M)[[\epsilon]] & \to & C^\infty(M)[[\epsilon]] \\ (f,g) & \mapsto & f \star g \end{array}$$

such that

i) $f \star g = f \cdot g + \sum_{i=1}^{\infty} B_i(f,g)\, \epsilon^i$,
ii) $(f \star g) \star h = f \star (g \star h)$ $\quad \forall f,g,h \in C^\infty(M)$ $\quad$ *(associativity)*,
iii) $1 \star f = f \star 1 = f$ $\quad \forall f \in C^\infty(M)$.

The B_i could in principle be just bilinear operators, but, in order to encode locality from a physical point of view, one requires them to be bidifferential operators on $C^\infty(M)$ of globally bounded order, that is, bilinear operators which moreover are differential operators w.r.t. each argument; writing the i-th term in local coordinates:

$$B_i(f,g) = \sum_{K,L} \beta_i^{KL}\, \partial_K f\, \partial_L g$$

where the sum runs over all multi-indices $K = (k_1, \ldots, k_m)$ and $L = (l_1, \ldots, l_n)$ of any length $m, n \in \mathbb{N}$ and the usual notation for higher order derivatives is applied; the β_i^{KL}'s are smooth functions, which are non-zero only for finitely many choices of the multi-indices K and L.

Example 2.2 The Moyal star product

We have already introduced the Moyal star product as the first example of a deformed product on the algebra of functions on $\mathbb{R}^{2d}$ endowed with the canonical symplectic form. Choosing Darboux coordinates $(q,p) = (q_1, \ldots, q_d, p_1, \ldots, p_d)$ we can now give an explicit formula for the product of two functions $f, g \in C^\infty(\mathbb{R}^{2d})$:

$$f \star g\,(q,p) := f(q,p)\ \exp\left(\mathrm{i}\frac{\hbar}{2}\left(\overleftarrow{\partial}_q \overrightarrow{\partial}_p - \overleftarrow{\partial}_p \overrightarrow{\partial}_q\right)\right)\ g(q,p),$$

where the $\overleftarrow{\partial}$'s operate on f and the $\overrightarrow{\partial}$'s on g; the parameter ϵ has been replaced by the expression $\mathrm{i}\frac{\hbar}{2}$ that usually appears in the physical literature.

More generally, given a constant skew-symmetric tensor $\{\alpha^{ij}\}$ on $\mathbb{R}^d$ with $i, j = 1, \ldots, d$, we can define a star product by:

$$f \star g\,(x) = \exp\left(\mathrm{i}\frac{\hbar}{2}\,\alpha^{ij}\frac{\partial}{\partial x^i}\frac{\partial}{\partial y^j}\right) f(x)\, g(y)\Big|_{y=x}. \tag{1}$$

We can easily check that such a star product is associative for any choice of α_{ij}

$$
\begin{aligned}
&((f \star g) \star h)\,(x) \\
&= e^{\left(\mathrm{i}\frac{\hbar}{2}\,\alpha^{ij}\,\frac{\partial}{\partial x^i}\frac{\partial}{\partial z^j}\right)}(f \star g)(x)h(z)\Big|_{x=z} \\
&= e^{\left(\mathrm{i}\frac{\hbar}{2}\,\alpha^{ij}\,(\frac{\partial}{\partial x^i}+\frac{\partial}{\partial y^i})\frac{\partial}{\partial z^j}\right)}e^{\left(\mathrm{i}\frac{\hbar}{2}\,\alpha^{kl}\,\frac{\partial}{\partial x^k}\frac{\partial}{\partial y^l}\right)}f(x)g(y)h(z)\Big|_{x=y=z} \\
&= e^{\left(\mathrm{i}\frac{\hbar}{2}\,\alpha^{ij}\,\frac{\partial}{\partial x^i}\frac{\partial}{\partial z^j}+\alpha^{kl}\,\frac{\partial}{\partial y^k}\frac{\partial}{\partial z^l}+\alpha^{mn}\,\frac{\partial}{\partial x^m}\frac{\partial}{\partial y^n}\right)}f(x)g(y)h(z)\Big|_{x=y=z} \\
&= e^{\left(\mathrm{i}\frac{\hbar}{2}\,\alpha^{ij}\,\frac{\partial}{\partial x^i}(\frac{\partial}{\partial y^j}+\frac{\partial}{\partial z^j})\right)}e^{\left(\mathrm{i}\frac{\hbar}{2}\,\alpha^{kl}\,\frac{\partial}{\partial y^k}\frac{\partial}{\partial z^l}\right)}f(x)g(y)h(z)\Big|_{x=y=z} \\
&= (f \star (g \star h))\,(x).
\end{aligned}
$$

Point i) and iii) in Definition (2.1) and the $\mathbb{R}[[\epsilon]]$-linearity can be checked as well directly from the formula (1).

We would like to emphasize that condition iii) in the Definition 2.1 implies that the degree 0 term in the r.h.s. of i) has to be the usual product and it moreover ensures that the B_i's are bidifferential operators in the strict sense, i.e. they have no term of order 0

$$B_i(f,1) = B_i(1,f) = 0 \qquad \forall i \in \mathbb{N}_0. \tag{2}$$

As another consequence of the previous requirements on the B_i's, it is straightforward to prove that the skew-symmetric part B_1^- of the first bidifferential operator, defined by

$$B_1^-(f,g) := \frac{1}{2}\Big(B_1(f,g) - B_1(g,f)\Big)$$

satisfies the following equations:

- $B_1^-(f,g) = -B_1^-(g,f)$,
- $B_1^-(f,g\cdot h) = g\cdot B_1^-(f,h) + B_1^-(f,g)\cdot h$,
- $B_1^-(B_1^-(f,g),h) + B_1^-(B_1^-(g,h),f) + B_1^-(B_1^-(h,f),g) = 0$.

A bilinear operator on $C^\infty(M)$ which satisfies these three identities is called a Poisson bracket. A smooth manifold M endowed with a Poisson bracket on the algebra of smooth functions is called a Poisson manifold (see also [9] and references therein).

It is therefore natural to look at the inverse problem: given a Poisson manifold M, can we define an associative, but possibly non commutative, product $\star$ on the algebra of smooth functions, which is a deformation of the pointwise product and such that

$$\frac{f \star g - g \star f}{\epsilon} \mod \epsilon = \{ f \, , g \}$$

for any pair of functions $f, g \in C^\infty(M)$?

In order to reduce an irrelevant multiplicity of solutions, the problem can be brought down to the study of equivalence classes of such products, where the equivalence is to be understood in the sense of the following

Definition 2.3 Two star products $\star$ and $\star'$ on $C^\infty(M)$ are said to be equivalent iff there exists a linear operator $\mathcal{D}\colon C^\infty(M)[[\epsilon]] \to C^\infty(M)[[\epsilon]]$ of the form

$$\mathcal{D}f := f + \sum_{i=1}^{\infty} D_i(f)\, \epsilon^i$$

such that

$$f \star' g = \mathcal{D}^{-1} \left(\mathcal{D}f \star \mathcal{D}g \right) \tag{3}$$

where $\mathcal{D}^{-1}$ has to be understood as the inverse in the sense of formal power series.

It follows from the very definition of star product that also the D_i's have to be differential operators which vanish on constants, as was shown in [27] (and without proof in [45]).

This notion of equivalence leads immediately to a generalization of the previously stated problem, according to the following

Lemma 2.4 *In any equivalence class of star products, there exists a representative whose first term B_1 in the ϵ expansion is skew-symmetric.*

Proof Given any star product

$$f \star g := f \cdot g + \epsilon\, B_1(f,g) + \epsilon^2 B_2(f,g) + \cdots$$

we can define an equivalent star product as in (3) with the help of a formal differential operator

$$\mathcal{D} = \mathrm{id} + \epsilon\, D_1 + \epsilon^2 D_2 + \cdots .$$

The condition for the first term of the new star product to be skew-symmetric $B_1'(f,g) + B_1'(g,f) = 0$ gives rise to an equation for the first term of the differential operator

$$D_1(f\, g) = D_1 f\, g + f\, D_1 g + \frac{1}{2}\Big(B_1(f,g) + B_1(g,f)\Big), \tag{4}$$

which can be used to define D_1 locally on polynomials and hence by completion on any smooth function. By choosing a partition of unity, we may finally apply D_1 to any smooth function on M.

We can start by choosing D_1 to vanish on linear functions. Then the equation (4) defines uniquely the action of D_1 on quadratic terms, given by the symmetric part B_1^+ of the bilinear operator B_1:

$$D_1(x^i x^j) = B_1^+(x^i, x^j) := \frac{1}{2}\Big(B_1(x^i, x^j) + B_1(x^j, x^i)\Big).$$

where $\{x^k\}$ are local coordinates on the manifold M. The process extends to any monomial and — as a consequence of the associativity of $\star$ — gives rise to a well defined operator since it does not depend on the way we group the factors. We check this on a cubic term:

$$\begin{aligned} D_1((x^i x^j)\, x^k) &= D_1(x^i x^j)\, x^k + x^i x^j\, D_1(x_k) + B_1^+(x^i x^j, x^k) \\ &= B_1^+(x^i, x^j)\, x^k + B_1^+(x^i x^j, x^k) \\ &= B_1^+(x^i, x^j x^k) + x^i\, B_1^+(x^j, x^k) \\ &= x^i\, D_1(x^j x^k) + D_1(x_i)\, x^j x^k + B_1^+(x^i, x^j x^k) \\ &= D_1(x^i\, (x^j x^k)). \end{aligned}$$

The equality between the second and the third lines is a consequence of the associativity of the star product: it is indeed the term of order ϵ in $(x^i \star x^j) \star x^k = x^i \star (x^j \star x^k)$ once we restrict the operators appearing on both sides to their symmetric part. □

The above proof is actually a particular case of the Hochschild–Kostant–Rosenberg theorem. Associativity implies in fact that B_1^+ is a Hochschild cocycle, while in (4) we want to express it as a Hochschild coboundary: the HKR theorem states exactly that this is always possible on $\mathbb{R}^d$ and thus locally on any manifold.

From this point of view, the natural subsequent step is to look for the existence and uniqueness of equivalence classes of star products which are deformations of a given Poisson structure on the smooth manifold M. As already mentioned in the introduction, the existence of such products was first proved by DeWilde and Lecomte [16] in the symplectic case, where the Poisson structure is defined via a symplectic form (a non degenerate closed 2-form). Independently of this previous result, Fedosov [19] gave an explicit geometric construction: the star product is obtained "glueing" together local expressions obtained via the Moyal formula.

As for the classification, the role played by the second de Rham cohomology of the manifold, whose occurrence in connection with this problem can be traced back to [5], has been clarified in subsequent works by different authors ([38], [6], [26], [48], [7], [15]) until it came out that equivalence classes of star products on a symplectic manifold are in one-to-one correspondence with elements in $H^2_{dR}(M)[[\epsilon]]$.

The general case was solved by Kontsevich in [34], who gave an explicit recipe for the construction of a star product starting from any Poisson structure on $\mathbb{R}^d$. This formula can thus be used to define locally a star product on any Poisson manifold; the local expressions can be once again glued together to obtain a global star product, as explained in Section 6. As already mentioned, this result is a straightforward consequence of the formality theorem, which was already announced as a conjecture in [33] and subsequently proved in [34]. In the following, we will review this stronger result which relates two apparently very different mathematical objects — multivector fields and multidifferential operators — and we will come to the explicit formula as a consequence in the end.

As a concluding act, we anticipate the Kontsevich formula even though we will fully understand its meaning only in the forthcoming Sections.

$$f \star g := f \cdot g + \sum_{n=1}^{\infty} \epsilon^n \sum_{\Gamma \in G_{n,2}} w_\Gamma \, B_\Gamma(f,g) \tag{5}$$

The bidifferential operators as well as the weight coefficients are indexed by the elements Γ of a suitable subset $G_{n,2}$ of the set of graphs on $n+2$ vertices, the so-called admissible graphs.

3

Rephrasing the main problem: the formality

In this Section we introduce the main tools that we will need to review Kontsevich's construction of a star product on a Poisson manifold.

The problem of classifying star products on a given Poisson manifold M is solved by proving that there is a one-to-one correspondence between equivalence classes of star products and equivalence classes of formal Poisson structures.

While the former were defined in the previous Section, the equivalence relation on the set of formal Poisson structures is defined as follows. First of all, to give a Poisson structure on M is the same as to choose a Poisson bivector field, i.e. a section π of $\bigwedge^2 \mathrm{T}M$ with certain properties that we will specify later, and define the Poisson bracket via the pairing between (exterior powers of the) tangent and cotangent space:

$$\{ f \, , g \} := \frac{1}{2} \, \langle \pi \, , \mathrm{d}f \wedge \mathrm{d}g \rangle \qquad \forall f, g \in C^\infty(M). \tag{6}$$

The set of Poisson structures is acted on by the group of diffeomorphisms of M, the action being given through the push-forward by

$$\pi_\phi := \phi_* \pi. \tag{7}$$

To extend this notion to formal power series, we can introduce a bracket on $C^\infty(M)[[\epsilon]]$ by:

$$\{ f \, , g \}_\epsilon := \sum_{m=0}^{\infty} \epsilon^m \sum_{\substack{i,j,k=0 \\ i+j+k=m}}^{m} \langle \pi_i \, , \mathrm{d}f_j \wedge \mathrm{d}g_k \rangle \tag{8}$$

where

$$f = \sum_{j=0}^{\infty} \epsilon^j f_j \qquad \text{and} \qquad g = \sum_{k=0}^{\infty} \epsilon^k g_k$$

One says that

$$\pi_\epsilon := \pi_0 + \pi_1 \epsilon + \pi_2 \epsilon^2 + \cdots$$

is a formal Poisson structure if $\{\ ,\ \}_\epsilon$ is a Lie bracket on $C^\infty(M)[[\epsilon]]$.

The gauge group in this case is given by formal diffeomorphisms, i.e. formal power series of the form

$$\phi_\epsilon := \exp(\epsilon \mathsf{X})$$

where $\mathsf{X} := \sum_{k=0}^{\infty} \epsilon^k X_k$ is a formal vector field, i.e. a formal power series whose coefficients are vector fields. This set is given the structure of a group defining the product of two such exponentials via the Baker–Campbell–Hausdorff formula:

$$\exp(\epsilon \mathsf{X}) \cdot \exp(\epsilon \mathsf{Y}) := \exp(\epsilon\, \mathsf{X} + \epsilon\, \mathsf{Y} + \frac{1}{2}\epsilon\, [\mathsf{X}, \mathsf{Y}] + \cdots). \qquad (9)$$

The action which generalizes (7) is then given via the Lie derivatives $\mathcal{L}$ on bivector fields by

$$\exp(\epsilon \mathsf{X})_* \pi := \sum_{m=0}^{\infty} \epsilon^m \sum_{\substack{i,j,k=0 \\ i+j+k=m}}^{m} (\mathcal{L}_{\mathsf{X}_i})^j \pi_k \qquad (10)$$

Kontsevich's main result in [34] was to find an identification between the set of star products modulo the action of the differential operators defined in (2.3) and the set of formal Poisson structure modulo this gauge group. (For further details the reader is referred to [1] and [36].)

3.1 DGLA's, L_∞- algebras and deformation functors

In the classical approach to deformation theory, (see e.g [2]) to each deformation is attached a DGLA via the solutions to the Maurer–Cartan equation modulo the action of a gauge group. The first tools we need to approach our problem are then contained in the following definitions.

Definition 3.1 A graded Lie algebra (briefly GLA) is a $\mathbb{Z}$-graded vector space $\mathfrak{g} = \bigoplus_{i\in\mathbb{Z}} \mathfrak{g}^i$ endowed with a bilinear operation

$$[\ ,\ ]\colon \mathfrak{g}\otimes\mathfrak{g}\to\mathfrak{g}$$

satisfying the following conditions:

a) $[a\,,b]\in\mathfrak{g}^{\alpha+\beta}$ (*homogeneity*)
b) $[a\,,b] = -(-)^{\alpha\beta}[b\,,a]$ (*skew-symmetry*)
c) $[a\,,[b\,,c]] = [[a\,,b]\,,c] + (-)^{\alpha\beta}[b\,,[a\,,c]]$ (*Jacobi identity*)

for any $a\in\mathfrak{g}^\alpha$, $b\in\mathfrak{g}^\beta$ and $c\in\mathfrak{g}^\gamma$

As an example we can consider any Lie algebra as a GLA concentrated in degree 0. Conversely, for any GLA $\mathfrak{g}$, its degree zero part $\mathfrak{g}^0$ (as well as the even part $\mathfrak{g}^{even} := \bigoplus_{i\in\mathbb{Z}}\mathfrak{g}^{2i}$) is a Lie algebra in the usual sense.

Definition 3.2 A differential graded Lie algebra is a GLA $\mathfrak{g}$ together with a differential, $\mathrm{d}\colon\mathfrak{g}\to\mathfrak{g}$, i.e. a linear operator of degree 1 ($\mathrm{d}\colon \mathfrak{g}^i\to\mathfrak{g}^{i+1}$) which satisfies the Leibniz rule

$$\mathrm{d}[a\,,b] = [\mathrm{d}a\,,b] + (-)^\alpha[a\,,\mathrm{d}b] \qquad a\in\mathfrak{g}^\alpha, b\in\mathfrak{g}^\beta$$

and squares to zero ($\mathrm{d}\circ\mathrm{d}=0$).

Again we can make any Lie algebra into a DGLA concentrated in degree 0 with trivial differential $\mathrm{d}=0$. More examples can be found for instance in [36]. In the next Section we will introduce the two DGLA's that play a role in deformation quantization.

The categories of graded and differential graded Lie algebras are completed with the natural notions of morphisms as graded linear maps which moreover commute with the differentials and the brackets[1]. Since we have a differential, we can form a cohomology complex out of any DGLA defining the cohomology of $\mathfrak{g}$ as

$$\mathcal{H}^i(\mathfrak{g}) := \mathrm{Ker}(\mathrm{d}\colon\mathfrak{g}^i\to\mathfrak{g}^{i+1})\Big/\mathrm{Im}(\mathrm{d}\colon\mathfrak{g}^{i-1}\to\mathfrak{g}^i).$$

The set $\mathcal{H} := \bigoplus_i \mathcal{H}^i(\mathfrak{g})$ has a natural structure of graded vector space and, because of the compatibility condition between the differential d and the bracket on $\mathfrak{g}$, it inherits the structure of a GLA, defined unambiguously on equivalence classes $|a|, |b|\in\mathcal{H}$ by:

$$[\,|a|\,,|b|\,]_{\mathcal{H}} := \Big|[a\,,b]_{\mathfrak{g}}\Big|.$$

[1] We recall that a graded linear map $\phi\colon\mathfrak{g}\to\mathfrak{h}$ of degree k is a linear map such that $\phi(\mathfrak{g}^i)\subset\mathfrak{h}^{i+k}\ \forall i\in\mathbb{N}$. We remark that, in the case of DGLA's, a morphism has to be a degree 0 linear map in order to commute with the other structures.

Finally, the cohomology of a DGLA can itself be turned into a DGLA with zero differential.

It is evident that every morphism $\phi\colon \mathfrak{g}_1 \to \mathfrak{g}_2$ of DGLA's induces a morphism $\mathcal{H}(\phi)\colon \mathcal{H}_1 \to \mathcal{H}_2$ between cohomologies. Among these, we are particularly interested in the so-called quasi-isomorphisms, i.e. morphisms of DGLA's inducing isomorphisms in cohomology. Such maps generate an equivalence relation: two DGLA's $\mathfrak{g}_1$ and $\mathfrak{g}_2$ are called quasi-isomorphic if they are equivalent under this relation.[2]

Definition 3.3 A differential graded Lie algebra $\mathfrak{g}$ is called formal if it is quasi-isomorphic to its cohomology, regarded as a DGLA with zero differential and the induced bracket.

The main result of Kontsevich's work — the formality theorem contained in [34] – was to show that the DGLA of multidifferential operators, which we are going to introduce in the next Section, is formal.

In order to achieve this goal, however, one has to rephrase the problem in a broader category, which we will define in this Section, though its structure will become clearer in Chapter 4, where it will be analyzed from a dual point of view.

To introduce the notation that will be useful throughout, we start from the very basic definitions.

Definition 3.4 A graded coalgebra (briefly GCA in the following) on the base ring $\mathbb{K}$ is a $\mathbb{Z}$-graded vector space $\mathfrak{h} = \bigoplus_{i\in\mathbb{Z}} \mathfrak{h}^i$ endowed with a comultiplication, i.e. a graded linear map

$$\Delta\colon \mathfrak{h} \to \mathfrak{h} \otimes \mathfrak{h}$$

such that

$$\Delta(\mathfrak{h}^i) \subset \bigoplus_{j+k=i} \mathfrak{h}^j \otimes \mathfrak{h}^k$$

and which moreover satisfies the coassociativity condition

$$(\Delta \otimes \mathrm{id})\Delta(a) = (\mathrm{id} \otimes \Delta)\Delta(a)$$

for every $a \in \mathfrak{h}$. It is said to be with counit if there exists a morphism

$$\epsilon\colon \mathfrak{h} \to \mathbb{K}$$

[2] We want to stress the fact that the existence of a quasi-isomorphism $\phi\colon \mathfrak{g}_1 \to \mathfrak{g}_2$ does not imply the existence of a "quasi-inverse" $\phi^{-1}\colon \mathfrak{g}_2 \to \mathfrak{g}_1$: therefore these maps do not define automatically an equivalence relation. This is the main reason why we have to consider the broader category of L_∞-algebras.

such that $\epsilon(\mathfrak{h}^i) = 0$ for any $i > 0$ and

$$(\epsilon \otimes \mathrm{id})\Delta(a) = (\mathrm{id} \otimes \epsilon)\Delta(a) = a$$

for every $a \in \mathfrak{h}$. It is said to be cocommutative if

$$\mathsf{T} \circ \Delta = \Delta$$

where $\mathsf{T} \colon \mathfrak{h} \otimes \mathfrak{h} \to \mathfrak{h} \otimes \mathfrak{h}$ is the twisting map, defined on a product $x \otimes y$ of homogeneous elements of degree respectively $|x|$ and $|y|$ by

$$\mathsf{T}(x \otimes y) := (-)^{|x||y|}\, y \otimes x$$

and extended by linearity.

Given a (graded) vector space V over $\mathbb{K}$, we can define new graded vector spaces over the same ground field by:

$$\begin{aligned} T(V) &:= \textstyle\bigoplus_{n=0}^{\infty} V^{\otimes n} \\ \overline{T}(V) &:= \textstyle\bigoplus_{n=1}^{\infty} V^{\otimes n} \end{aligned} \qquad V^{\otimes n} := \begin{cases} \underbrace{V \otimes \cdots \otimes V}_{n} & n \geq 1 \\ \mathbb{K} & n = 0 \end{cases}, \tag{11}$$

and turn them into associative algebras w.r.t. the tensor product. $T(V)$ has also a unit given by $1 \in \mathbb{K}$. They are called respectively the tensor algebra and the reduced tensor algebra. As a graded vector space, $T(V)$ can be endowed with a coalgebra structure defining the comultiplication Δ_T on homogeneous elements by:

$$\begin{aligned} \Delta_T(v_1 \otimes \cdots \otimes v_n) := {} & 1 \otimes (v_1 \otimes \cdots \otimes v_n) \\ & + \sum_{j=1}^{j=n-1} (v_1 \otimes \cdots \otimes v_j) \otimes (v_{j+1} \otimes \cdots \otimes v_n) \\ & + (v_1 \otimes \cdots \otimes v_n) \otimes 1 \end{aligned}$$

and the counit ϵ_T as the canonical projection $\epsilon_T \colon T(V) \to V^{\otimes 0} = \mathbb{K}$. The projection $T(V) \overset{\overline{\pi}}{\to} \overline{T}(V)$ and the inclusion $\overline{T}(V) \overset{i}{\hookrightarrow} T(V)$ induce a comultiplication also on the reduced algebra, which gives rise to a coalgebra without counit.

The tensor algebra gives rise to two other special algebras, the symmetric $S(V)$ and exterior $\Lambda(V)$ algebras, defined as vector spaces as the quotients of $T(V)$ by the two-sided ideals — respectively $\mathcal{I}_S$ and $\mathcal{I}_\Lambda$ — generated by homogeneous elements of the form $v \otimes w - \mathsf{T}(v \otimes w)$ and $v \otimes w + \mathsf{T}(v \otimes w)$. These graded vector spaces inherit the structure of associative algebras w.r.t. the tensor product. The reduced versions $\overline{S}(V)$ and $\overline{\Lambda}(V)$ are defined replacing $T(V)$ by the reduced algebra $\overline{T}(V)$.

Also in this case, the underlying vector spaces can be endowed with a comultiplication which gives them the structure of coalgebras (without counit in the reduced cases). In particular on $S(V)$ the comultiplication is given on homogeneous elements $v \in V$ by

$$\Delta_S(v) := 1 \otimes v + v \otimes 1,$$

and extended as an algebra homomorphism w.r.t. the tensor product.

All the usual additional structures that can be put on an algebra can be dualized to give a dual version on coalgebras. Having in mind the structure of DGLA's we introduce the analog of a differential by defining first coderivations.

Definition 3.5 A coderivation of degree k on a GCA $\mathfrak{h}$ is a graded linear map $\delta\colon \mathfrak{h}^i \to \mathfrak{h}^{i+k}$ which satisfies the (co–)Leibniz identity:

$$\Delta\delta(v) = (\delta \otimes \mathrm{id})\Delta(v) + ((-)^{k|v|}\,\mathrm{id} \otimes \delta)\Delta(v) \qquad \forall v \in \mathfrak{h}^{|v|}$$

A differential Q on a coalgebra is a coderivation of degree one that squares to zero.

With these premises, we can give the definition of the main object we will deal with.

Definition 3.6 An L_∞-algebra is a graded vector space $\mathfrak{g}$ on $\mathbb{K}$ endowed with a degree 1 coalgebra differential Q on the reduced symmetric space $\overline{S}(\mathfrak{g}[1])$.[3] An L_∞-morphism $F\colon (\mathfrak{g}, Q) \to (\tilde{\mathfrak{g}}, \widetilde{Q})$ is a morphism

$$F\colon \overline{S}(\mathfrak{g}[1]) \longrightarrow \overline{S}(\tilde{\mathfrak{g}}[1])$$

of graded coalgebras (sometimes called a pre-L_∞-morphism), which moreover commutes with the differentials ($FQ = \widetilde{Q}F$).

As in the dual case an algebra morphism $f\colon S(\mathsf{A}) \to S(\mathsf{A})$ (resp. a derivation $\delta\colon S(\mathsf{A}) \to S(\mathsf{A})$) is uniquely determined by its restriction to the algebra $\mathsf{A} = S^1(\mathsf{A})$ because of the homomorphism condition $f(ab) = f(a)f(b)$ (resp. the Leibniz rule), an L_∞-morphism F and a coderivation Q are uniquely determined by their projection onto the first

[3] We recall that given any graded vector space $\mathfrak{g}$, we can obtain a new graded vector space $\mathfrak{g}[k]$ by shifting each component by k, i.e.

$$\mathfrak{g}[k] = \bigoplus_{i \in \mathbb{Z}} \mathfrak{g}[k]^i \qquad \text{where} \qquad \mathfrak{g}[k]^i := \mathfrak{g}^{i+k}.$$

component F^1 resp. Q^1. It is useful to generalize this notation introducing the symbol F^i_j (resp. Q^i_j) for the projection to the i-th component of the target vector space restricted to the j-th component of the domain space.[4] With this notation, we can express in a more explicit way the condition which F (resp. Q) has to satisfy to be an L_∞-morphism (resp. a differential). Since, with the above notation, QQ, FQ and $\widetilde{Q}F$ are coderivations (as it can be checked by a straightforward computation), it is sufficient to verify these conditions on their projection to the first component.

We deduce that a coderivation Q is a differential iff

$$\sum_{i=1}^{n} Q^1_i Q^i_n = 0 \qquad \forall n \in \mathbb{N}_0 \tag{12}$$

while a morphism F of graded coalgebras is an L_∞-morphism iff

$$\sum_{i=1}^{n} F^1_i Q^i_n = \sum_{i=1}^{n} \widetilde{Q}^1_i F^i_n \qquad \forall n \in \mathbb{N}_0. \tag{13}$$

In particular, for $n = 1$ we have

$$Q^1_1 Q^1_1 = 0 \qquad \text{and} \qquad F^1_1 Q^1_1 = \widetilde{Q}^1_1 F^1_1;$$

therefore every coderivation Q induces the structure of a complex of vector spaces on $\mathfrak{g}$ and every L_∞-morphism restricts to a morphism of complexes F^1_1. We can thus generalize the definitions given for a DGLA to this case, defining a quasi-isomorphism of L_∞-algebras to be an L_∞-morphism F such that F^1_1 is a quasi-isomorphism of complexes. The notion of formality can be extended in a similar way. We quote a result on L_∞-quasi-isomorphisms we will need later, which follows from a classification theorem on L_∞-algebras.

Lemma 3.7 *Let $F\colon (\mathfrak{g}, Q) \to (\tilde{\mathfrak{g}}, \widetilde{Q})$ be an L_∞-morphism. If F is a quasi isomorphism it admits a* quasi-inverse, *i.e. there exists an L_∞-morphism $G\colon (\tilde{\mathfrak{g}}, \widetilde{Q}) \to (\mathfrak{g}, Q)$ which induces the inverse isomorphism in the corresponding cohomologies.*

For a complete proof of this Lemma together with an explicit expression of the quasi-inverse and a discussion of the above mentioned classification theorem we refer the reader to [10].

[4] With the help of this decomposition, it can be showed that for any given j, only finitely many F^i_j (and analogously Q^i_j) are non trivial, namely $F^i_j = 0$ for $i > j$. For an explicit formula we refer the reader to [24] and [10].

In particular, Lemma 3.7 implies that L_∞-quasi-isomorphisms define an equivalence relation, i.e. two L_∞-algebras are L_∞-quasi-isomorphic iff there is an L_∞-quasi-isomorphism between them. This is considerably simpler than in the case of DGLA's, where the equivalence relation is only generated by the corresponding quasi-isomorphisms, and explains finally why L_∞-algebras are a preferred tool in the solution of the problem at hand.

Example 3.8 To clarify in what sense we previously introduced L_∞-algebras as a generalization of DGLA's, we will show how to induce an L_∞-algebra structure on any given DGLA $\mathfrak{g}$.

We have already a suitable candidate for Q_1^1, since we know that it fulfills the same equation as the differential d: we may then define Q_1^1 to be a multiple of the differential. If we write down explicitly (12) for $n = 2$, we get:

$$Q_1^1\, Q_2^1 + Q_2^1\, Q_2^2 = 0;$$

since every Q_j^i can be expressed in term of a combination of products of some Q_k^1, Q_2^2 must be a combination of Q_1^1 acting on the first or on the second argument of Q_2^1 (for an explicit expression of the general case see [24]). Identifying Q_1^1 with d (up to a sign), the above equation has thus the same form as the compatibility condition between the bracket $[\;,\;]$ and the differential and suggests that Q_2^1 should be defined in terms of the Lie bracket. A simple computation points out the right signs, so that the coderivation is completely determined by

$$\begin{aligned} &Q_1^1(a) := (-)^\alpha \mathrm{d}a && a \in \mathfrak{g}^\alpha, \\ &Q_2^1(b\,c) := (-)^{\beta(\gamma-1)}[\,b\,,\,c\,] && b \in \mathfrak{g}^\beta, c \in \mathfrak{g}^\gamma, \\ &Q_n^1 = 0 && \forall n \geq 3. \end{aligned}$$

The only other equation involving non trivial terms follows from (12) when $n = 3$:

$$Q_1^1\, Q_3^1 + Q_2^1\, Q_3^2 + Q_3^1\, Q_3^3 = 0.$$

Inserting the previous definition and expanding Q_3^2 in terms of Q_2^1 we get

$$\begin{aligned} &(-)^{(\alpha+\beta)(\gamma-1)}\Big[(-)^{\alpha(\beta-1)}[\,a\,,\,b\,]\,,\,c\Big]+ \\ (-)^{(\alpha+\gamma)(\beta-1)}&(-)^{(\gamma-1)(\beta-1)}\Big[(-)^{\alpha(\gamma-1)}[\,a\,,\,c\,]\,,\,b\Big]+ \\ (-)^{(\beta+\gamma)(\alpha-1)}&(-)^{(\beta+\gamma)(\alpha-1)}\Big[(-)^{\beta(\gamma-1)}[\,b\,,\,c\,]\,,\,a\Big] = 0, \end{aligned} \tag{14}$$

which, after a rearrangement of the signs, turns out to be the (graded) Jacobi identity.

According to the same philosophy, a DGLA morphism $F\colon \mathfrak{g} \to \tilde{\mathfrak{g}}$ induces an L_∞-morphism $\overline{F}$ which is completely determined by its first component $\overline{F}^1_1 := F$. In fact, the only two non trivial conditions on $\overline{F}$ coming from (13) with $n = 0$ resp. $n = 1$ are:

$$\begin{aligned} \overline{F}^1_1 Q^1_1(f) &= \widetilde{Q}^1_1 \overline{F}^1_1(f) \Leftrightarrow F(d\,f) = \tilde{d}\,F(f) \\ \overline{F}^1_1 Q^1_2(fg) + \overline{F}^2_1 Q^2_2(fg) &= \widetilde{Q}^1_1 \overline{F}^2_1(fg) + \widetilde{Q}^1_2 \overline{F}^2_2(fg) \\ &\Leftrightarrow F\Big([\,f\,,\,g\,]\Big) = [\,F(f)\,,\,F(g)\,] \end{aligned}$$

If we had chosen Q^1_3 not to vanish, the identity (14) would have been fulfilled up to homotopy, i.e. up to a term of the form

$$\mathrm{d}\rho(g,h,k) \pm \rho(\mathrm{d}g,h,k) \pm \rho(g,\mathrm{d}h,k) \pm \rho(g,h,\mathrm{d}k),$$

where $\rho\colon \Lambda^3\mathfrak{g} \to \mathfrak{g}[-1]$; in this case $\mathfrak{g}$ is said to have the structure of a homotopy Lie algebra.

This construction can be generalized, introducing the canonical isomorphism between the symmetric and exterior algebra (usually called décalage isomorphism[5]) to define for each n a multibracket of degree $2-n$

$$[\cdot,\cdots,\cdot]_n\colon \Lambda^n\mathfrak{g} \to \mathfrak{g}[2-n]$$

starting from the corresponding Q^1_n. Equation (12) gives rise to an infinite family of condition on these multibracket. A graded vector space $\mathfrak{g}$ together with such a family of operators is a strong homotopy Lie algebra (SHLA).

To conclude this overview of the main tools we will need in the following — and to give an account of the last term in the title of this Section — we introduce now the Maurer–Cartan equation of a DGLA $\mathfrak{g}$:

$$\mathrm{d}\,a + \frac{1}{2}[\,a\,,\,a\,] = 0 \qquad a \in \mathfrak{g}^1, \tag{15}$$

[5] More precisely, the décalage isomorphism is given on the n-symmetric power of $\mathfrak{g}$ shifted by one by

$$\begin{aligned} \mathrm{dec}_n : S^n(\mathfrak{g}[1]) &\to \Lambda^n(\mathfrak{g})[n] \\ x_1\cdots x_n &\mapsto (-1)^{\sum_{i=1}^n (n-i)(|x_i|-1)} x_1 \wedge \ldots \wedge x_n, \end{aligned}$$

where the sign is chosen precisely to compensate for the graded antisymmetry of the wedge product.

which plays a central role in deformation theory, as will be exemplified in next Section, in (17) and (22).

It is a straightforward application of the definition 3.1 to show that the set of solutions to this equation is preserved under the action of any morphism of DGLA's and — as we will see in the next Section — of any L_∞-morphism between the corresponding L_∞-algebras.

There is another group which preserve the solutions to the Maurer–Cartan equation, namely the gauge group that can be defined canonically starting from the degree zero part of any formal DGLA.

It is a basic result of Lie algebra theory that there exists a functor exp from the category of nilpotent Lie algebras to the category of groups. For every such Lie algebra $\mathfrak{g}$, the set defined formally as $\exp(\mathfrak{g})$ can be endowed with the structure of a group defining the product via the Baker–Campbell–Hausdorff formula as in (9); the definition is well-posed since the nilpotency ensures that the infinite sum reduces to a finite one.

In the case at hand, generalizing what was somehow anticipated in (9), we can introduce the formal counterpart $\mathfrak{g}[[\epsilon]]$ of any DGLA $\mathfrak{g}$ defined as a vector space by $\mathfrak{g}[[\epsilon]] := \mathfrak{g} \otimes \mathbb{K}[[\epsilon]]$ and show that it has the natural structure of a DGLA. It is clear that the degree zero part $\mathfrak{g}^0[[\epsilon]]$ is a Lie algebra, although non–nilpotent. Nevertheless, we can define the gauge group formally as the set $\mathsf{G} := \exp(\epsilon\, \mathfrak{g}^0[[\epsilon]])$ and introduce a well–defined product taking the Baker–Campbell–Hausdorff formula as the definition of a formal power series. Finally, the action of the group on $\epsilon\, \mathfrak{g}^1[[\epsilon]]$ can be defined generalizing the adjoint action in (9). Namely:

$$\begin{aligned}\exp(\epsilon\, g)\mathsf{a} &:= \sum_{n=0}^{\infty} \frac{(\mathrm{ad}\, g)^n}{n!}(\mathsf{a}) - \sum_{n=0}^{\infty} \frac{(\mathrm{ad}\, g)^n}{(n+1)!}(\mathrm{d}g) \\ &= \mathsf{a} + \epsilon\,[\,g\,,\,\mathsf{a}\,] - \epsilon\, \mathrm{d}g + o(\epsilon^2)\end{aligned}$$

for any $g \in \mathfrak{g}^0[[\epsilon]]$ and $\mathsf{a} \in \mathfrak{g}^1[[\epsilon]]$.

It is a straightforward computation to show that this action preserves the subset $\mathrm{MC}(\mathfrak{g}) \subset \epsilon\, \mathfrak{g}^1[[\epsilon]]$ of solutions to the (formal) Maurer–Cartan equation.

3.2 Multivector fields and multidifferential operators

As we already mentioned, a Poisson structure is completely defined by the choice of a bivector field satisfying certain properties; on the other hand a star product is specified by a family of bidifferential operators. In order to work out the correspondence between these two objects, we are finally going to introduce the two DGLA's they belong to: multivector fields $\mathcal{V}$ and multidifferential operators $\mathcal{D}$.

3.2.1 The DGLA $\mathcal{V}$

A k-multivector field X is a Section of the k-th exterior power $\bigwedge^k \mathrm{T}M$ of the tangent space $\mathrm{T}M$; choosing local coordinates $\{x^i\}_{i=1,\ldots,\dim M}$ and denoting by $\{\partial_i\}_{i=1,\ldots,\dim M}$ the corresponding basis of the tangent space:

$$\mathsf{X} = \sum_{i_1,\ldots,i_k=1}^{\dim M} X^{i_1\cdots i_k}(x)\,\partial_{i_1}\wedge\cdots\wedge\partial_{i_k}.$$

The direct sum of such vector spaces has thus the natural structure of a graded vector space

$$\widetilde{\mathcal{V}} := \bigoplus_{i=0}^{\infty}\widetilde{\mathcal{V}}^i \qquad\qquad \widetilde{\mathcal{V}}^i := \begin{cases} C^\infty(M) & i=0 \\ \Gamma(\bigwedge^i \mathrm{T}M) & i\geq 1 \end{cases},$$

having added smooth functions in degree 0.

The most natural way to define a Lie structure on $\widetilde{\mathcal{V}}$ is by extending the usual Lie bracket on vector fields given in terms of the Lie derivative w.r.t. the first vector field:

$$[\,\mathsf{X},\,\mathsf{Y}\,] := \mathcal{L}_{\mathsf{X}}\mathsf{Y}.$$

The same definition can be applied to the case when the second argument is a function, setting:

$$[\,\mathsf{X},\,f\,] := \mathcal{L}_{\mathsf{X}}(f) = \sum_{i=1}^{\dim M} X^i\frac{\partial f}{\partial x^i}.$$

where we have given also an explicit expression in local coordinates. Setting then the Lie bracket of any two functions to vanish makes $\widetilde{\mathcal{V}}^0\oplus\widetilde{\mathcal{V}}^1$ into a GLA.

Then we define the bracket between a vector field X and a homogeneous element $\mathsf{Y}_1\wedge\ldots\wedge\mathsf{Y}_k\in\widetilde{\mathcal{V}}^k$ with $k>1$ by the following formula:

$$[\,\mathsf{X},\,\mathsf{Y}_1\wedge\ldots\wedge\mathsf{Y}_k\,] := \sum_{i=1}^{k}(-)^{i+1}[\,\mathsf{X},\,\mathsf{Y}_i\,]\wedge\mathsf{Y}_1\wedge\ldots\wedge\widehat{\mathsf{Y}}_i\wedge\ldots\wedge\mathsf{Y}_k,$$

where the bracket on the r.h.s. is just the usual bracket on $\widetilde{\mathcal{V}}^1$; we can then extend it to the case of two generic multivector fields by requiring it to be linear, graded commutative and such that for any $\mathsf{X} \in \widetilde{\mathcal{V}}^k$, $\mathrm{ad}_{\mathsf{X}} := [\mathsf{X}, \cdot]$ is a derivation of degree $k-1$ w.r.t. the wedge product.

Finally, by iterated application of the Leibniz rule, we can find also an explicit expression for the case of a function and a k-vector field:

$$[\mathsf{X}_1 \wedge \cdots \wedge \mathsf{X}_k , f] := \sum_{i=1}^{k} (-)^{k-i}\, \mathcal{L}_{\mathsf{X}_i}(f)\, \mathsf{X}_1 \wedge \cdots \wedge \widehat{\mathsf{X}}_i \wedge \cdots \wedge \mathsf{X}_k$$

and two homogeneous multivector fields of degree greater than 1:

$$\begin{aligned}
&[\mathsf{X}_1 \wedge \cdots \wedge \mathsf{X}_k , \mathsf{Y}_1 \wedge \cdots \wedge \mathsf{Y}_l] := \\
&\qquad \sum_{i=1}^{k}\sum_{j=1}^{l} (-)^{i+j}\, [\mathsf{X}_i , \mathsf{Y}_j] \wedge \mathsf{X}_1 \wedge \cdots \wedge \widehat{\mathsf{X}}_i \wedge \cdots \wedge \mathsf{X}_k \wedge \\
&\qquad\qquad \wedge \mathsf{Y}_1 \wedge \cdots \wedge \widehat{\mathsf{Y}}_j \wedge \cdots \wedge \mathsf{Y}_l .
\end{aligned}$$

With the help of these formulae, we can finally check that the bracket defined so far satisfies also the Jacobi identity. We give here a sketch proof: to simplify the notation the wedge product has not been explicitly written, a small caret $\hat{\mathsf{V}}_i$ represents the i-th component of the missing vector field V and θ^a_b is equal to 1 if $a > b$ and zero otherwise.

Given any three multivector fields X, Y and Z of positive degree n, l and m respectively:

$$\begin{aligned}
[\mathsf{X}, [\mathsf{Y}, \mathsf{Z}]] &= \sum_{i,j}^{l,m} (-)^{i+j} \Big[\mathsf{X}, [\mathsf{Y}_i , \mathsf{Z}_j]\, \underset{i}{\hat{\mathsf{Y}}}\, \underset{j}{\hat{\mathsf{Z}}}\Big] \\
&= \sum_{i,j,k}^{l,m,n} (-)^{i+j+k+1} [\mathsf{X}_k, [\mathsf{Y}_i, \mathsf{Z}_j]]\, \underset{k}{\hat{\mathsf{X}}}\, \underset{i}{\hat{\mathsf{Y}}}\, \underset{j}{\hat{\mathsf{Z}}} \\
&\qquad + \sum_{i,j,k,r\neq i}^{l,m,n} (-)^{i+j+k+r+\theta^r_i} [\mathsf{X}_k, \mathsf{Y}_r][\mathsf{Y}_i, \mathsf{Z}_j]\, \underset{k}{\hat{\mathsf{X}}}\, \underset{i,r}{\hat{\mathsf{Y}}}\, \underset{j}{\hat{\mathsf{Z}}} \\
&\qquad\quad + \sum_{i,j,k,s\neq j}^{l,m,n} (-)^{i+j+k+s+l-1+\theta^s_l} [\mathsf{X}_k, \mathsf{Z}_s][\mathsf{Y}_i, \mathsf{Z}_j]\, \underset{k}{\hat{\mathsf{X}}}\, \underset{i}{\hat{\mathsf{Y}}}\, \underset{j,s}{\hat{\mathsf{Z}}} \\
&= \sum_{i,j,k}^{l,m,n} (-)^{i+j+k+1} \Big([[\mathsf{X}_k, \mathsf{Y}_i], \mathsf{Z}_j]\, \underset{k}{\hat{\mathsf{X}}}\, \underset{i}{\hat{\mathsf{Y}}}\, \underset{j}{\hat{\mathsf{Z}}} \\
&\qquad + (-)^{(n+1)(l+1)} [\mathsf{Y}_i, [\mathsf{X}_k, \mathsf{Z}_j]]\, \underset{i}{\hat{\mathsf{Y}}}\, \underset{k}{\hat{\mathsf{X}}}\, \underset{j}{\hat{\mathsf{Z}}} \Big) + \cdots \\
&= [[\mathsf{X}, \mathsf{Y}], \mathsf{Z}] + (-)^{(n+1)(l+1)} [\mathsf{Y}, [\mathsf{X}, \mathsf{Z}]]
\end{aligned}$$

Analogous computations show that the Jacobi identity is fulfilled also in the case when one or two of the multivector fields is of degree 0, while in the case of three functions the identity becomes trivial.

This inductive recipe to construct a Lie bracket out of its action on the components of lowest degree of the GLA together with its defining properties completely determines the bracket on the whole algebra, as the following proposition summarizes.

Proposition 3.9 *There exists a unique extension of the Lie bracket on* $\widetilde{\mathcal{V}}^0 \oplus \widetilde{\mathcal{V}}^1$ *— called* Schouten–Nijenhuis bracket *— onto the whole* $\widetilde{\mathcal{V}}$

$$[\ ,\ ]_{\mathsf{SN}} : \widetilde{\mathcal{V}}^k \otimes \widetilde{\mathcal{V}}^l \to \widetilde{\mathcal{V}}^{k+l-1}$$

for which the following identities hold:

i) $[\mathsf{X}, \mathsf{Y}]_{\mathsf{SN}} = -(-)^{(x+1)(y+1)} [\mathsf{Y}, \mathsf{X}]_{\mathsf{SN}}$
ii) $[\mathsf{X}, \mathsf{Y} \wedge \mathsf{Z}]_{\mathsf{SN}} = [\mathsf{X}, \mathsf{Y}]_{\mathsf{SN}} \wedge \mathsf{Z} + (-)^{(y+1)z}\, \mathsf{Y} \wedge [\mathsf{X}, \mathsf{Z}]_{\mathsf{SN}}$
iii) $\big[\mathsf{X}, [\mathsf{Y}, \mathsf{Z}]_{\mathsf{SN}}\big]_{\mathsf{SN}} = \big[[\mathsf{X}, \mathsf{Y}]_{\mathsf{SN}}, \mathsf{Z}\big]_{\mathsf{SN}} + (-)^{(x+1)(y+1)} \big[\mathsf{Y}, [\mathsf{X}, \mathsf{Z}]_{\mathsf{SN}}\big]_{\mathsf{SN}}$

for any triple X,Y *and* Z *of degree resp.* x, y *and* z.

The sign convention adopted thus far is the original one, as can be found for instance in the seminal paper [5]. In order to recover the signs we introduced in 3.1, we have to shift the degree of each element by one, defining the graded Lie algebra of multivector fields $\mathcal{V}$ as

$$\mathcal{V} := \bigoplus_{i=-1}^{\infty} \mathcal{V}^i \qquad \mathcal{V}^i := \widetilde{\mathcal{V}}^{i+1} \qquad i = -1, 0, \ldots, \tag{16}$$

which in a shorthand notation is indicated by $\mathcal{V} := \widetilde{\mathcal{V}}[1]$, together with the above defined Schouten–Nijenhuis bracket.

The GLA $\mathcal{V}$ is then turned into a differential graded Lie algebra setting the differential $\mathrm{d} \colon \mathcal{V} \to \mathcal{V}$ to be identically zero.

We now turn our attention to the particular class of multivector fields we are most interested in: Poisson bivector fields. We recall that given a bivector field $\pi \in \mathcal{V}^1$, we can uniquely define a bilinear bracket $\{\ ,\ \}$ as in (6), which is by construction skew-symmetric and satisfies Leibniz rule. The last condition for $\{\ ,\ \}$ to be a Poisson bracket — the Jacobi identity — translates into a quadratic equation on the bivector field, which in local coordinates is:

$$\{\{f\,,g\}\,,h\}+\{\{g\,,h\}\,,f\}+\{\{h\,,f\}\,,g\}=0$$
$$\Updownarrow$$
$$\pi^{ij}\,\partial_j\pi^{kl}\,\partial_i f\,\partial_k g\,\partial_l h+\pi^{ij}\,\partial_j\pi^{kl}\,\partial_i g\,\partial_k h\,\partial_l f+\pi^{ij}\,\partial_j\pi^{kl}\,\partial_i h\,\partial_k f\,\partial_l g=0$$
$$\Updownarrow$$
$$\pi^{ij}\,\partial_j\pi^{kl}\,\partial_i\wedge\partial_k\wedge\partial_l=0$$

The last line is nothing but the expression in local coordinates of the vanishing of the Schouten–Nijenhuis bracket of π with itself. If we finally recall that we defined $\mathcal{V}$ to be a DGLA with zero differential, we see that Poisson bivector fields are exactly the solutions to the Maurer–Cartan equation (15) on $\mathcal{V}$

$$\mathrm{d}\pi+\frac{1}{2}[\,\pi\,,\,\pi\,]_{\mathsf{SN}}=0,\qquad \pi\in\mathcal{V}^1. \tag{17}$$

Finally, formal Poisson structures $\{\ ,\ \}_\epsilon$ are associated to a formal bivector $\pi\in\epsilon\,\mathcal{V}^1[[\epsilon]]$ as in (8) and the action defined in (10) is exactly the gauge group action in the sense of Section 3.1, since the formal diffeomorphisms acting on $\{\ ,\ \}_\epsilon$ are generated by elements of $\mathcal{V}^0[[\epsilon]]$.

3.2.2 The DGLA $\mathcal{D}$

The second DGLA that plays a role in the formality theorem is a subalgebra of the Hochschild DGLA, whose definition and main properties we are going to review in what follows.

To any associative algebra with unit A on a field $\mathbb{K}$ we can associate the complex of multilinear maps from A to itself.

$$\mathcal{C}:=\sum_{i=-1}^{\infty}\mathcal{C}^i\qquad \mathcal{C}^i:=\mathrm{Hom}_{\mathbb{K}}(A^{\otimes(i+1)},A)$$

In analogy to what we have done for the case of multivector fields, we shifted the degree by one in order to match our convention for the signs that will appear in the definition of the bracket.

Having the case of linear operators in mind, on which the Lie algebra structure arises from the underlying associative structure given by the composition of operators, we try to extend this notion to multilinear operators. Clearly, when composing an $(m+1)$-linear operator ϕ with an $(n+1)$-linear operator ψ we have to specify an inclusion $A\hookrightarrow A^{\otimes(m+1)}$ to identify the target space of ψ with one of the component of the domain of ϕ: loosely speaking we have to know where to plug in the output of ψ

into the inputs of ϕ. We therefore define a whole family of compositions $\{\circ_i\}$ such that for ϕ and ψ as above

$$(\phi \circ_i \psi)(f_0, \dots, f_{m+n}) := \phi(f_0, \dots, f_{i-1}, \psi(f_i, \dots, f_{i+n}), f_{i+n+1}, \dots, f_{m+n})$$

for any $(m+n+1)$-tuple of elements of A; this operation can be better understood through the pictorial representation in Fig. 3.1.

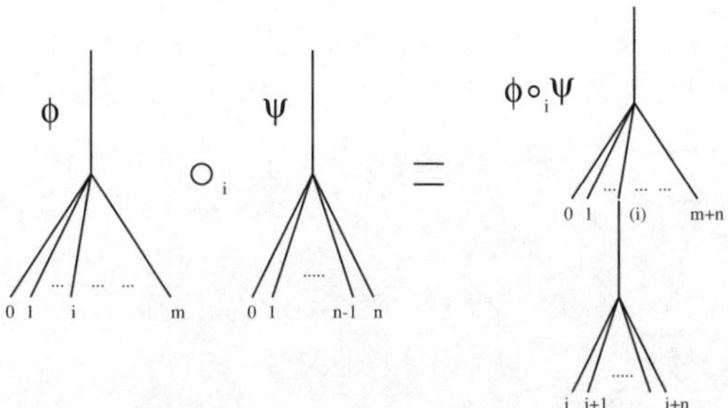

Figure 3.1 The i- composition.

We can further sum up with signs all the possible partial compositions to find an "almost associative" product on $\mathcal{C}$ — in fact a pre-Lie structure — given by

$$\phi \circ \psi := \sum_{i=0}^{m} (-)^{ni} \phi \circ_i \psi$$

with the help of which we can give $\mathcal{C}$ the structure of a GLA.

Proposition 3.10 *The graded vector space $\mathcal{C}$ together with the Gerstenhaber bracket $[\ ,\ ]_{\mathsf{G}} : \mathcal{C}^m \otimes \mathcal{C}^n \to \mathcal{C}^{m+n}$ defined (on homogeneous elements) by*

$$[\phi, \psi]_{\mathsf{G}} := \phi \circ \psi - (-)^{mn} \psi \circ \phi \tag{18}$$

is a graded Lie algebra, called the Hochschild GLA

Proof Since this bracket, introduced by Gerstenhaber in [23], is defined as a linear combination of terms of the form $\phi \circ_i \psi$ and $\psi \circ_i \phi$, it is clearly linear and homogeneous by construction. The presence of the sign $(-)^{mn}$ ensures that it is also (graded) skew-symmetric, since clearly

$$[\phi, \psi]_{\mathsf{G}} = -(-)^{mn}\Big(\psi \circ \phi - (-)^{mn} \phi \circ \psi\Big) = -(-)^{mn}[\psi, \phi]_{\mathsf{G}}$$

for any $\phi \in \mathcal{C}^m$ and $\psi \in \mathcal{C}^n$.

As for the Jacobi identity, we have to prove that the following holds:

$$[\phi, [\psi, \chi]_{\mathsf{G}}]_{\mathsf{G}} = [[\phi, \psi]_{\mathsf{G}}, \chi]_{\mathsf{G}} + (-)^{mn}[\psi, [\phi, \chi]_{\mathsf{G}}]_{\mathsf{G}} \tag{19}$$

for any triple ϕ, ψ, χ of multilinear operator of degree resp. m, n and p. Expanding the first term on r.h.s. of (19) we get

$$\begin{aligned}
&\left(\phi \circ \psi - (-)^{mn} \psi \circ \phi\right) \circ \chi - (-)^{(m+n)p} \chi \circ \left(\phi \circ \psi - (-)^{mn} \psi \circ \phi\right) \\
&= \sum_{i,k=0}^{m,m+n} (-)^{ni+kp} \, (\phi \circ_i \psi) \circ_k \chi \\
&\quad - \sum_{j,k=0}^{n,m+n} (-)^{m(j+n)+kp} \, (\psi \circ_j \phi) \circ_k \chi \\
&\quad - \sum_{i,k=0}^{m,p} (-)^{(m+n)(k+p)+ni} \, \chi \circ_k (\phi \circ_i \psi) \\
&\quad + \sum_{j,k=0}^{n,p} (-)^{(m+n)(k+p)+m(j+n)} \, \chi \circ_k (\psi \circ_j \phi)
\end{aligned}$$

The first sum can be decomposed according to the following rule for iterated partial compositions

$$(\phi \circ_i \psi) \circ_k \chi = \begin{cases} (\phi \circ_k \chi) \circ_i \psi & k < i \\ \phi \circ_i (\psi \circ_{k-i} \chi) & i \leq k \leq i+n \\ (\phi \circ_{k-n} \chi) \circ_i \psi & i+n < k \end{cases}$$

in a term of the form

$$\sum_{\substack{i \\ i \leq k \leq i+n}}^{m} (-)^{ni+kp} \, \phi \circ_i (\psi \circ_{k-i} \chi) = \sum_{i,k=0}^{m,n} (-)^{(n+p)i+kp} \, \phi \circ_i (\psi \circ_k \chi),$$

whose sign matches the one of the corresponding term coming from $(\phi \circ \psi) \circ \chi$ on the l.h.s, plus those terms in which the i-th and k-th composition commute, which cancel with the corresponding terms coming from the expansion of the second term of the r.h.s. of (19).

Upon application of the same procedure to the remaining terms, the claim follows. □

For a different approach refer to [44], where, after having identified multilinear maps on A with graded coderivations of the free cocommutative coalgebra cogenerated by A as a module, the bracket is interpreted as the commutator w.r.t. the composition of coderivations.

Before introducing a differential on $\mathcal{C}$, we have to pick out a particular class of degree one linear operators. It is clear from the above definitions that associative multiplications are elements of $\mathcal{C}^1$ which moreover satisfy the associativity condition. Writing this equation explicitly in terms of such an element $\mathfrak{m}$

$$(f \cdot g) \cdot h = f \cdot (g \cdot h) \Leftrightarrow \mathfrak{m}(\mathfrak{m}(f,g),h) - \mathfrak{m}(f,\mathfrak{m}(g,h)) = 0 \qquad (20)$$

we realize immediately that this is — up to a multiplicative factor — the requirement that the Gerstenhaber bracket of $\mathfrak{m}$ with itself vanishes, since

$$\begin{aligned} [\mathfrak{m}\,,\,\mathfrak{m}]_{\mathsf{G}}(f,g,h) &= \sum_{i=0}^{1}(-)^i(\mathfrak{m}\circ_i \mathfrak{m})(f,g,h) \\ &\qquad - (-)^1\sum_{i=0}^{1}(-)^i(\mathfrak{m}\circ_i \mathfrak{m})(f,g,h) \\ &= 2\Big(\mathfrak{m}(\mathfrak{m}(f,g),h) - \mathfrak{m}(f,\mathfrak{m}(g,h))\Big), \end{aligned} \qquad (21)$$

as is shown in a pictorial way in Fig. 3.2

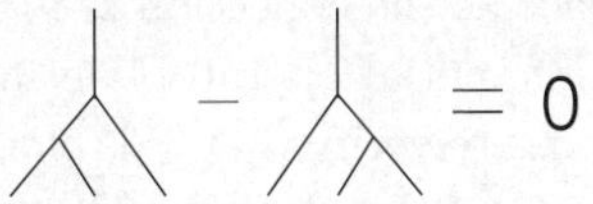

Figure 3.2 The associativity constraint

Now, for each element ϕ of degree k of a (DG) Lie algebra $\mathfrak{g}$, $\mathrm{ad}_\phi := [\phi\,,\ ]$ is a derivation (of degree k), since the Jacobi identity can also be written as:

$$\mathrm{ad}_\phi\,[\psi\,,\,\xi] = [\mathrm{ad}_\phi\,\psi\,,\,\xi] + (-)^{km}[\psi\,,\,\mathrm{ad}_\phi\,\xi]$$

for any $\psi \in \mathfrak{g}^m$ and $\xi \in \mathfrak{g}^n$. It is therefore natural to introduce the Hochschild differential

$$\begin{aligned} \mathrm{d}_{\mathfrak{m}} : \mathcal{C}^i &\rightarrow \mathcal{C}^{i+1} \\ \psi &\mapsto \mathrm{d}_{\mathfrak{m}}\psi := [\mathfrak{m}\,,\,\psi]_{\mathsf{G}}. \end{aligned}$$

The only thing that we still have to check is that $\mathrm{d}_{\mathfrak{m}}$ squares to zero, which follows immediately from the Jacobi identity and the associativity constraint on $\mathfrak{m}$ expressed in terms of the Gerstenhaber bracket as shown in (20) and (21):

$$(\mathrm{d_m} \circ \mathrm{d_m})\psi = \left[\mathfrak{m}, [\mathfrak{m}, \psi]_{\mathsf{G}}\right]_{\mathsf{G}} = \left[[\mathfrak{m}, \mathfrak{m}]_{\mathsf{G}}, \psi\right]_{\mathsf{G}} - \left[\mathfrak{m}, [\mathfrak{m}, \psi]_{\mathsf{G}}\right]_{\mathsf{G}} = \\ = -\left[\mathfrak{m}, [\mathfrak{m}, \psi]_{\mathsf{G}}\right]_{\mathsf{G}} \quad \Leftrightarrow \quad \mathrm{d}_{\mathfrak{m}}^2 = 0$$

So we have proved the following

Proposition 3.11 *The GLA $\mathcal{C}$ together with the differential* $\mathrm{d_m}$ *is a differential graded Lie algebra.*

We can also give an explicit expression of the action of the differential on an element $\psi \in \mathcal{C}^n$:

$$(\mathrm{d_m}\psi)(f_0, \ldots, f_{n+1}) = \sum_{i=0}^{n} (-)^{i+1}\psi(f_0, \ldots, f_{i-1}, f_i \cdot f_{i+1}, \ldots, f_{n+1}) + \\ + f_0 \cdot \psi(f_1, \ldots, f_{n+1}) + (-)^{(n+1)}\psi(f_0, \ldots, f_n) \cdot f_{n+1}.$$

As we already mentioned, in the case $A = C^\infty(M)$, what we are actually interested in is not the whole Hochschild DGLA, but rather a subalgebra of $\mathcal{C}$: the DGLA of multidifferential operators $\widetilde{\mathcal{D}}$. It is defined as a (graded) vector space as the collection $\widetilde{\mathcal{D}} := \bigoplus \widetilde{\mathcal{D}}^i$ of the subspaces $\widetilde{\mathcal{D}}^i \subset \mathcal{C}^i$ consisting of differential operators acting on smooth functions on M. It is an easy exercise to verify that $\widetilde{\mathcal{D}}$ is closed under Gerstenhaber bracket and the action of $\mathrm{d_m}$ and thus is a DGL subalgebra.

We stress the fact that $\widetilde{\mathcal{D}}$ also includes operators of order 0, i.e. loosely speaking operators which "do not differentiate": this way also the associative product $\mathfrak{m}$ is still an element of $\widetilde{\mathcal{D}}^1$.

Having in mind the defining properties of the star product given in Section 2 and in particular the requirement that $B_i(1, f) = 0 \quad \forall i \in \mathbb{N}, f \in C^\infty(M)$, which ensures that the unity is preserved through deformation, we restrict our choice further, considering only differential operators which vanish on constant functions; they build a new DGL subalgebra $\mathcal{D} \subset \widetilde{\mathcal{D}}$. We remark, however, that $\mathrm{d_m}$ is no longer an inner derivation when restricted to $\mathcal{D}$, since clearly the multiplication does not vanish on constants.

Finally, we want to work out also for this DGLA the role played by the Maurer–Cartan equation: we will show that in this case this equation encodes the associativity of the product.

Given an element $\mathsf{B} \in \mathcal{D}^1$, we can interpret $\mathfrak{m} + \mathsf{B}$ as a deformation of the original product. As shown in (20) and (21), the associativity

constraint on $\mathfrak{m} + \mathsf{B}$ translates into

$$[\mathfrak{m} + \mathsf{B}\,,\, \mathfrak{m} + \mathsf{B}]_{\mathsf{G}} = 0$$

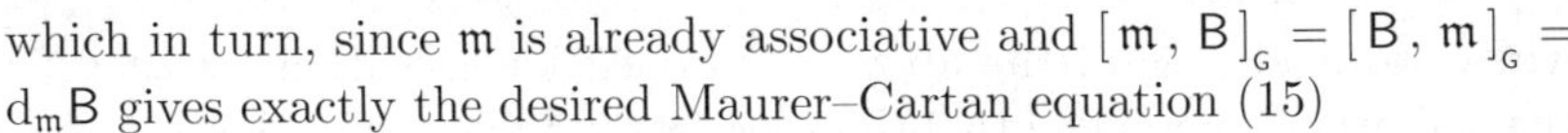

which in turn, since $\mathfrak{m}$ is already associative and $[\mathfrak{m}\,,\, \mathsf{B}]_{\mathsf{G}} = [\mathsf{B}\,,\, \mathfrak{m}]_{\mathsf{G}} = \mathrm{d}_{\mathfrak{m}}\mathsf{B}$ gives exactly the desired Maurer–Cartan equation (15)

$$\mathrm{d}_{\mathfrak{m}}\mathsf{B} + \frac{1}{2}[\mathsf{B}\,,\, \mathsf{B}]_{\mathsf{G}} = 0. \tag{22}$$

Introducing the formal counterpart of $\mathcal{D}$, it is clear that the deformed product turns out to be nothing but a star product as in Definition 2.1, since now $\mathsf{B} \in \epsilon\,\mathcal{D}^1[[\epsilon]]$ is a formal sum of bidifferential operators. Analogously, the gauge group is given exactly by formal differential operators and the action on the star product is the one given in (3), since the adjoint action, due to the definition of the Gerstenhaber bracket, is nothing but the composition of D_i with B_j.

3.3 The first term: U_1

In this last Section we will give an account for the structures we had to introduce and for the two particular cases of DGLA we defined above.

As we already mentioned, our main goal is to prove the formality of the DGLA $\mathcal{D}$ of multidifferential operators. This approach relies on the existence of a previous result by Hochschild, Kostant and Rosenberg [28] which, for any given smooth manifold M, establishes an isomorphism between the cohomology of the algebra of multidifferential operators and the algebra of multivector fields which, according to our previous definition, coincides with its cohomology.

$$\text{HKR}: \mathcal{H}(\widetilde{\mathcal{D}}) \xrightarrow{\sim} \widetilde{\mathcal{V}} = \mathcal{H}(\widetilde{\mathcal{V}})$$

Actually the original result concerned smooth affine algebraic varieties, but it can be extended to smooth manifolds, as it is shown for instance in [34]. This isomorphism is induced by the natural map

$$U_1^{(0)} : \widetilde{\mathcal{V}} \longrightarrow \widetilde{\mathcal{D}}$$

which extends the usual identification between vector fields and first order differential operators, mapping a homogeneous element of the form $\xi_0 \wedge \cdots \wedge \xi_n$ to the multidifferential operator whose action on functions

$f_0, \ldots, f_n$ is given by

$$\frac{1}{(n+1)!} \sum_{\sigma \in S_{n+1}} \mathrm{sgn}(\sigma)\, \xi_{\sigma(0)}(f_0) \cdots \xi_{\sigma(n)}(f_n),$$

where we made use of the above mentioned identification for each ξ_i; the definition is extended to 0-th order vector fields as the identity map. Unfortunately this map, which can be easily checked to be a chain map, fails to preserve the Lie structure, as can be easily verified already at order 2. Given two homogeneous bivector fields $\chi_1 \wedge \chi_2$ and $\xi_1 \wedge \xi_2$, we can verify explicitly that in general

$$U_1^{(0)}\left([\chi_1 \wedge \chi_2\,, \xi_1 \wedge \xi_2]\right) \neq \left[U_1^{(0)}(\chi_1 \wedge \chi_2)\,, U_1^{(0)}(\xi_1 \wedge \xi_2)\right].$$

Omitting the subscripts SN and G and the wedge products to ease the notation, the l.h.s. applied to a triple of functions gives

$$\begin{aligned} U_1^{(0)}\Big([\chi_1\,, \xi_1]\,\chi_2\,\xi_2 - [\chi_1\,, \xi_2]\chi_2\xi_1 \\ - [\chi_2\,, \xi_1]\chi_1\xi_2 + [\chi_2\,, \xi_2]\chi_1\xi_1\Big)(f \otimes g \otimes h) \\ = \frac{1}{6}\Big(\chi_1\xi_1 f\, \chi_2 g\, \xi_2 h - \xi_1\chi_1 f\, \chi_2 g\, \xi_2 h - \chi_1\xi_2 f\, \chi_2 g\, \xi_1 h \\ + \xi_2\chi_1 f\, \chi_2 g\, \xi_1 h + -\chi_2\xi_1 f\, \chi_1 g\, \xi_2 h + \xi_1\chi_2 f\, \chi_1 g\, \xi_2 h \\ + \chi_2\xi_2 f\, \chi_1 g\, \xi_1 h + \xi_2\chi_2 f\, \chi_1 g\, \xi_1 h\Big) + \text{perm.} \end{aligned}$$

while the r.h.s. is

$$\begin{aligned} \left[\frac{1}{2}(\chi_1 \cdot \chi_2 - \chi_2 \cdot \chi_1)\,, \frac{1}{2}(\xi_1 \cdot \xi_2 - \xi_2 \cdot \xi_1)\right](f \otimes g \otimes h) = \\ = \frac{1}{4}\Big(\chi_1(\xi_1 f\, \xi_2 g)\, \chi_2 h + \cdots\Big). \end{aligned}$$

However the difference between the two terms is the image of a closed term in the cohomology of $\mathcal{D}$. We have therefore a way to control the defect of this map in being a Lie algebra morphism and we can hope to find a way to extend it somehow to a morphism whose first order approximation is this isomorphism of complexes. This is exactly the role played by the L_∞-morphism U we will define in the next Sections: in order to give a geometric interpretation of this approximation we will look at the same problem from a dual perspective.

4

Digression: what happens in the dual

The whole machinery of the Kontsevich's construction can be better understood by looking at the mathematical objects and structures we previously introduced from a dual point of view.

Given a vector space V, polynomials on V can be naturally identified with symmetric functions on the dual space V^* defining

$$f(v) := \sum \frac{1}{k!} f_k(v \cdots v) \qquad \forall v \in V$$

where the coefficients f_k are elements of $S^k(V^*)$.

To extend this construction to the case when V is a graded vector space we have to consider the exterior algebra instead. If we introduce the completion $\overline{\Lambda}(V^*)$ of this algebra[1], we can define in a similar way a function in a formal neighborhood of 0 to be given by the formal Taylor expansion in the parameter ϵ

$$f(\epsilon v) := \sum \frac{\epsilon^k}{k!} f_k(v \cdots v) \qquad \forall v \in V.$$

Following this recipe, a vector field X on V can be identified with a derivation on $\overline{\Lambda}(V^*)$ and the Leibniz rule ensures that X is completely determined by its restriction on V^*. In an analogous way an algebra homomorphism

$$\phi \colon \overline{\Lambda}(W^*) \to \overline{\Lambda}(V^*),$$

determines a map $f = \phi^* \colon \overline{\Lambda}(V) \to \overline{\Lambda}(W)$ whose components f_k are

[1] To be more precise, we should specify the topology w.r.t. which we define this completion. This can be done in a natural way considering $\overline{S}(V^*)$ (resp. $\overline{\Lambda}(V^*)$) as the injective limit of the $S^k(V^*)$ (resp. $\Lambda^k(V^*)$) with the induced topology, as in the case of formal power series.

completely determined by their projection on W as the ϕ_k are determined by their restriction on W^*.

In the following we will need the pointed version of these objects, namely we will consider the pair $(V, 0)$ as a pointed manifold and define a (formal) pointed map to be an algebra homomorphism between the reduced symmetric algebras (as introduced in 11)

$$\phi\colon \overline{\Lambda}(W^*)_{>0} \to \overline{\Lambda}(V^*)_{>0},$$

where the subscript "$_{>0}$" indicates that we are considering the two coalgebras as the (completion of the) quotients of $\overline{T}(W^*)$ (resp. $\overline{T}(V^*)$). Analogously, a pointed vector field X is a vector field which has zero as a fixed point, i.e. such that

$$\mathsf{X}(f)(0) = 0 \qquad \forall f$$

or equivalently such that $(Xf)_0 = 0$ for every map f.

We will further call a pointed vector field cohomological — or Q-field — iff it commutes with itself, i.e. iff $\mathsf{X}^2 = \frac{1}{2}[\mathsf{X}, \mathsf{X}] = 0$ and pointed Q-manifold a (formal) pointed manifold together with a cohomological vector field.

We turn now our attention to the non commutative case, taking a Lie algebra $\mathfrak{g}$. The bracket $[\ ,\ ]\colon \Lambda^2\mathfrak{g} \to \mathfrak{g}$ gives rise to a linear map

$$[\ ,\ ]^*\colon \mathfrak{g}^* \to \Lambda^2(\mathfrak{g})^*.$$

We can extend it to whole exterior algebra to

$$\delta\colon \Lambda^\bullet(\mathfrak{g})^* \to \Lambda^{\bullet+1}(\mathfrak{g})^*$$

requiring that $\delta|_{\mathfrak{g}^*} \equiv [\ ,\ ]^*$ and imposing the Leibniz rule to get a derivation.

The exterior algebra can now be interpreted as some odd analog of a manifold, on which δ plays the role of a (pointed) vector field. Since the Jacobi identity on $[\ ,\ ]$ translates to the equation $\delta^2 = 0$, δ is a cohomological pointed vector field.

If we now consider two Lie algebras $\mathfrak{g}$ and $\mathfrak{h}$ and endow their exterior algebras with differentials $\delta_\mathfrak{g}$ and $\delta_\mathfrak{h}$, a Lie algebra homomorphism $\phi\colon \mathfrak{g} \to \mathfrak{h}$ will correspond in this case to a chain map $\phi^*\mathfrak{h}^* \to \mathfrak{g}^*$, since

$$\phi\left([\cdot, \cdot]_\mathfrak{g}\right) = [\phi(\cdot), \phi(\cdot)]_\mathfrak{h} \qquad \Longleftrightarrow \qquad \delta_\mathfrak{g} \circ \phi^* = \phi^* \circ \delta_\mathfrak{h}$$

This is the first glimpse of the correspondence between L_∞-algebras and pointed Q-manifolds: a Lie algebra is a particular case of DGLA, which in turn can be endowed with an L_∞-structure; from this point of view the map ϕ satisfies the same equation of the first component of an L_∞-morphism as given in (13) for $n = 1$.

To get the full picture, we have to extend the previous construction to the case of a graded vector space Z which has odd and even parts. Functions on such a space can be identified with elements in the tensor product $S(Z^*) := S(V^*) \otimes \Lambda(W^*)$, where $Z = V \oplus \Pi W$ is the natural decomposition of the graded space in even and odd subspaces.[2]

The conditions for a vector field $\delta\colon S^\bullet(Z^*) \to S^{\bullet+1}(Z^*)$ to be cohomological can now be expressed in terms of its coefficients

$$\delta_k\colon S^k(Z^*) \to S^{k+1}(Z^*)$$

expanding the equation $\delta^2 = 0$. This gives rise to an infinite family of equations:

$$\begin{cases} \delta_0\,\delta_0 = 0 \\ \delta_1\,\delta_0 + \delta_0\,\delta_1 = 0 \\ \delta_2\,\delta_0 + \,\delta_1\delta_1 + \,\delta_0\delta_2 = 0 \\ \dots \end{cases}$$

If we now define the dual coefficients $m_k := \left(\delta_k|_{Z^*}\right)^*$ and introduce the natural pairing $\langle\ ,\ \rangle : Z^* \otimes Z \to \mathbb{C}$, we can express the same condition in terms of the maps

$$m_k\colon S^{k+1}(Z) \to Z,$$

paying attention to the signs we have to introduce for δ to be a (graded) derivation.

The first equation ($m_0\,m_0 = 0$) tells us that m_0 is a differential on Z and defines therefore a cohomology $\mathcal{H}_{m_0}(Z)$.

For $k = 1$, with an obvious notation, we get

$$\langle\, \delta_1\,\delta_0\,f\,,\,xy\,\rangle = \langle\, \delta_0\,f\,,\,m_1(xy)\,\rangle = \langle\, f\,,\,m_0(m_1(xy))\,\rangle$$

and

$$\begin{aligned} &\langle\, \delta_0\,\delta_1\,f\,,\,xy\,\rangle = \langle\, \delta_1\,f\,,\,m_0(x)\,y\,\rangle + (-)^{|x|}\,\langle\, \delta_1\,f\,,\,x\,m_0(y)\,\rangle = \\ &= \langle\, f\,,\,m_1(m_0(x)\,y)\,\rangle + (-)^{|x|}\,\langle\, f\,,\,m_1(x\,m_0(y))\,\rangle\,, \end{aligned}$$

[2] In the following we will denote by ΠW the (odd) space defined by a parity reversal on the vector space W, which can be also written as $W[1]$, using the notation introduced in Section 3.1.

i.e. m_0 is a derivation w.r.t. the multiplication defined by m_1.

If we now write Z as $\mathfrak{g}[1]$ and identify the symmetric and exterior algebras with the décalage isomorphism $S^n(\mathfrak{g}[1]) \xrightarrow{\sim} \Lambda^n(V[n])$, m_1 can be interpreted as a bilinear skew-symmetric operator on $\mathfrak{g}$.

The next equation, which involves m_1 composed with itself, tells us exactly that this operator is indeed a Lie bracket for which the Jacobi identity is satisfied up to terms containing m_0, i.e. — since m_0 is a differential — up to homotopy.

Putting the equations together, this gives rise to a strong homotopy Lie algebra structure on $\mathfrak{g}$, thus establishing a one-to-one correspondence between pointed Q-manifolds and SHLA's, which in turn are equivalent to L_∞-algebras, as we already observed in Section 3.1.

Finally, to complete this equivalence and to express the formality condition (13) more explicitly, we spell out the equations for the coefficients of a Q-map, i.e. a (formal) pointed map between two Q-manifolds Z and $\widetilde{Z}$ which commutes with the Q-fields; namely:

$$\begin{array}{rcl} \phi\colon S\left(\widetilde{Z}^*_{>0}\right) & \longrightarrow & S(Z^*_{>0}) \\ & \text{s. t.} & \\ \phi\circ\tilde{\delta} & = & \delta\circ\phi. \end{array} \tag{23}$$

As for the case of the vector field δ, we consider only the restriction of this map to the original space $\widetilde{Z}$ and define the coefficients of the dual map as

$$U_k := \left(\phi_k|_{\widetilde{Z}^*}\right)^* : S^k(Z) \to \widetilde{Z}.$$

With the same notation as above, we can express the condition (23) on the dual coefficients with the help of the natural pairing. The first equation reads:

$$\begin{array}{rcl} \left\langle \phi\,\tilde{\delta}\, f\,,\, x \right\rangle & = & \langle \delta\,\phi\, f\,,\, x\rangle \\ & \Downarrow & \\ \left\langle \tilde{\delta}_0\, f\,,\, U_1(x) \right\rangle & = & \langle \phi\, f\,,\, m_0(x)\rangle \\ & \Downarrow & \\ \langle f\,,\, \widetilde{m}_0(U_1(x))\rangle & = & \langle f\,,\, U_1(m_0(x))\rangle\,. \end{array}$$

As we could have guessed from the discussion in Section 3.1, the first coefficient U_1 is a chain map w.r.t. the differential defined by the first coefficient of the Q-structures.

$$[U_1] : \mathcal{H}_{m_0}(Z) \to \mathcal{H}_{\widetilde{m}_0}(\widetilde{Z}).$$

An analogous computation gives the equation for the next coefficient:

$$\begin{aligned}\widetilde{m}_1(U_1(x)\,U_1(y)) + \widetilde{m}_1(U_2(x\,y))\\ = U_2(m_0(x)\,y) + (-)^{|x|}U_2(x\,m_0(y)) + U_1(m_1(x\,y)),\end{aligned}$$

which shows that U_1 preserves the Lie structure induced by m_1 and $\widetilde{m}_1$ up to terms containing m_0 and $\widetilde{m}_0$, i.e. up to homotopy.

This is exactly what we were looking for: as the map $U_1^{(0)}$ defined in Section 3.3 is a chain map which fails to be a DGLA morphism, a Q-map U (or equivalently an L_∞-morphism) induces a map U_1 which shares the same property.

We restrict thus our attention to DGLA's, considering now a pair of pointed Q-manifolds Z and $\widetilde{Z}$ such that $m_k = \widetilde{m}_k = 0$ for $k > 1$. Equivalently, we consider two L_∞-algebras as in Example 3.8, whose coderivation have only two non-vanishing components.

A straightforward computation which follows the same steps as above for $k = 1, 2$, leads in this case to the following condition on the n-th coefficient of U:

$$\begin{aligned}\widetilde{m}_0\left(U_n(x_1\cdots x_n)\right) + \frac{1}{2}\sum_{\substack{I\sqcup J=\{1,\dots n\}\\ I,J\neq\emptyset}} \varepsilon_x(I,J)\,\widetilde{m}_1\left(U_{|I|}(x_I)\cdot U_{|J|}(x_J)\right)\\ = \sum_{k=1}^{n}\varepsilon_x^k U_n\left(m_0(x_k)\cdot x_1\cdots\widehat{x}_k\cdots x_n\right)\\ + \frac{1}{2}\sum_{k\neq l}\varepsilon_x^{kl}U_{n-1}\left(m_1(x_k\cdot x_l)\cdot x_1\cdots\widehat{x}_k\cdots\widehat{x}_l\cdots x_n\right) \qquad (24)\end{aligned}$$

To avoid a cumbersome expression involving lots of signs, we introduced a shorthand notation $\varepsilon_x(I,J)$ for the Koszul sign associated to the $(|I|,|J|)$-shuffle permutation associated to the partition $I\sqcup J = \{1,\dots,n\}$[3] and ε_x^k (resp. ε_x^{kl}) for the particular case $I = \{k\}$ (resp. $I = \{k,l\}$); we further simplified the expression adopting the multiindex notation $x_I := \prod_{i\in I} x_i$.

[3] Whenever a vector space V is endowed with a graded commutative product, the Koszul sign $\varepsilon(\sigma)$ of a permutation σ is the sign defined by

$$x_1\cdots x_n = \varepsilon(\sigma)\,x_{\sigma(1)}\cdots x_{\sigma(n)} \qquad x_i\in V.$$

An $(l, n-l)$-shuffle permutation is a permutation σ of $(1,\dots,n)$ such that $\sigma(1) < \cdots < \sigma(l)$ and $\sigma(l+1) < \cdots \sigma(n)$. The shuffle permutation associated to a partition $I_1\sqcup\cdots\sqcup I_k = \{1,\dots,n\}$ is the permutation that takes first all the elements indexed by the subset I_1 in the given order, then those indexed by I_2 and so on.

This expression will be specialized in next Section to the case of the L_∞-morphism introduced by Kontsevich to give a formula for the star product on $\mathbb{R}^d$: we will choose as Z the DGLA $\mathcal{V}$ of multivector fields and as $\widetilde{Z}$ the DGLA $\mathcal{V}$ of multidifferential operators and derive the equation that the coefficients U_n must satisfy to determine the required formality map.

As a concluding act of this digression, we will establish once and for all the relation between the formality of $\mathcal{D}$ and the solution of the problem of classifying all possible star products on $\mathbb{R}^d$.

As we already worked out in Section 3.1, the associativity of the star product as well as the Jacobi identity for a bivector field are encoded in the Maurer–Cartan equations 22 resp. 17. In order to translate these equations in the language of pointed Q-manifolds, we have first to introduce the generalized Maurer–Cartan equation on an (formal) L_∞-algebra $(\mathfrak{g}[[\epsilon]], Q)$:

$$Q(\exp \epsilon\, x) = 0 \qquad x \in \mathfrak{g}^1[[\epsilon]],$$

where the exponential function exp maps an element of degree 1 to a formal power series in $\epsilon\mathfrak{g}[[\epsilon]]$.

From a dual point of view, this amounts to the request that x is a fixed point of the cohomological vector field δ, i.e. that for every f in $S(\mathfrak{g}^*[[\epsilon]][1])$

$$\delta\, f(\epsilon\, x) = 0.$$

Since $(\delta f)_k = \delta_{k-1} f$, expanding the previous equation in a formal Taylor series and using the pairing as above to get $\langle\, \delta_{k-1} f\,,\, x \cdots x\,\rangle = \langle\, f\,,\, m_{k-1}(x \cdots x)\,\rangle$, the generalized Maurer–Cartan equation can be written in the form

$$\sum_{k=1}^{\infty} \frac{\epsilon^k}{k!} m_{k-1}(x \cdots x) = \epsilon\, m_0(x) + \frac{\epsilon^2}{2}\, m_1(x\, x) + o(\epsilon^3) = 0. \tag{25}$$

It is evident that (the formal counterpart of) equation 15 is recovered as a particular case when $m_k = 0$ for $k > 1$.

Finally, as a morphism of DGLA's preserves the solutions of the Maurer–Cartan equation, since it commutes both with the differential and with the Lie bracket, an L_∞-morphism ϕ: $S((\mathfrak{h}^*[[\epsilon]][1])) \to S((\mathfrak{g}^*[[\epsilon]][1]))$, according to (23), preserves the solutions of the above generalization; with the usual notation, if x is a solution to (25) on $\mathfrak{g}[[\epsilon]]$,

$$U(\epsilon x) = \sum_{k=1} \frac{\epsilon^k}{k!} U_k(x \cdots x)$$

is a solution of the same equation on $\mathfrak{h}$.

The action of the gauge group on the set MC($\mathfrak{g}$) can analogously be generalized to the case of L_∞-algebras and a similar computation shows that, if x and x' are equivalent modulo this generalized action, their images under U are still equivalent solutions.

In conclusion, reducing the previous discussion to the specific case we are interested in, namely when $\mathfrak{g} = \mathcal{V}$ and $\mathfrak{h} = \mathcal{D}$, given an L_∞-morphism U we have a formula to construct out of any (formal) Poisson bivector field π an associative star product given by

$$U(\pi) = \sum_{k=0} \frac{\epsilon^k}{k!} U_k(\pi \cdots \pi) \tag{26}$$

where we reinserted the coefficient of order 0 corresponding to the original non deformed product. If moreover U is a quasi-isomorphism, the correspondence between (formal) Poisson structures on M and formal deformations of the pointwise product on $C^\infty(M)$ is one-to-one: in other terms once we give a formality map, we have solved the problem of existence and classification of star products on M.

This is exactly the procedure followed by Kontsevich to give his formula for the star product on $\mathbb{R}^d$.

5

The Kontsevich formula

In this Section we will finally give an explicit expression of Kontsevich's formality map from $\mathcal{V}$ to $\mathcal{D}$ which induces the one-to-one map from (formal) Poisson structures on $\mathbb{R}^d$ to star products on $C^\infty(\mathbb{R}^d)$.

The main idea is to introduce a pictorial way to describe how a multivector field can be interpreted as a multidifferential operator and to rewrite the equations introduced in (23) in terms of graphs.

As a toy model we can consider the Moyal star product introduced in Section 2 and give a pictorial version of formula (1) as follows:

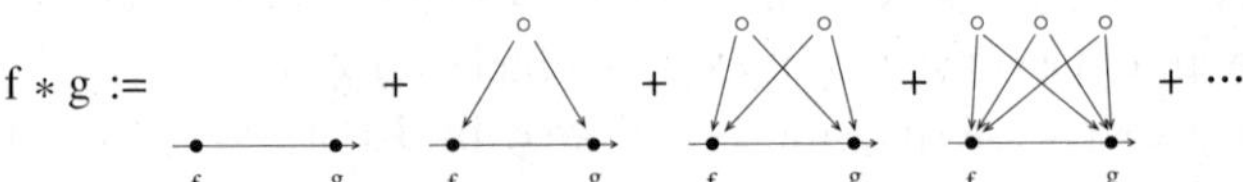

Figure 5.1 A pictorial representation of the first terms of the Moyal star product.

To the n-th term of the series we associate a graph with n "unfilled" vertices – which represent the n copies of the Poisson tensor π – and two "filled" vertices – which stand for the two functions that are to be differentiated; the left (resp. right) arrow emerging from the vertex corresponding to π^{ij} represent ∂_i (resp. ∂_j) acting on f (resp. g) and the sum over all indices involved is understood.

This setting can be generalized introducing vertices of higher order, i.e. with more outgoing arrows, to represent multivector fields and letting arrows point also to "unfilled" vertices, to represent the composition of differential operators: in the Moyal case, since the Poisson tensor is constant such graphs do not appear.

The main intuition behind the Kontsevich formula for the star product is that one can introduce an appropriate set of graphs and assign to each graph Γ a multidifferential operator B_Γ and a weight w_Γ in such a way that the map that sends an n-tuple of multivector fields to the corresponding weighted sum over all possible graphs in this set of multidifferential operators is an L_∞-morphism.

This procedure will become more explicit in the next Section, where we will go into the details of Kontsevich's construction.

5.1 Admissible graphs, weights and B_Γ's

First of all, we have to introduce the above mentioned set of graphs we will deal with in the following.

Definition 5.1 The set $\mathcal{G}_{n,\bar{n}}$ of admissible graphs consists of all connected graphs Γ which satisfy the following properties:

- the set of vertices $V(\Gamma)$ is decomposed in two ordered subsets $V_1(\Gamma)$ and $V_2(\Gamma)$ isomorphic to $\{1,\ldots,n\}$ resp. $\{\bar{1},\ldots,\bar{n}\}$ whose elements are called vertices of the first resp. second type;
- the following inequalities involving the number of vertices of the two types are fulfilled: $n \geq 0$, $\bar{n} \geq 0$ and $2n+\bar{n}-2 \geq 0$;
- the set of edges $E(\Gamma)$ is finite and does not contain small loops, i.e. edges starting and ending at the same vertex;
- all edges in $E(\Gamma)$ are oriented and start from a vertex of the first type;
- the set of edges starting at a given vertex $v \in V_1(\Gamma)$, which will be denoted in the following by $\mathrm{Star}(v)$, is ordered.

Example 5.2 Admissible graphs

Graphs *i)* and *ii)* in Fig. 5.2 are admissible, while graphs *iii)* and *iv)* are not.

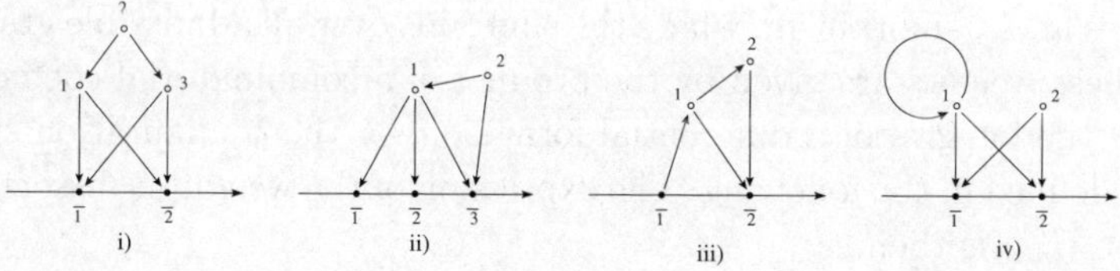

Figure 5.2 Some examples of admissible and non-admissible graphs.

We now introduce the procedure to associate to each pair $(\Gamma, \xi_1 \otimes \cdots \otimes \xi_n)$ consisting of a graph $\Gamma \in \mathcal{G}_{n,\bar{n}}$ with $2n+m-2$ edges and of a tensor product of n multivector fields on $\mathbb{R}^d$ a multidifferential operator $B_\Gamma \in \mathcal{D}^{\bar{n}-1}$.

- We associate to each vertex v of the first type with k outgoing arrows the skew-symmetric tensor $\xi_i^{j_1,\dots,j_k}$ corresponding to a given ξ_i via the natural identification.
- We place a function at each vertex of the second type.
- We associate to the l-th arrow in $\mathrm{Star}(v)$ a partial derivative w.r.t. the coordinate labeled by the l-th index of ξ_i acting on the function or the tensor appearing at its endpoint.
- We multiply such elements in the order prescribed by the labeling of the graph.

As an example, the multidifferential operator corresponding to the first graph in Fig.5.2 and to the triple (α, β, γ) of bivector fields is given by

$$U_{\Gamma_1}(\alpha,\beta,\gamma)(f,g) := \beta^{b_1 b_2}\,\partial_{b_1}\alpha^{a_1 a_2}\,\partial_{b_2}\gamma^{c_1 c_2}\,\partial_{a_1}\partial_{c_1} f\,\partial_{a_2}\partial_{c_2} g,$$

while the operator corresponding to the second graph and the pair $(\pi, \rho)$is

$$U_{\Gamma_1}(\pi,\rho)(f,g,h) := \pi^{p_1 p_2}\partial_{p_1}\rho^{r_1 r_2 r_3}\partial_{r_1} f \partial_{r_2} g \partial_{r_3}\partial_{p_2} h$$

This construction gives rise for each Γ to a linear map $U_\Gamma \colon T^n(\mathcal{V}) \to \mathcal{D}$ which is equivariant w.r.t. the action of the symmetric group, i.e. permuting the order in which we choose the edges we get a sign equal to the signature of the permutation. The main point in Kontsevich's formality theorem was to show that there exist a choice of weights w_Γ such that the linear combination

$$U := \sum_\Gamma w_\Gamma B_\Gamma$$

defines an L_∞-morphism, where the sum runs over all admissible graphs.

These weights are given by the product of a combinatorial coefficient times the integral of a differential form ω_Γ over the configuration space $C_{n,\bar{n}}$ defined in the following. The expression of the weight w_Γ associated to $\Gamma \in \mathcal{G}_{n,\bar{n}}$ is then:

$$w_\Gamma := \prod_{k=1}^{n} \frac{1}{(\#\,\mathrm{Star}(k))!}\,\frac{1}{(2\pi)^{2n+\bar{n}-2}} \int_{\bar{C}^+_{n,\bar{n}}} \omega_\Gamma \qquad (27)$$

if Γ has exactly $2n + \bar{n} - 2$ edges, while the weight is set to vanish otherwise. The definition of ω_Γ and of the configuration space can be better understood if we imagine embedding the graph Γ in the upper half plane $\mathcal{H} := \{z \in \mathbb{C} | \, \Im(z) \geq 0\}$ binding the vertices of the second type to the real line.

We can now introduce the open configuration space of the $n + \bar{n}$ distinct vertices of Γ as the smooth manifold:

$$\mathrm{Conf}_{n,\bar{n}} := \Big\{ (z_1, \dots, z_n, z_{\bar{1}}, \dots, z_{\bar{n}}) \in \mathbb{C}^{n+\bar{n}} \Big| \, z_i \in \mathcal{H}^+, z_{\bar{i}} \in \mathbb{R}, \\ z_i \neq z_j \text{ for } i \neq j, \; z_{\bar{i}} \neq z_{\bar{j}} \text{ for } \bar{i} \neq \bar{j} \Big\}.$$

In order to get the right configuration space we have to quotient $\mathrm{Conf}_{n,\bar{n}}$ by the action of the 2-dimensional Lie group G consisting of translations in the horizontal direction and rescaling, whose action on a given point $z \in \mathcal{H}$ is given by:

$$z \mapsto az + b \qquad\qquad a \in \mathbb{R}^+, b \in \mathbb{R}.$$

In virtue of the condition imposed on the number of vertices in (5.1), the action of G is free; therefore the quotient space, which will be denoted by $C_{n,\bar{n}}$, is again a smooth manifold, of (real) dimension $2n + \bar{n} - 2$.

Particular care has to be devoted to the case when the graph has no vertices of the second type. In this situation, having no points on the real line, the open configuration space can be defined as a subset of $\mathbb{C}^n$ instead of $\mathcal{H}^n$ and we can introduce a more general Lie group G', acting by rescaling and translation in any direction; the quotient space $C_n := \mathrm{Conf}_{n,0} /_{G'}$ for $n \geq 2$ is again a smooth manifold, of dimension $2n - 3$.

In order to get a connected manifold, we restrict further our attention to the component $C^+_{n,\bar{n}}$ in which the vertices of the second type are ordered along the real line in ascending order, namely:

$$C^+_{n,\bar{n}} := \Big\{ (z_1, \dots, z_n, z_{\bar{1}}, \dots, z_{\bar{n}}) \in C_{n,\bar{n}} \Big| \, z_{\bar{i}} < z_{\bar{j}} \text{ for } \bar{i} < \bar{j} \Big\}.$$

On these spaces we can finally introduce the differential form ω_Γ. We first define an angle map

$$\phi \colon C_{2,0} \longrightarrow S^1$$

which associates to each pair of distinct points z_1, z_2 in the upper half plane the angle between the geodesics w.r.t. the Poincaré metric connecting z_1 to $+i\infty$ and to z_2, measured in the counterclockwise direction (cfr. Fig. 5.3).

The differential of this function is now a well-defined 1-form on $C_{2,0}$ which we can pull-back to the configuration space corresponding to the

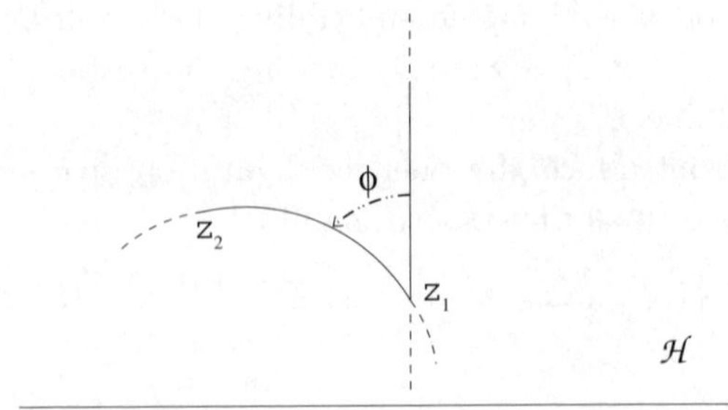

Figure 5.3 The angle map ϕ

whole graph with the help of the natural projection π_e associated to each edge $e = (z_i, z_j)$ of Γ

$$\begin{array}{rccc} \pi_e : & C_{n,\bar{n}} & \longrightarrow & C_{2,0} \\ & (z_1, \dots, z_{\bar{n}}) & \mapsto & (z_i, z_j) \end{array}$$

to obtain $d\phi_e := \pi_e^* \, d\phi \in \Omega^1(C_{n,\bar{n}})$. The form that appears in the definition of the weight w_Γ can now be defined as

$$\omega_\Gamma := \bigwedge_{e \in \Gamma} d\phi_e$$

where the ordering of the 1-forms in the product is the one induced on the set of all edges by the ordering on the (first) vertices and the ordering on the set Star(v) of edges emerging from the vertex v. We want to remark hereby that, as long as we consider graphs with $2n + \bar{n} - 2$ edges, the degree of the form matches exactly the dimension of the space over which it has to be integrated, which gives us a real valued weight.

This geometric construction has a more natural interpretation if one derives the Kontsevich formula for the star product from a path integral approach, as it was done for the first time in [11].

For the weights to be well-defined, we also have to require that the integrals involved converge. However, as the geometric construction of ϕ suggests, as soon as two points approach each other, the differential form $d\phi$ is not defined. The solution to this problem has already been given implicitly in (27): the differential form is not integrated over the open configuration space, but on a suitable compact space whose definition and properties are contained in the following

Lemma 5.3 *For any configuration space $C_{n,\bar{n}}$ (resp. C_n) there exists a compact space $\bar{C}_{n,\bar{n}}$ (resp. $\bar{C}_n$) whose interior is the open configuration space and such that the projections π_e, the angle map ϕ and thus the*

differential form ω_Γ extend smoothly to the corresponding compactifications.

The compactified configuration spaces are (compact) smooth manifolds with corners. We recall that a smooth manifold with corner of dimension m is a topological Hausdorff space M which is locally homeomorphic to $\mathbb{R}^{m-n} \times \mathbb{R}^n_+$ with $n = 0, \dots, m$. The points $x \in M$ whose local expression in some (and thus any) chart has the form $x_1, \dots, x_{m-n}, 0, \dots, 0)$ are said to be of type n and form submanifolds of M called strata of codimension n.

The general idea behind such a compactification is that the naive approach of considering the closure of the open space in the cartesian product would not take into account the different speeds with which two or more points "collapse" together on the boundary of the configuration space.

For a more detailed description of the compactification we refer the reader to [22] for an algebraic approach and to [4] and [8] for an explicit description in local coordinates. More recently Sinha [42] gave a simplified construction in the spirit of Kontsevich's original ideas. In [3] the orientation of such spaces and of their codimension one strata – whose relevance will be clarified in the following – is discussed.

Finally, the integral in (27) is well-defined and yields a weight $w_\Gamma \in \mathbb{R}$ for any admissible graph Γ, since we defined w_Γ to be non zero only when Γ has exactly $2n + m - 2$ edges, i.e. when the degree of ω_Γ matches the dimension of the corresponding configuration space.

5.2 The proof: Lemmas, Stokes' theorem, Vanishing theorems

Having defined all the tools we will need, we can now give a sketch of the proof.

In order to verify that U defines the required L_∞ morphism we have to check that the following conditions hold:

I The first component of the restriction of U to $\mathcal{V}$ is – up to a shift in the degrees of the two DGLAs – the natural map introduced in Section (3.3).

II U is a graded linear map of degree 0.

III U satisfies the equations for an L_∞-morphism defined in Section (4).

Lemma 5.4 I *The map*

$$U_1 \colon \mathcal{V} \longrightarrow \mathcal{D}$$

is the natural map that identifies each multivector field with the corresponding multiderivation.

Proof The set $\mathcal{G}_{1,\bar{n}}$ consist of only one element, namely the graph $\Gamma_{\bar{n}}$ with one vertex of the first type with $2 \cdot 1 + \bar{n} - 2 = \bar{n}$ arrows with an equal number of vertices of the second type as endpoints.

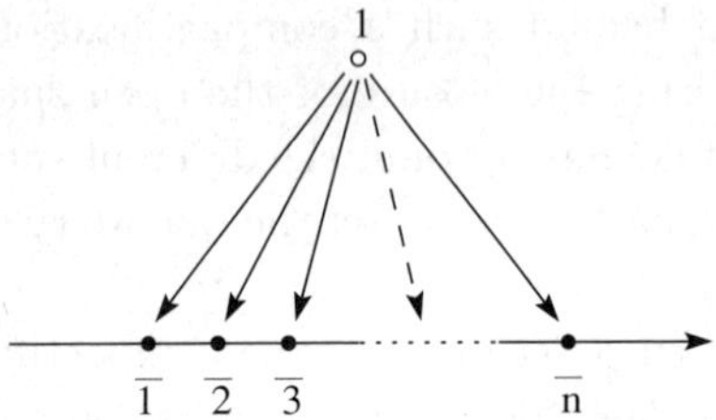

Figure 5.4 The admissible graph $\Gamma_{\bar{n}}$

To each k-vector field ξ we associate thus the multidifferential operator given by

$$U_{\Gamma_{\bar{n}}}(\xi)(f_{\bar{1}}, \ldots, f_{\bar{n}}) := w_{\Gamma_{\bar{n}}}\, \xi^{i_{\bar{1}},\ldots,i_{\bar{n}}}\, \partial_{i_{\bar{1}}} f_{\bar{1}} \cdots \partial_{i_{\bar{n}}} f_{\bar{n}}.$$

An easy computation shows that the integral of $\omega_{\Gamma_{\bar{n}}}$ over $\bar{C}_{1,\bar{n}}$ cancels the power of $\frac{1}{2\pi}$ and leaves us with the right weight

$$w_{\Gamma_{\bar{n}}} = \frac{1}{\bar{n}!}$$

we expect for U_1 to be the natural map that induces the HKR isomorphism. □

Lemma 5.5 II *The n-th component*

$$U_n := \sum_{\bar{n}=1}^{\infty} \sum_{\Gamma \in \mathcal{G}_{n,\bar{n}}} w_\Gamma B_\Gamma$$

has the right degree for U to be an L_∞-morphism.

Proof To each vertex v_i with $\#\operatorname{Star}(v_i)$ outgoing arrows corresponds an element of $\mathcal{V}^{r_i} = \widetilde{\mathcal{V}}^{r_i+1}$ where $r_i = \#\operatorname{Star}(v_i)$. On the other side, each graph with $\bar{n}$ vertices of the second type together with an n-tuple of multivector fields gives rise to a differential operator of degree $s = \bar{n} - 1$.

Since we consider only graphs with $2n + \bar{n} - 2$ edges and this is equal by construction to

$$\sum_{i=1}^{n} \#\,\mathrm{Star}(v_i),$$

the degree of $U_n(\xi_1, \ldots, \xi_n)$ can be written as

$$s = (2n + \bar{n} - 2) + 1 - n = \sum_{i=1}^{n} r_i + 1 - n$$

which is exactly the prescribed degree for the n-th component of an L_∞-morphism.

Although the construction we gave in the previous section involves a tensor product of multivector fields, the signs and weights in U_n are chosen in such a way that, upon symmetrization, it descends to the symmetric algebra. □

We come now to the main part of Kontsevich's construction: the geometric proof of the formality.

First of all we have to extend our morphism U to include also a 0-th component which represents the usual multiplication between smooth functions — the associative product we want to deform via the higher order corrections. We can now specialize the L_∞ condition (24) to the case at hand, where m_0 $\widetilde{m}_0$ can be expressed in terms of of the Taylor coefficients U_n as:

$$\sum_{l=0}^{n} \sum_{k=-1}^{m} \sum_{i=0}^{m-k} \varepsilon_{kim} \sum_{\sigma \in S_{l,n-l}} \varepsilon_\xi(\sigma)\, U_l\Big(\xi_{\sigma(1)}, \ldots, \xi_{\sigma(l)}\Big)$$
$$\Big(f_0 \otimes \cdots \otimes f_{i-1} \otimes U_{n-l}(\xi_{\sigma(l+1)}, \ldots, \xi_{\sigma(n)})(f_i \otimes \cdots \otimes f_{i+k}) \otimes f_{i+k+1} \otimes \cdots \otimes f_m\Big)$$
$$= \sum_{i \neq j = 1}^{n} \varepsilon_\xi^{ij}\, U_{n-1}(\xi_i \circ \xi_j, \xi_1, \ldots, \widehat{\xi_i}, \ldots, \widehat{\xi_j}, \ldots, \xi_n)(f_0 \otimes \cdots \otimes f_n), \quad (28)$$

where

- $\{\xi_j\}_{j=1,\ldots,n}$ are multivector fields;
- $f_0, \ldots, f_m$ are the smooth functions on which the multidifferential operator is acting;
- $S_{l,n-l}$ is the subset of S_n consisting of $(l, n-l)$-shuffles
- the product $\xi_i \circ \xi_j$ is defined in such a way that the Schouten–Nijenhuis bracket can be expressed in terms of this composition by

a formula similar to the one relating the Gerstenhaber bracket to the analogous composition $\circ$ on $\mathcal{D}$ given in 3.2.2;
- the signs involved are defined as follows: $\varepsilon_{kim} := (-1)^{k(m+i)}$, $\varepsilon_{\xi}(\sigma)$ is the Koszul sign associated to the permutation σ and ε_{ξ}^{ij} is defined as in (24).

This equation encodes the formality condition since the l.h.s. corresponds to the Gerstenhaber bracket between multidifferential operators while the r.h.s. contains "one half" of the Schouten–Nijenhuis bracket; the differentials do not appear explicitly since on $\mathcal{V}$ we defined d to be identically zero, while on $\mathcal{D}$ it is expressed in terms of the bracket with the multiplication $\mathfrak{m}$, which we included in the equation as U_0.

For a detailed explanation of the signs involved we refer again to [3].

We can now rewrite equation (28) in a form that involves again admissible graphs and weights to show that it actually holds. It should be clear from the previous construction of the coefficients U_k that the difference between the l.h.s. and the r.h.s. of equation (28) can be written as a linear combination of the form

$$\sum_{\Gamma \in \mathcal{G}_{n,\bar{n}}} c_\Gamma U_\Gamma(\xi_1, \ldots, \xi_n)(f_0 \otimes \cdots \otimes f_n) \tag{29}$$

where the the sum runs in this case over the set of admissible graphs with $2n + \bar{n} - 3$ edges. Equation (28) is thus fulfilled for every n if these coefficients c_Γ vanish for every such graph.

The main tool to prove the vanishing of these coefficients is the Stokes Theorem for manifolds with corners, which ensures that also in this case the integral of an exact form $\mathrm{d}\Omega$ on a manifold M can be expressed as the integral of Ω on the boundary ∂M. In the case at hand, this implies that if we choose as Ω the differential form ω_γ corresponding to an admissible graph, since each $d\phi_e$ is obviously closed and the manifolds $\bar{C}^+_{n,\bar{n}}$ are compact by construction, the following holds:

$$\int_{\partial \bar{C}^+_{n,\bar{n}}} \omega_\Gamma = \int_{\bar{C}^+_{n,\bar{n}}} d\,\omega_\Gamma = 0. \tag{30}$$

We will now expand the l.h.s. of (30) to show that it gives exactly the coefficient c_Γ occurring in (29) for the corresponding admissible graph.

First of all, we want to give an explicit description of the manifold $\partial \bar{C}^+_{n,\bar{n}}$ on which the integration is performed. Since the weights w_Γ involved in (28) are set to vanish identically if the degree of the differential form does not match the dimension of the space on which we integrate, we can restrict our attention to codimension 1 strata of $\partial \bar{C}^+_{n,\bar{n}}$, which have the required dimension $2n + \bar{n} - 3$ equal to the number of edges and thus of the 1-forms $d\phi_e$.

In an intuitive description of the configuration space $\bar{C}_{n,\bar{n}}$, the boundary represents the degenerate configurations in which some of the $n + \bar{n}$ points "collapse together". The codimension 1 strata of the boundary can thus be classified as follows:

- strata of type S1, in which $i \geq 2$ points in the upper half plane $\mathcal{H}^+$ collapse together to a point still lying above the real line. Points in such a stratum can be locally described by the product
$$C_i \times C_{n-i+1,\bar{n}}. \tag{31}$$
where the first term stand for the relative position of the collapsing points as viewed "through a magnifying glass" and the second is the space of the remaining points plus a single point toward which the first i collapse.
- strata of type S2, in which $i > 0$ points in $\mathcal{H}^+$ and $j > 0$ points in $\mathbb{R}$ with $2i + j \geq 2$ collapse to a single point on the real line. The limit configuration is given in this case by
$$C_{i,j} \times C_{n-i,\bar{n}-j+1}. \tag{32}$$

These strata have a pictorial representation in Figure 5.5. In both cases the integral of ω_Γ over the stratum can be split into a product of two integrals of the form (27): the product of those $d\phi_e$ for which the edge e connects two collapsing points is integrated over the first component in the decomposition of the stratum given by (31) resp. (32), while the remaining 1-forms are integrated over the second.

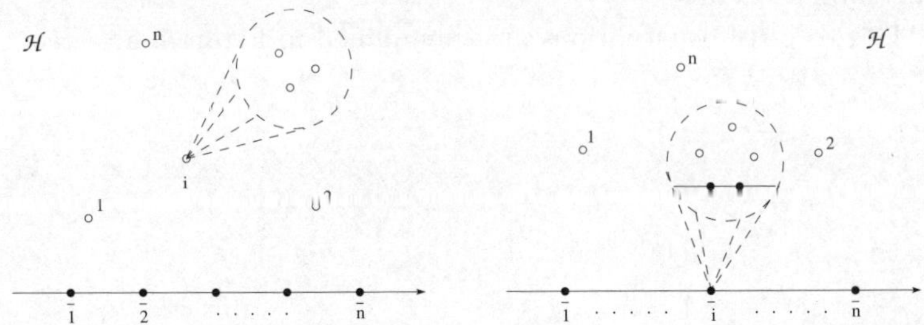

Figure 5.5 Looking at codimension 1 strata "through a magnifying glass".

According to this description, we can split the integral in the l.h.s. of (30) into a sum over different terms coming from strata of type S1 and S2. Now we are going to list all the possible configurations leading to such strata to show that most of these terms vanish and that the only remaining terms are exactly those required to give rise to (28). We will not check directly that the signs we get by the integration match with those in (28), since we did not give explicitly the orientation of the configuration spaces and of their boundaries, but we refer once again the reader to the only paper completely devoted to the careful computation of all signs involved in Kontsevich's construction [3].

Among the strata of type S1, we distinguish two subcases, according to the number i of vertices collapsing. Since the integrals are set to vanish if the degree of the form does not match the dimension of the domain, a simple dimensional argument shows that the only contributions come from those graphs Γ whose subgraph Γ_1 spanned by the collapsing vertices contains exactly $2i-3$ edges.

If $i = 2$ there is only an edge e involved and in the first integral coming from the decomposition (31) the differential of the angle function is integrated over $C_2 \cong S^1$ and we get (up to a sign) a factor 2π which cancels the coefficient in (27). The remaining integral represents the weight of the corresponding quotient graph Γ_2 obtained from the original graph after the contraction of e: to the vertex j of type I resulting from this contraction is now associated the j-composition of the two multivector fields that were associated to the endpoints of e. Therefore, summing over all graphs and all strata of this subtype we get the r.h.s. of the desired equation (28).

If $i \geq 3$, the integral corresponding to this stratum involves the product of $2i-3$ angle forms over C_i and vanishes according to the following Lemma, which contains the most technical result among Kontsevich's "vanishing theorems".

The two possible situations are exemplified in Figure 5.6.

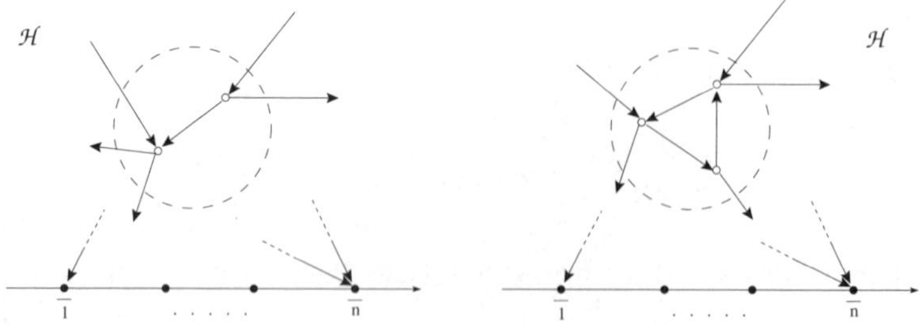

Figure 5.6 Example of a non vanishing and of a vanishing term.

Lemma 5.6 *The integral over the configuration space C_n of $n \geq 3$ points in the upper half plane of any $2n-3$ $(= \dim C_n)$ angle forms $d\phi_{e_i}$ with $i = 1, \ldots n$ vanishes for $n \geq 3$*

Proof The first step consists in restricting the integration to an even number of angle forms. This is achieved by identifying the configuration space C_n with the subset of $\mathcal{H}^n$ where one of the endpoints of e_1 is set to be the origin and the second is forced to lie on the unit circle (this particular configuration can always be achieved with the help of the action of the Lie group G'). The integral decomposes then into a product of $d\phi_{e_1}$ integrated over S^1 and the remaining $2n-4 =: 2N$ forms integrated over the resulting complex manifold U given by the isomorphism $C_n \cong S^1 \times U$. The claim is then a consequence of the following chain of equalities:

$$\int_U \bigwedge_{j=1}^{2N} d\arg(f_j) = \int_U \bigwedge_{j=1}^{2N} d\log|f_j| = \int_{\overline{U}} \mathcal{I}\Big(d\big(\log|f_1| \bigwedge_{j=2}^{2N} d\log|z_j|\big)\Big) =$$
$$= \int_{\overline{U}} d\mathcal{I}\Big(\big(\log|f_1| \bigwedge_{j=2}^{2N} d\log|z_j|\big)\Big) = 0 \tag{33}$$

where we gave an expression for the angle function ϕ_{e_j} in terms of the argument of the (holomorphic) function f_j (which is nothing but the difference of the coordinates of the endpoints of e_j).

The first equality is what Kontsevich calls a "trick using logarithms" and follows from the decompositions

$$d\arg(f_j) = \frac{1}{2i}\left(d\log(f_j) - d\log(\overline{f}_j)\right)$$

and

$$d\log|f_j| = \frac{1}{2}\left(d\log(f_j) + d\log(\overline{f}_j)\right).$$

The product of $2N$ such expressions is thus a linear combination of products of k holomorphic and $2N-k$ anti-holomorphic forms. A basic result in complex analysis ensures that, upon integration over the complex manifold U, the only terms that do not vanish are those with $k = N$. It is a straightforward computation to check that the non vanishing terms coming from the first decomposition match with those coming from the second.

In the second equality the integral of the differential form is replaced by the integration of a suitable 1-form with values in the space of distributions over the compactification $\overline{U}$ of U. A final Lemma in [34] shows that this map $\mathcal{I}$ from standard to distributional 1-forms commutes with the differential, thus proving the last step in (33). In [30], Khovanskii gave a more elegant proof of this result in the category of complete complex algebraic varieties, deriving the first equality rigorously on the set of non singular points of X and resolving the singularities with the help of a local representation in polar coordinates. □

Finally, turning our attention to the strata of type S2, the same dimensional argument introduced for the previous case restricts the possible non vanishing terms to the condition that the subgraph Γ_1 spanned by the $i+j$ collapsing vertices (resp. of the first and of the second type) contains exactly $2i+j-2$ edges.

With the same definition as before for the quotient graph Γ_2 obtained by contracting Γ_1, we claim that the only non vanishing contributions come from those graphs for which both graphs obtained from a given Γ are admissible. In this case the weight w_Γ will decompose into the product $w_{\Gamma_1} \cdot w_{\Gamma_2}$ which in general, by the conditions on the number of edges of Γ and Γ_1, does not vanish.

Since all other properties required by Definition 5.1 are inherited from Γ, we have only to check that we do not get "bad edges" by contraction. The only such possibility is depicted in the graph on the right in Figure 5.7 and occurs when Γ_2 contains an edge which starts from a vertex of the second type: in this case the corresponding integral vanishes because it contains the differential of an angle function evaluated on the pair (z_1, z_2), where the first point is constrained to lie on the real line and such a function vanishes for every z_2 because the angle is measured w.r.t. the Poincaré metric (as it can be inferred intuitively from Figure 5.3).

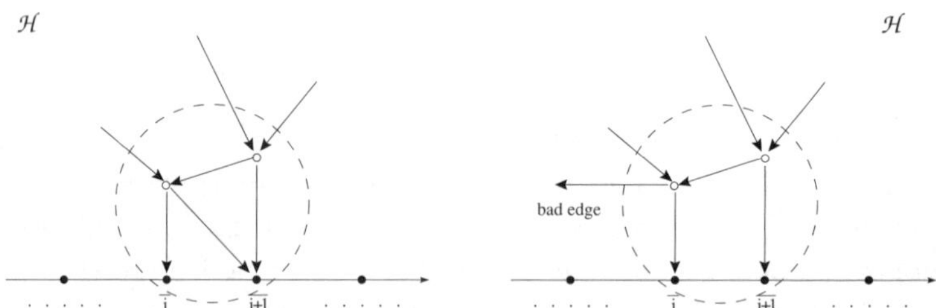

Figure 5.7 Example of a collapse leading to an admissible quotient graph and of a collapse corresponding to a vanishing term because of a bad edge.

The only non vanishing terms thus correspond to the case when we plug the differential operator corresponding to the subgraph Γ_1 as k-th argument of the one corresponding to Γ_2, where k is the vertex of the second type emerging from the collapse. Summing over all such possibilities and having checked (up to a sign as usual) that we get the right weights, it should be clear that the contribution due to the strata of type S2 accounts for the l.h.s. of (28).

In conclusion, we have proved that the morphism U is an L_∞-morphism and since its first coefficient U_1 coincides with the map $U_1^{(0)}$ given in Section 3.3 it is also a quasi-isomorphism and thus determines uniquely a star product given by (26) for any given bivector field π on $\mathbb{R}^d$.

6
From local to global deformation quantization

The content of this last section is based mainly on the work of Cattaneo, Felder and Tomassini [13] (see also [14] and [12]), who gave a direct construction of the quantization of a general Poisson manifold.

The Kontsevich formula, in fact, gives a quantization only for the case $M = \mathbb{R}^d$ for any Poisson bivector field π and can thus be adopted in the general case to give only a local expression of the star product.

The globalization Kontsevich sketched in [34] was carried through in [35] by abstract arguments, extending the formality theorem to the general case.

The works of Cattaneo, Felder and Tomassini instead give an explicit recipe to define the star product globally, in a similar way to what Fedosov has done in the symplectic category [19]. Also in their approach, the main tool is a flat connection $\overline{D}$ on a vector bundle over M such that the algebra of the horizontal sections w.r.t. to $\overline{D}$ is a quantization of the Poisson algebra of the manifold.

We give now an outline of the construction, addressing the reader to [13] for details and proofs.

In the first step, we introduce the vector bundle $E_0 \to M$ of infinite jets of functions together with the canonical flat connection D_0. The fiber E_0^x over $x \in M$ is naturally a commutative algebra and inherits the Poisson structure induced fiberwise by the Poisson structure on $C^\infty(M)$. The canonical map which associates to any globally defined function its infinite jet at each point x is a Poisson isomorphism onto the Poisson algebra of horizontal sections of E_0 w.r.t. D_0.

As the star product yields a deformation of the pointwise product on $C^\infty(M)$, we need also a "quantum version" of the vector bundle and of the flat connection in order to find an analogous isomorphism. The vector bundle $E \to M$ is defined in terms of a section ϕ^{aff} of the fiber bundle $M^{aff} \to M$, where M^{aff} is the quotient of the manifold M^{coor} of jets of coordinates systems on M by the action of the group $\mathrm{GL}(d, \mathbb{R})$ of linear diffeomorphisms, namely $E := (\phi^{aff})^* \widetilde{E}$ where $\widetilde{E}$ is the bundle of $\mathbb{R}[[\epsilon]]$-modules

$$M^{coor} \times_{\mathrm{GL}(d,\mathbb{R})} \mathbb{R}[[y^1, \ldots, y^d]][[\epsilon]] \to M^{aff}.$$

Since the section ϕ^{aff} can be realized explicitly by a collection of infinite jets at 0 of maps $\phi_x \colon \mathbb{R}^d \to M$ such that $\phi_x(0) = x$ for every $x \in M$ (defined modulo the action of $GL(d, \mathbb{R})$), we can suppose for simplicity that we have fixed a representative ϕ_x of the equivalence class for each open set of a given covering, thus realizing a trivialization of the bundle E. Therefore, from now on we will identify E with the trivial bundle with fiber $\mathbb{R}[[y^1, \ldots, Y^d]][[\epsilon]]$; in this way E realizes the desired quantization, since it is isomorphic (as a bundle of $\mathbb{R}[[\epsilon]]$-modules) to the bundle $E_0[[\epsilon]]$ whose elements are formal power series with infinite jets of functions as coefficients.

In order to define the star product and the connection on E, we have first to introduce some new objects whose existence and properties are byproducts of the formality theorem. Given a Poisson bivector field π and two vector fields ξ and η on $\mathbb{R}^d$, we define:

$$\begin{aligned} P(\pi) &:= \sum_{k=0}^{\infty} \frac{\epsilon^k}{k!} U_k(\pi, \ldots, \pi), \\ A(\xi, \pi) &:= \sum_{k=0}^{\infty} \frac{\epsilon^k}{k!} U_{k+1}(\xi, \pi, \ldots, \pi), \\ F(\xi, \eta, \pi) &:= \sum_{k=0}^{\infty} \frac{\epsilon^k}{k!} U_{k+2}(\xi, \eta, \pi, \ldots, \pi). \end{aligned} \tag{34}$$

A straightforward computation of the degree of the multidifferential operators on the r.h.s. of (34) shows that $P(\pi)$ is a (formal) bidifferential operator, $A(\xi, \pi)$ a differential operator and $F(\xi, \eta, \pi)$ a function. Indeed $P(\pi)$ is nothing but the star product associated to π as introduced at the end of Section 4.

More precisely, P, A and F are elements of degree resp. 0, 1 and 2 of the Lie algebra cohomology complex of (formal) vector fields with values in the space of local polynomial maps, i.e. multidifferential operators depending polynomially on π: an element of degree k of this complex is a map that sends $\xi_1 \wedge \cdots \wedge \xi_k$ to a multidifferential operator $S(\xi_1, \ldots, \xi_k, \pi)$ (we refer the reader to [13] for details). The differential δ on this complex is then defined by

$$\begin{aligned}\delta\, S(\xi_1,\ldots,\xi_{k+1},\pi) \;:=&\\ &\sum_{i=1}^{k+1}(-)^i \frac{\partial}{\partial t}\Big|_{t=0} S(\xi_1,\ldots,\hat{\xi}_i,\ldots,\xi_{k+1},(\Phi^t_\xi)_*\,\pi)\\ &\quad+\sum_{i<j}(-)^{i+j}\; S([\,\xi_i\,,\,\xi_j\,],\xi_1,\ldots,\hat{\xi}_i,\ldots,\hat{\xi}_j,\ldots,\xi_{k+1},\pi)\end{aligned}$$

where a caret denotes as usual the omission of the corresponding argument and Φ^t_ξ is the flow of the vector field xi.

As the associativity condition on the star product, which can now be written in the form $P\circ(P\otimes\mathrm{id}-\mathrm{id}\otimes P)=0$, follows from the formality theorem, the following equations are a corollary of the same result and can be proved with analogous computations:

$$P(\pi)\circ(A(\xi,\pi)\otimes\mathrm{id}+\mathrm{id}\otimes A(\xi,\pi))=A(\xi,\pi)\circ P(\pi)+\delta P(\xi,\pi)$$

$$\begin{aligned}P(\pi)\circ(F(\xi,\eta,\pi)\otimes\mathrm{id}-\mathrm{id}\otimes F(\xi,\eta,\pi))&\\ =-A(\xi,\pi)\circ A(\eta,\pi)+A(\eta,\pi)\circ A(\xi,\pi)&\\ +\,\delta A(\xi,\eta,\pi)&\end{aligned}\tag{35}$$

$$\begin{aligned}-\,A(\xi,\pi)\circ F(\eta,\zeta,\pi)-A(\eta,\pi)\circ F(\zeta,\xi,\pi)-A(\zeta,\pi)\circ F(\xi,\eta,\pi)&\\ =\delta F(\xi,\eta,\zeta,\pi)&\end{aligned}$$

The first of these equations describes the fact that under the coordinate transformation induced by ξ the star product $P(\pi)$ is changed to an equivalent one up to higher order terms. The last two equations will be used in the construction of the connection and its curvature, since they represent an analogous of the defining relations between a connection 1-form A and its curvature F_A.

Upon explicit computation of the configuration space integrals involved in the definition of the Taylor coefficients U_k, we can also give the lowest order terms in the expansion of P, A and F and their action on functions:

(i) $P(\pi)(f\otimes g)=f\,g+\epsilon\,\pi(df,dg)+O(\epsilon^2)$;
(ii) $A(\xi,\pi)=\xi+O(\epsilon)$, where we identify ξ with a first order differential operator on the r.h.s.;
(iii) $A(\xi,\pi)=\xi$, if ξ is a linear vector field;
(iv) $F(\xi,\eta,\alpha)=O(\epsilon)$;
(v) $P(\pi)(1\otimes f)=P(\pi)(f\otimes 1)=f$;

(vi) $A(\xi, \pi)1 = 0$.

Equations i) and v) where already introduced in Definition 2.1 as two of the defining conditions of a star product, while the ones involving A are used to construct a connection D on sections of E.

A section $f \in \Gamma(E)$ is given locally by a map $x \to f_x$ where for every y, $f_x(y)$ is a formal power series whose coefficients are infinite jets. On the space of such sections we can introduce a deformed product $\star$ which will give us the desired star product on $C^\infty(M)$ once we identify horizontal sections with ordinary functions. Denoting analogously by π_x the push-forward by ϕ_x^{-1} of the Poisson bivector π on $\mathbb{R}^d$, we can define the deformed product through the formal bidifferential operator $P(\pi_x)$ in the same way as $P(\pi)$ represents the usual star product:

$$(f \star g)_x(y) := f_x(y)\, g_x(y) + \epsilon\, \pi_x^{ij}(y) \frac{\partial f_x(y)}{\partial y^i} \frac{\partial g_x(y)}{\partial y^j} + O(\epsilon^2).$$

We can define the connection D on $\Gamma(E)$ by

$$(D\, f)_x = d_x f + A_x^M f$$

where $d_x f$ is the de Rham differential of f regarded as a function with values in $\mathbb{R}[[y^1, \ldots, y^d]][[\epsilon]]$ and the formal connection 1-form is specified by its action on a tangent vector ξ by

$$A_x^M(\xi) = A(\hat{\xi}_x, \pi_x)$$

where A is the operator defined in (34) evaluated on the multivector fields ξ and π expressed in the local coordinate system given by ϕ_x.

The important point is that since the coefficients U_k of the formality map that appear in the definition of P and A are polynomial in the derivatives of the coordinate of the arguments ξ and π, all results holding for $P(\pi)$ and $A(\xi, \pi)$ are inherited by their formal counterparts. In particular equalities i) and v) above (together with the formality theorem from which they are derived) ensure that $\star$ is an associative deformation of the pointwise product on sections and equalities ii) and iii) can be used to prove that D is indeed independent of the choice of ϕ and therefore induces a global connection on E.

We can finally extend D and $\star$ by the (graded) Leibniz rule to the whole complex of formal differential forms $\Omega^\bullet(E) = \Omega M \otimes_{C^\infty(M)} \Gamma(E)$ and use (35) to verify the following

Lemma 6.1 *Let F^M be the E-valued 2-form given by $x \to F_x^M$ where $F_x^M(\xi, \eta) = F(\hat{\xi}_x, \hat{\eta}_x, \pi_x)$ for any pair of vector fields ξ, η. Then F^M*

represent the curvature of D *and the two are related to each other and to the star product by the usual identities:*

a) $D(f \star g) = D(f) \star g + f \star D(g)$*;*
b) $D^2(\cdot) = [\, F^M \overset{\star}{,} \cdot \,]$*;*
c) $D\, F^M = 0$

Proof The identities follow directly from the relations (35), in which the star commutator $[\, f \overset{\star}{,} g \,] = f \star g - g \star f$ is already implicitly defined, once we identify the complex of formal multivector fields endowed with the differential δ with the complex of formal multidifferential operators with the de Rham differential. The map that realizes this isomorphism is explicitly defined in [13]. □

A connection D satisfying the above relations on a bundle E of associative algebras is called a Fedosov connection with Weyl curvature F: it is the kind of connection Fedosov introduced to give a global construction in the symplectic case. Following Fedosov, the last step to the required globalization is to deform D into a new connection $\overline{D}$ which enjoys the same properties and moreover has zero Weyl curvature, so that we can define the complex $H^k(E, \overline{D})$ and in particular the (sub)algebra of horizontal sections $H^0(E, \overline{D})$.

The construction of $\overline{D}$ relies on the following Lemmata.

Lemma 6.2 *Let* D *be a Fedosov connection on* E *with Weyl curvature* F *and* γ *an* E*-valued* 1*-form, then*

$$\overline{D} := D + [\, \gamma \overset{\star}{,} \cdot \,]$$

is also a Fedosov connection whose Weyl curvature is $\overline{F} = F + D\,\gamma + \gamma \star \gamma$*.*

Proof For any given section f, a direct computation shows

$$\begin{aligned}\overline{D}^2 f &= [\, F \overset{\star}{,} f \,] + D[\, \gamma \overset{\star}{,} f \,] + [\, \gamma \overset{\star}{,} Df \,] + [\, \gamma \overset{\star}{,} [\, \gamma \overset{\star}{,} f \,]] = \\ &= [\, F \overset{\star}{,} f \,] + [\, D\gamma \overset{\star}{,} f \,] + [\, \gamma \overset{\star}{,} [\, \gamma \overset{\star}{,} f \,]] = \left[\, F + D\,\gamma + \frac{1}{2}[\, \gamma \overset{\star}{,} \gamma \,] \overset{\star}{,} f \,\right]\end{aligned}$$

where the last equality follows from the Jacobi identity for the star commutator, since every associative product induces a Lie bracket given by the commutator.

Applying $\overline{D}$ on the new curvature, we can check explicitly that

$$\begin{aligned}\overline{D}\Big(F + D\gamma + \frac{1}{2}[\gamma \stackrel{\star}{,} \gamma]\Big) \\ &= D^2\gamma + \frac{1}{2}[D\gamma \stackrel{\star}{,} \gamma] - \frac{1}{2}[\gamma \stackrel{\star}{,} D\gamma] + [\gamma \stackrel{\star}{,} F + D\gamma] \\ &= [F \stackrel{\star}{,} \gamma] + [\gamma \stackrel{\star}{,} F] \\ &= 0\end{aligned}$$

where we made use again of the (graded) Jacobi identity and of the (graded) skew-symmetry of $[\ \stackrel{\star}{,}\ ]$. □

Lemma 6.3 *Let D be a Fedosov connection on a bundle $E = E_0[[\epsilon]]$ and F its Weyl curvature and let*

$$D = D_0 + \epsilon D_1 + \cdots \qquad \text{and} \qquad F = F_0 + \epsilon F_1 + \cdots$$

be their expansions as formal power series. If $F_0 = 0$ and the second cohomology of E_0 w.r.t. D_0 is trivial, there exist a 1-form γ such that $\overline{D}$ has zero Weyl curvature.

Proof By the previous Lemma, the claim is equivalent to the existence of a solution to the equation

$$\overline{F} = F + D\gamma + \frac{1}{2}[\gamma \stackrel{\star}{,} \gamma] = 0.$$

A solution can be explicitly constructed by induction on the order in ϵ, starting from $\gamma_0 = 0$ and assuming that $\gamma^{(k)}$ is a solution mod ϵ^{k+1}. We can thus add to $\overline{F}^k = F + D\gamma^{(k)} + \frac{1}{2}\big[\gamma^{(k)} \stackrel{\star}{,} \gamma^{(k)}\big]$ the next term $\epsilon^{k+1} D_0\, \gamma_{k+1}$ to get $\overline{F}^{(k+1)}$ modulo higher terms. From $D\overline{F}^{(k)} + \Big[\gamma^{(k)} \stackrel{\star}{,} \overline{F}^{(k)}\Big] = 0$ and the induction hypothesis $\overline{F}^{(k)} = 0$ mod ϵ^{k+1} we get $D_0\, \overline{F}^{(k)} = 0$. Since now $H^2(E_0, D_0) = 0$, we can invert D_0 to define γ_{k+1} in terms of the lower order terms $\overline{F}^{(k)}$ in such a way that $\overline{F}^{(k+1)} = 0$ is satisfied mod ϵ^{k+2}, thus completing the induction step. □

Since in our case D is a deformation of the natural flat connection D_0 on sections of the bundle of infinite jets, the hypothesis of the previous Lemma are satisfied and we can actually find a flat connection $\overline{D}$ which is still a good deformation of D_0.

A last technical Lemma gives us an isomorphism between the algebra of the horizontal sections $H^0(E, \overline{D})$ and its non-deformed counterpart

$H^0(E_0, D_0)$, which in turn is isomorphic to the Poisson algebra $C^\infty(M)$: this concludes the globalization procedure.

Only recently, Dolgushev [18] gave a new proof of Kontsevich's formality theorem for a general manifold. The main difference in this approach is that it is based on the use of covariant tensors unlike Kontsevich's original proof, which is based on ∞-jets of multidifferential operators and multivector fields and is therefore intrinsically local. In particular, he gave a solution of the deformation quantization problem for an arbitrary Poisson orbifold.

Bibliography

1. E. Arbarello, "Introduction to Kontsevich's Result on Deformation–Quantization of Poisson Structures," in: *Seminari di geometria algebrica*, Pisa (1999), 5–20.
2. M. Artin, "Deformation of singularities," Tata Institute of Fundamental Research, Bombay (1976).
3. D. Arnal, D. Manchon, M. Masmoudi, "Choix de signes pour la formalité de M. Kontsevich," `math.QA/0003003`.
4. S. Axelrod and I. Singer, "Chern-Simons perturbation theory II," *J. Diff. Geom.* **39** (1994), no. 1, 173–213.
5. F. Bayen, M. Flato, C. Fronsdal, A. Lichnerowicz and D. Sternheimer, "Quantum mechanics as a deformation of classical mechanics," *Lett. Math. Phys.* **1** (1977), 521–530.
6. M. Bertelson, M. Cahen and S. Gutt, "Equivalence of star products," *Class. Quantum Grav.* **14** (1997), A93–A107.
7. P. Bonneau, "Fedosov star products and one-differentiable deformations". *Lett. Math. Phys.* **45** (1998), 363–376.
8. R. Bott and C. Taubes, "On the self-linking of knots. Topology and physics," *J. Math. Phys* **35** (1994), no. 10, 5247–5287.
9. H. Bursztyn and A. Weinstein, "Poisson geometry and Morita equivalence," in this volume.
10. A. Canonaco, "L_∞-Algebras and Quasi–Isomorphisms," in: *Seminari di geometria algebrica*, Pisa (1999), 67–86.
11. A. S. Cattaneo and G. Felder, "A Path Integral Approach to the Kontsevich Quantization Formula," *C*ommun. Math. Phys. **212** (2000), 591–611.
12. A. S. Cattaneo and G. Felder, "On the globalization of Kontse-

vich's star product and the perturbative Poisson sigma model," *Prog. Theor. Phys. Suppl.* **144** (2001), 38–53.

13. A. S. Cattaneo, G. Felder and L. Tomassini, "From local to global deformation quantization of Poisson manifolds," *Duke Math. J.* **115** (2002), 329–352.
14. A. S. Cattaneo, G. Felder and L. Tomassini, "Fedosov connections on jet bundles and deformation quantization," in: *Deformation Quantization* (ed. G. Halbout), IRMA Lectures in Mathematics and Theoretical Physics (ed. V. Turaev), de Gruyter, Berlin, (2002), 191–202.
15. P. Deligne, "Déformations de l'Algèbre des Fonctions d'une Variété Symplectique : Comparaison entre Fedosov et DeWilde, Lecomte," *Selecta Math. N.S.* **1** (1995), 667–697.
16. M. De Wilde and P. B. A. Lecomte, "Existence of star-products and of formal deformations of the Poisson Lie algebra of arbitrary symplectic manifolds," *Lett. Math. Phys.* **7** (1983), no. 6, 487–496.
17. G. Dito and D. Sternheimer, "Deformation quantization: genesis, developments and metamorphoses," in *Deformation quantization*, Strasbourg (2001), 9–54, IRMA Lect. Math. Theor. Phys., 1, de Gruyter, Berlin (2002).
18. V. Dolgushev, "Covariant and Equivariant Formality Theorems," `math.QA/0307212`.
19. B. Fedosov, "Deformation quantization and index theory," Mathematical Topics 9, Akademie Verlag, Berlin (1996).
20. M. Flato, A. Lichnerowicz and D. Sternheimer, "Déformations 1-différentiables des algèbres de Lie attachées à une variété symplectique ou de contact,". *Compositio Math.* **31** (1975), no. 1, 47–82.
21. M. Flato, A. Lichnerowicz and D. Sternheimer, "Crochet de Moyal-Vey et quantification," *C. R. Acad. Sci. Paris Sér. A-B* **283** (1976), no. 1, Aii, A19–A24.
22. W. Fulton and R. MacPherson, "Compactification of configuration spaces," *Annals of Mathematics* **139** (1994), 183–225.
23. M. Gerstenhaber, "The cohomology structure of an associative ring," *Ann. Math.(2)* **78** (1963), 267–288.
24. M. Grassi, "DG (Co)Algebras, DG Lie Algebras and L_∞ Algebras," in: *Seminari di geometria algebrica*, Pisa (1999), 49–66.
25. H. J. Groenewold,"On the principles of elementary quantum mechanics," *Physics* **12** (1946), 405–460.
26. S. Gutt, *Déformations formelles de l'algèbre des fonctions différentiables sur une variété symplectique*, Thesis, Université Libre de Bruxelles (1980).

27. S. Gutt and J. Rawnsley, "Equivalence of star-products on a symplectic manifold: an introduction to Deligne's Čech cohomology class," *J. Geom. Phys.* **29** (1999), 347–392.
28. G. Hochschild, B. Kostant and A. Rosenberg, "Differential forms on regular affine algebras," *Trans. Amer. Math. Soc.* **02** (1962), 383–408.
29. N. Ikeda, "Two-dimensional gravity and nonlinear gauge theory," *Ann. Phys.* **235**, (1994) 435–464.
30. A. G. Khovanskii, "On a Lemma of Kontsevich," *Funktsional. Anal. i Prilozhen.* **31** (1997), no. 4, 89–91; translation in *Funct. Anal. Appl.* **31** (1997), no. 4, 296–298.
31. A. A. Kirillov, "Unitary representations of nilpotent Lie groups", Russian Math. Surveys 17 (4), (1962), 53–104; "Elements of the theory of representations", Springer, Berlin, (1976).
32. B. Kostant, "Quantization and unitary representations", Lects. in Mod. Anal. and Appl. III, Lecture Notes in Mathematics 170, Springer, New York (1970), 87–208.
33. M. Kontsevich, "Formality conjecture," pp. 139–156 in: D. Sternheimer, J. Rawnsley and S. Gutt (eds.) *Deformation theory and symplectic geometry*, Ascona (1996), Math. Phys. Stud. **20**, Kluwer Acad. Publ., Dordrecht (1997).
34. M. Kontsevich, "Deformation quantization of Poisson manifolds I," `math.QA/9709040`.
35. M. Kontsevich, "Deformation quantization of algebraic varieties," EuroConférence Moshé Flato 2000, Part III, Dijon. *Lett. Math. Phys.* **56** (2001), no. 3, 271–294.
36. M. Manetti, "Deformation Theory Via Differential Graded Lie Algebras," in: *Seminari di geometria algebrica*, Pisa (1999), 21–48.
37. J. E. Moyal, "Quantum mechanics as a statistical theory," *Proc. Cambridge Phyl. Soc.* (1949), 99–124.
38. R. Nest and R. Tsygan, "Algebraic index theorem," *Comm. Math. Phys.* **172** (1995), 223–262; "Algebraic index theorem for families," *Adv. Math.* **113** (1995), 151–205; "Formal deformations of symplectic manifolds with boundary," *J. Reine Angew. Math.* **481** (1996), 27–54.
39. P. Schaller and T. Strobl, "Poisson structure induced (topological) field theories," *Modern Phys. Lett. A* **9** (1994), no. 33, 3129–3136.
40. M. Schlessinger and J. Stasheff, "The Lie algebra structure on tangent cohomology and deformation theory," *J. Pure Appl. Algebra* **38** (1985), 313–322.

41. I. E. Segal, "Quantization of non-linear systems", *J. Math. Phys.* **1** (1960), 468–488.
42. D. Sinha, "Manifold theoretic compactifications of configuration spaces," math.GT/0306385.
43. J. M. Souriau, "Structure des Systèmes Dynamiques," Dunod, Paris (1969)
44. J. Stasheff, "The intrinsic bracket on the deformation complex of an associative algebra," *J. Pure Appl. Algebra,* **89** (1993), 231–235.
45. J. Vey, "Déformation du crochet de Poisson sur une variété symplectique," *Comment. Math. Helv.* **50** (1975), 421–454.
46. H. Weyl, "The theory of groups and quantum mechanics," Dover, New York (1931), translated from "Gruppentheorie und Quantenmechanik," Hirzel Verlag, Leipzig (1928); "Quantenmechanik und Gruppentheorie," *Z. Physik* **46** (1927), 1–46.
47. E. P. Wigner, "Quantum corrections for thermodynamic equilibrium," *Phys Rev.* **40** (1932), 749–759.
48. P. Xu, "Fedosov $\star$-Products and Quantum Momentum Maps," *Comm. Math. Phys.* **197** (1998), 167–197.

PART THREE

Lie groupoids, sheaves and cohomology

I. Moerdijk
moerdijk@math.uu.nl

Mathematical Institute
Utrecht University
P.O. Box 80.010
3508 TA Utrecht
The Netherlands

J. Mrčun
janez.mrcun@fmf.uni-lj.si

Department of Mathematics
University of Ljubljana
Jadranska 19
1000 Ljubljana
Slovenia

1

Introduction

The purpose of these lectures is to provide a quick introduction to a part of the theory of Lie groupoids, and to present some of their algebraic invariants, in particular the fundamental group and the equivariant sheaf cohomology.

In many different areas of mathematics, an increasingly important role is played by groupoids with some extra structure (smooth groupoids, groupoids in schemes, simplicial groupoids, symplectic groupoids, quantum groupoids, etc.). Many of the constructions and results to be presented here translate easily from one context to the other, and we hope that our introductory text will be of some use to students in different fields. For a detailed discussion of the role of groupoids in symplectic and Poisson geometry, we refer the reader to the contribution to this volume by Bursztyn and Weinstein [9], and the references cited there.

Groupoids often represent not-so-nice 'quotients' of nice structures, such as leaf spaces of foliations, stacks or orbifolds. Here the representing groupoids are usually only defined up to a weak 'Morita' equivalence, and it is this kind of equivalence which is playing a leading role in our lectures. The invariants we introduce (various cohomologies and the fundamental group) are all functorial and stable under this Morita equivalence. Thus it is natural to try and modify the category of Lie groupoids in such a way that the equivalences are turned into isomorphisms. It is known from various contexts (leaf spaces, toposes, stacks, cf. the introduction to Chapter 2) that it is possible to do so, and we give a precise treatment of it in the last two sections of Chapter 2.

In Chapter 3 we introduce the category of equivariant sheaves over a groupoid, as well as associated 'derived' categories, and prove some of the basic properties of these categories. This is also the natural context to discuss the fundamental group of a Lie groupoid, since it represents the category of those equivariant sheaves which are locally constant. Chapter 4 then discusses general sheaf cohomology, while sheaf cohomology with compact supports is discussed in Chapter 5.

Much of the material presented here appeared earlier in various other sources mentioned explicitly below, and our only purpose was to give a more systematic and hopefully more easily accessible exposition. Explicit references are given at the beginning of each chapter. The exposition in Chapters 4 and 5 closely follows the one outlined in [45].

There are three things in this paper, however, which are new, or at least have not been published earlier in this form. First, in our treatment of the category of generalized morphisms between Lie groupoids, we give an explicit proof that the category whose morphisms are principal bundles is isomorphic to the category obtained by universally inverting the weak equivalences. Secondly, as far as we know, there is some novelty in our treatment of the fundamental groupoid of an arbitrary Lie groupoid, together with the proof that this provides a left adjoint into the category of those Lie groupoids which are weakly equivalent to discrete groupoids. It is a good illustration of the principle that the category of Lie groupoids is large enough to contain manifolds as well as discrete groups, and hence should be a good setting to adequately express the universal properties of the fundamental group functor. We also give a concrete application to codimension 1 foliations of our approach. Thirdly, the material in Section 4.4 is new. Here, for a foliated manifold $(M, \mathcal{F})$, we compare the cohomology of the underlying manifold with that of the holonomy (or monodromy) groupoid of the foliation. It answers a question of Haefliger at the Boulder conference[1] in 1999, which could be paraphrased as: when does the manifold M itself behave like the classifying space $\mathrm{B}(\mathrm{Hol}(M, \mathcal{F}))$ of the foliation? We give an answer based on change-of-base properties of the cohomology of étale groupoids.

We would like to thank the Organizers of the Euroschool for inviting us to give the series of lectures and write this contribution. In our work on subjects related to this exposition over the years, we have benefited from our discussions with many colleagues, including H. Bursztyn, M. Crainic, W.T. van Est, A. Haefliger, A. Kumjian, N.P. Landsman, K.C.H. Mackenzie, D. Pronk and A. Weinstein. We acknowledge financial support from the Dutch Science Foundation (NWO) and the Slovenian Ministry of Science (MŠZŠ grant J1-3148).

[1] see http://pompeiu.imar.ro/~ramazan/groupoid/open_prb/gop.html

2
Lie groupoids

In this chapter we give a short introduction to the theory of Lie groupoids. We also recall some of the most important constructions and examples, such as the holonomy groupoid of a foliation [26, 61]. Groupoids were first mentioned in the work of Brandt from the beginning of the twentieth century. Haefliger [23] already used étale Lie groupoids and pseudogroups to describe the transversal structure of foliations, and later Lie groupoids turned out to be one of the most adequate geometric models for non-commutative geometry (Connes [12]).

This chapter closely follows our presentation at the Euroschool, and much of it is taken directly from [48] where the reader can find more details. Sections 2.5 and 2.6 on the category of generalized morphisms and principal bundles were originally written (several years ago) as part of [48], but not included in the final version.

In our presentation, we have emphasized the notion of weak equivalence between Lie groupoids. A Lie groupoid may be viewed as a formal quotient of a manifold, such as the formal space of leaves of a foliation, where all the information can be lost if one takes the usual topological quotient. Weak equivalence is the equivalence relation between Lie groupoids, which, intuitively speaking, identifies Lie groupoids which represent the same quotient. For example, for an action of a discrete group on a manifold, a representation of the 'space of orbits' which is more refined than the topological orbit space is provided by the action groupoid of the action, as well as by any groupoid weakly equivalent to this action groupoid. However, if the action is free and proper, there is nothing wrong with the usual quotient, which is in this case weakly equivalent to the action groupoid. Another example, to be discussed in Section 2.2, is the étale holonomy groupoid of a foliation, which is determined by a choice of a complete transversal section, but different choices of this section give us weakly equivalent groupoids.

This idea leads to the consideration of a category of groupoids in which weakly equivalent groupoids are identified – more precisely, a category where weak equivalences are turned into isomorphisms. The 'generalized' morphisms in this new category can be described more explicitly in terms of certain types of principal bundles. These bundles will be discussed in Section 2.5, while the fact that the resulting category is precisely the one obtained by turning weak equivalences into isomorphisms will be proved in Section 2.6.

Historically, the explicit construction of this category of Lie groupoids and generalized morphisms (and of similar categories, of topological groupoids or groupoids in schemes, for example), came up naturally and independently in various contexts. For example, Hilsum and Skandalis [30] considered such generalized morphisms between leaf spaces of foliations, and in a closely related context, so did Haefliger [26] between étale groupoids or pseudogroups, and Muhly-Renault-Williams [52] considered generalized (iso)morphisms in the context of groupoids and C^*-algebras. In a different context, based on the observation that weakly equivalent groupoids define the same 'orbit topos', Moerdijk [41, 42] introduced a category of groupoids and generalized morphisms and proved that this category was equivalent to the category of toposes. Here, one also finds the first explicit statement of the universal property of this category, in terms of 'categories of fractions'. A systematical treatment of the generalized morphisms as principal bundles in the context of topological groupoids, and in particular the proof that groupoids are weakly equivalent if and only if there is a generalized isomorphism between them, appears in Mrčun [50, 51].

The category of groupoids and generalized morphisms appears in yet another way in algebraic geometry, notably in the context of Deligne-Mumford stacks [17]. In fact, the category of such stacks can again be described in terms of groupoids and generalized morphisms. A more refined 'bicategorical' version of the universal property of generalized morphisms between groupoids is proved in Pronk [54], who also elaborates the relation to differentiable stacks.

The viewpoint that generalized morphisms are the 'correct ones' between groupoids, and the use of principal bundles as a tool to study these morphisms, is by now very common, and discussed and exploited in many sources, see e.g. [8, 13, 28, 34, 35, 40, 59].

2.1 Lie groupoids and weak equivalences

A *groupoid* is a small category in which every arrow is invertible. A groupoid G thus consists of two sets, a set of objects G_0 and a set of arrows G_1. Each arrow g of G has two objects assigned to it, its source $\mathrm{s}(g)$ and its target $\mathrm{t}(g)$. We write $g\colon x \to x'$ to indicate that $x = \mathrm{s}(g)$ and $x' = \mathrm{t}(g)$. There is an associative multiplication of such arrows for which source and target match, giving an arrow $g'g\colon x \to x''$ for any two arrows $g : x \to x'$ and $g' : x' \to x''$. For any object x there is the unit $1_x\colon x \to x$, and each arrow $g\colon x \to x'$ has its inverse arrow $g^{-1}\colon x' \to x$.

These operations in a groupoid G can be viewed as *structure maps* of G relating the sets G_0 and G_1, namely the source map $\mathrm{s}\colon G_1 \to G_0$, the target map $\mathrm{t}\colon G_1 \to G_0$, the (partial) multiplication map $G_1 \times_{G_0} G_1 \to G_1$, $(g', g) \mapsto g'g$ (defined for the arrows g', g of G with $\mathrm{s}(g') = \mathrm{t}(g)$), the unit map $G_0 \to G_1$, $x \mapsto 1_x$, and the inverse map $G_1 \to G_1$, $g \mapsto g^{-1}$.

Sometimes we say that G is a groupoid *over* G_0. For any $x, x' \in G_0$ we denote

$$G(x, x') = \{g \in G_1 \,|\, \mathrm{s}(g) = x,\ \mathrm{t}(g) = x'\}\ .$$

For any arrow $g\colon x \to x'$ we may say that g is an arrow *from x to x'*. Next, we denote the fibers of s and t by $G(x, \text{-}\,) = \mathrm{s}^{-1}(x)$ and $G(\,\text{-}\,, x') = \mathrm{t}^{-1}(x')$. The set of arrows from x to x is a group, called the *isotropy group* of G at x, and denoted by

$$G_x = G(x, x)\ .$$

A well-known example of a groupoid is the fundamental groupoid of a manifold M. The set of objects of this groupoid is M, the arrows from $x \in M$ to $x' \in M$ are the homotopy classes of paths (relative to end-points) in M from x to x', and the partial multiplication is induced by the concatenation of paths.

A homomorphism between groupoids H and G is a functor $\phi\colon H \to G$. It is given by a map on objects $H_0 \to G_0$ and a map on arrows $H_1 \to G_1$, both denoted again by ϕ, which together preserve the groupoid structure, i.e. commute with all the structure maps.

A *Lie groupoid* is a groupoid G together with the structure on G_0 of a smooth Hausdorff second-countable manifold and the structure on G_1 of a (perhaps non-Hausdorff, non-second countable) smooth manifold, such that the source map of G is a smooth submersion with Hausdorff fibers and all the structure maps of G are smooth. Note that the domain of the multiplication map, $G_2 = G_1 \times_{G_0} G_1$, has a natural smooth manifold structure because the source map is a submersion. Also note that it follows that the target map of G is a submersion as well.

Here, we allowed explicitly that G_1 may be non-Hausdorff and non-second countable, as this situation arises in our main examples. A Lie groupoid G is called *Hausdorff* if the manifold of arrows G_1 is also Hausdorff. (On some rare occasions, we will use constructions which will lead to groupoids having a non-Hausdorff base manifold.)

A homomorphism between Lie groupoids H and G is a functor $\phi : H \to G$ which is smooth both on objects and on arrows. We say that ϕ is a submersion if $\phi : H_1 \to G_1$ is a submersion; this implies that $\phi\colon H_0 \to G_0$ is also a submersion. Lie groupoids and homomorphisms between them form a category, which we shall denote by Gpd.

A homomorphism $\phi\colon H \to G$ between Lie groupoids is a *weak equivalence* if the map $\mathrm{t} \circ \mathrm{pr}_1 : G_1 \times_{G_0} H_0 \to G_0$, sending a pair (g, y) with $\mathrm{s}(g) = \phi(y)$ to $\mathrm{t}(g)$, is a surjective submersion, and the square

$$\begin{array}{ccc} H_1 & \xrightarrow{\phi} & G_1 \\ {\scriptstyle (\mathrm{s},\mathrm{t})}\downarrow & & \downarrow{\scriptstyle (\mathrm{s},\mathrm{t})} \\ H_0 \times H_0 & \xrightarrow{\phi\times\phi} & G_0 \times G_0 \end{array}$$

is a fibered product of manifolds. Two Lie groupoids G and G' are *weakly equivalent* if there exist weak equivalences $\phi\colon H \to G$ and $\phi' : H \to G'$ for some third Lie groupoid H. This defines an equivalence relation between Lie groupoids. If G and G' are weakly equivalent Lie groupoids, we may in fact find a Lie groupoid H and weak equivalences $H \to G$ and $H \to G'$ which are surjective submersions on objects (see [48]). As we will see, many important properties of Lie groupoids are stable under weak equivalence.

Let G be a Lie groupoid, and $x \in G_0$. The source and the target map of G are submersions, therefore the fibers $G(x, \text{-}) = \mathrm{s}^{-1}(x)$ and $G(\text{-}, x) = \mathrm{t}^{-1}(x)$ are closed submanifolds of G_1. The isotropy group G_x is a Lie group (see for example [37, 48]), and acts freely and transitively on $G(x, \text{-})$ from the right along the fibers of $\mathrm{t}_x = \mathrm{t}|_{G(x,\text{-})}$. The *orbit* of G passing through x is by definition

$$Gx = \mathrm{t}(G(x, \text{-})) \subset G_0$$

with the quotient topology and the smooth structure of $G(x, \text{-})/G_x$, which makes it into an immersed submanifold of G_0. With this smooth structure on Gx, the map $\mathrm{t}_x : G(x, \text{-}) \to Gx$ becomes a smooth principal G_x-bundle.

The orbits of G form a pair-wise disjoint partition of the manifold G_0. The quotient space of G_0 with respect to this partition is called the *space of orbits* of G, and denoted by G_0/G or by $|G|$.

Examples 2.1 (1) Any manifold M can be viewed as a Lie groupoid over M in which all the arrows are units, i.e. the manifold of arrows is also M. We denote this Lie groupoid again by M, and refer to as the *unit groupoid* associated to M.

(2) Any manifold M gives rise to another Lie groupoid $\mathrm{Pair}(M)$ over M, called the *pair groupoid* of M, with arrows $\mathrm{Pair}(M)_1 = M \times M$. The source and the target map are the first and the second projection. The multiplication is unique, because for any $x, x' \in M$ there is exactly one arrow from x to x'.

The homomorphism $\mathrm{Pair}(M) \to 1$ to the trivial one-point groupoid consisting of one object and one arrow, is a weak equivalence.

Note that any smooth map $p \colon N \to M$ induces a homomorphism of pair groupoids $p \colon \mathrm{Pair}(N) \to \mathrm{Pair}(M)$ in the obvious way. Furthermore, if p is a submersion we may define the *kernel groupoid* $\mathrm{Ker}(p)$ over N, which is a Lie subgroupoid of $\mathrm{Pair}(N)$, consisting of all pairs $(y, y') \in N \times N$ with $p(y) = p(y')$, i.e. $\mathrm{Ker}(p)_1 = N \times_M N$.

Suppose that $p \colon N \to M$ is a surjective submersion. The map p then induces a weak equivalence $\mathrm{Ker}(p) \to M$, where we view M as the unit groupoid (example (1)). A particular case of this is where $N = \coprod_i U_i$ is the disjoint union of an open cover $\{U_i\}$ of M, and p is the evident map. Then $\mathrm{Ker}(p)$ takes the form

$$\coprod_{i,j} U_i \cap U_j \rightrightarrows \coprod_i U_i \,.$$

(3) Any Lie group G can be viewed as a Lie groupoid over a one-point space, and with G as the manifold of arrows. We shall denote this Lie groupoid again by G.

(4) If G is a Lie group acting smoothly from the left on a manifold M, we define the associated *action groupoid* $G \ltimes M$ over M in which $(G \ltimes M)_1 = G \times M$. The source map is the second projection, the target is given by the action map, and the multiplication is defined by

$$(g', x')(g, x) = (g'g, x) \,.$$

The semi-direct product symbol is used because this construction is a special case of the semi-direct product construction described in Section 2.4.

(5) Let M be a manifold. The *fundamental groupoid* $\Pi(M)$ of M is a Lie groupoid over M in which the arrows from $x \in M$ to $y \in M$ are the homotopy classes of paths (relative end-points) in M from x to y, while the multiplication is induced by the concatenation of paths. It is not difficult to see that $\Pi(M)_1$ has indeed a natural smooth structure such that $\Pi(M)$ is a Lie groupoid.

(6) Let E be a vector bundle over a manifold M. One can define a Lie groupoid $\mathrm{GL}(E)$ over M such that the arrows from $x \in M$ to $y \in M$ are the linear isomorphisms $E_x \to E_y$ between the fibers of E.

(7) A Lie groupoid G is said to be *transitive* if the map $(\mathrm{s}, \mathrm{t})\colon G_1 \to G_0 \times G_0$ is a surjective submersion. This notion is obviously stable under weak equivalences. An example of a transitive Lie groupoid over M is the *gauge groupoid* associated to a (right) principal H-bundle $\pi : P \to M$, for a Lie group H. The manifold of arrows of this groupoid is the orbit space of the diagonal action of H on $P \times P$, and the source and the target map are induced by the composition of the first and the second projection with π.

For any object x of a transitive Lie groupoid G over M, the inclusion

$$G_x \longrightarrow G$$

of the isotropy (Lie) group at x into G is a weak equivalence, the map $\mathrm{t} : G(x, \text{-}) \to M$ is a principal G_x-bundle, and the gauge groupoid of this principal G_x-bundle is isomorphic to G.

(8) A Lie groupoid G is *proper* if the map $(\mathrm{s}, \mathrm{t}) : G_1 \to G_0 \times G_0$ is a proper map between Hausdorff manifolds. Propriety is stable under weak equivalence. Proper groupoids are closely related to orbifolds, see Example 2.5 (6).

2.2 The monodromy and holonomy groupoids of a foliation

In this section we will recall the important construction of the holonomy groupoid of a foliation, as well as that of the related monodromy groupoid [23, 53, 61]. These groupoids play a central role in many of the constructions of invariants of foliations, such as the characteristic classes of (transversal) bundles, K-theory of C^*-algebras of foliations, and cyclic homology of foliations.

Throughout this section $(M, \mathcal{F})$ denotes a fixed foliated manifold. Recall that each leaf L of $\mathcal{F}$ has a natural smooth structure for which the inclusion $L \to M$ is an immersion. We now describe the two Lie groupoids associated to the foliated manifold $(M, \mathcal{F})$.

First, the monodromy groupoid $\mathrm{Mon}(M, \mathcal{F})$ is a groupoid over M with the following arrows:

(a) if $x, y \in M$ lie on the same leaf L of $\mathcal{F}$, then the arrows in $\mathrm{Mon}(M, \mathcal{F})$ from x to y are the homotopy classes (relative end-points) of paths in L from x to y, while
(b) if $x, y \in M$ lie on different leaves of $\mathcal{F}$, there are no arrows between them.

The multiplication is induced by the concatenation of paths. In particular, the isotropy groups of the monodromy groupoid are the fundamental groups of the leaves, i.e. $\mathrm{Mon}(M, \mathcal{F})_x = \pi_1(L, x)$ for any point x on a leaf L.

The holonomy groupoid $\mathrm{Hol}(M, \mathcal{F})$ is defined analogously, except that one takes the holonomy classes of paths [48, p. 23] as arrows instead of the homotopy classes. The isotropy group $\mathrm{Hol}(M, \mathcal{F})_x$ at a point x of M on a leaf L is the holonomy group $\mathrm{Hol}(L, x)$ of L.

There are natural Lie groupoid structures on the monodromy and the holonomy groupoid of a foliation such that the following proposition holds true (cf. [15, 48]).

Proposition 2.2 *Let $\mathcal{F}$ be a foliation of a manifold M. We have:*

(i) The orbits of the monodromy and the holonomy groupoids of $(M, \mathcal{F})$ are exactly the leaves of $\mathcal{F}$ (with the leaf smooth structure).

(ii) The isotropy groups of the monodromy and the holonomy groupoid of $(M, \mathcal{F})$ are discrete.

(iii) For a point x on a leaf L, the target map of the monodromy groupoid restricts to the universal covering $\mathrm{Mon}(M, \mathcal{F})(x, \text{-}) \to L$.

(iv) For a point x on a leaf L, the target map of the holonomy groupoid restricts to the covering $\mathrm{Hol}(M, \mathcal{F})(x, \text{-}) \to L$ *corresponding to the kernel of the holonomy homomorphism* $\pi_1(L, x) \to \mathrm{Hol}(L, x)$.

(v) The quotient homomorphism of Lie groupoids $\mathrm{Mon}(M, \mathcal{F}) \to \mathrm{Hol}(M, \mathcal{F})$ *is a local diffeomorphism, and restricts to a covering projection* $\mathrm{Mon}(M, \mathcal{F})(x, \text{-}) \to \mathrm{Hol}(M, \mathcal{F})(x, \text{-})$ *for any $x \in M$.*

Examples 2.3 (1) Let $f : M \to N$ be a surjective submersion with connected fibers, and $\mathcal{F}$ the associated foliation of M. Then all the leaves have trivial holonomy, and the holonomy groupoid of $(M, \mathcal{F})$ is

$\mathrm{Ker}(f) = M \times_N M$. If the fibers of f are simply connected, then this groupoid is also the monodromy groupoid of $(M, \mathcal{F})$.

(2) Let $\mathcal{F}$ be the Reeb foliation of S^3. The compact leaf of $\mathcal{F}$ has $\mathbb{R}^2$ for its holonomy cover, while any other leaf is itself diffeomorphic to $\mathbb{R}^2$ and has trivial holonomy group. Since the fiber of the source map is the holonomy cover of the corresponding leaf, the holonomy groupoid is the same as the monodromy groupoid, and, as a set, it is the product $S^3 \times \mathbb{R}^2$. However, the topology of this space is not the product topology. In fact, one can see that this groupoid is not Hausdorff.

(3) Let $\mathcal{F}$ be a Riemannian foliation on M. Then the derivative $\mathrm{Hol}(M, \mathcal{F}) \to GL(N(\mathcal{F}))$ is an injective homomorphism. In particular, the holonomy groupoid of a Riemannian foliation is Hausdorff.

2.3 Etale groupoids and foliation groupoids

An *étale groupoid* is a Lie groupoid G with $\dim G_1 = \dim G_0$. Equivalently, a Lie groupoid is étale if its source map, and therefore all its structure maps, are étale (i.e. local diffeomorphisms).

It follows that for an étale groupoid, the fibers of the source map, the fibers of the target map, the isotropy groups and the orbits are discrete. Any weak equivalence between étale groupoids $G \to H$ is an étale map on objects and on arrows [54].

Lie groupoids which are weakly equivalent to étale groupoids are called *foliation groupoids*. They can be characterized more intrinsically as follows (cf. [15]):

Proposition 2.4 *A Lie groupoid is a foliation groupoid if and only if it has discrete isotropy groups.*

Examples 2.5 (1) The unit groupoid of a smooth manifold is étale.

(2) A discrete group is an étale groupoid over a one-point space.

(3) If G is a discrete group acting on a manifold M, the associated action groupoid $G \ltimes M$ is étale.

(4) Let M be a manifold. The germs of locally defined diffeomorphisms $f\colon U \to V$ between open subsets of M form a groupoid over M: the germ of f at x is an arrow from x to $f(x)$, and the multiplication in $\Gamma(M)$ is induced by the composition of diffeomorphisms. There is a natural sheaf topology and a smooth structure on $\Gamma(M)_1$ such that the structure maps of $\Gamma(M)$ are étale. Thus $\Gamma(M)$ is an étale groupoid. In particular, the étale groupoid $\Gamma(\mathbb{R}^q)$ is referred to as the Haefliger group-

oid, and denoted by Γ^q. The étale groupoid $\Gamma(M)$ is weakly equivalent to the Haefliger groupoid Γ^q with $q = \dim M$.

(5) Let $\mathcal{F}$ be a foliation of a manifold M. From Proposition 2.4 and Proposition 2.2 (ii) it follows that the monodromy and the holonomy groupoid of $(M, \mathcal{F})$ are foliation groupoids. This can be seen explicitly as follows. We can choose a *complete transversal section* S of $(M, \mathcal{F})$, i.e. an immersed submanifold of M of dimension equal to the codimension of $\mathcal{F}$ which is transversal to the leaves of $\mathcal{F}$ and intersects any leaf in at least one point. Denote by $\iota \colon S \to M$ the inclusion.

We can now define a Lie groupoid $\mathrm{Mon}_S(M, \mathcal{F})$ over S as the restriction of the monodromy groupoid $\mathrm{Mon}(M, \mathcal{F})$ to S: the arrows of $\mathrm{Mon}_S(M, \mathcal{F})$ are those arrows of $\mathrm{Mon}(M, \mathcal{F})$ which start and end in the submanifold S. It is easy to see that $\mathrm{Mon}_S(M, \mathcal{F})$ is indeed a Lie groupoid because the composition $\mathrm{t} \circ \mathrm{pr}_1 : \mathrm{Mon}(M, \mathcal{F})_1 \times_M S \to M$ is a surjective local diffeomorphism. (This groupoid can also be seen as the induced groupoid of $\mathrm{Mon}(M, \mathcal{F})$ along ι, i.e. $\mathrm{Mon}_S(M, \mathcal{F}) = \iota^*(\mathrm{Mon}(M, \mathcal{F}))$, see Section 2.4.)

In particular, note that $\dim \mathrm{Mon}_S(M, \mathcal{F})_1 = \dim S$, so the groupoid $\mathrm{Mon}_S(M, \mathcal{F})$ is étale. It is referred to as the *étale monodromy groupoid* over S associated to $(M, \mathcal{F})$. The inclusion $\mathrm{Mon}_S(M, \mathcal{F}) \to \mathrm{Mon}(M, \mathcal{F})$ is a weak equivalence. For any point $x \in S$ on a leaf L of $\mathcal{F}$ we have

$$\mathrm{Mon}_S(M, \mathcal{F})_x = \mathrm{Mon}(M, \mathcal{F})_x = \pi_1(L, x) .$$

In a completely analogous way we define the *étale holonomy groupoid* $\mathrm{Hol}_S(M, \mathcal{F})$ over S, weakly equivalent to $\mathrm{Hol}(M, \mathcal{F})$, as the restriction of the holonomy groupoid $\mathrm{Hol}(M, \mathcal{F})$ to S. Note that for any $x \in S$ we have

$$\mathrm{Hol}_S(M, \mathcal{F})_x = \mathrm{Hol}(M, \mathcal{F})_x = \mathrm{Hol}(L, x) ,$$

where L is the leaf of $\mathcal{F}$ through x.

(6) Proper foliation groupoids (Example 2.1 (8)) are called *orbifold groupoids*. Orbifolds can be represented as weak equivalence classes of orbifold groupoids (for details and further references, see [28, 36, 46]).

Remark. If G is a foliation groupoid, its orbits form a foliation $\mathcal{F}$ of G_0, and one says that G 'integrates' $\mathcal{F}$. The holonomy and monodromy groupoids of $\mathcal{F}$ are extreme integrals of $\mathcal{F}$, in the sense that if the fibers of the source map of G are connected, there are canonical surjective local diffeomorphisms $\mathrm{Mon}(G_0, \mathcal{F}) \to G \to \mathrm{Hol}(G_0, \mathcal{F})$ (cf. [15]).

Let G be an étale groupoid. There is a canonical homomorphism of Lie groupoids

$$\mathrm{Eff}\colon G \longrightarrow \Gamma(G_0)\,,$$

which is the identity on objects and is given on arrows by

$$\mathrm{Eff}(g) = \mathrm{germ}_{\mathrm{s}(g)}(\mathrm{t}\circ(\mathrm{s}|_U)^{-1})\,,$$

where $g \in G_1$ and U is any open neighbourhood of g in G_1 such that both $\mathrm{s}|_U$ and $\mathrm{t}|_U$ are injective. The map $\mathrm{Eff}: G_1 \to \Gamma(G_0)_1$ is a local diffeomorphism. An *effective groupoid* is an étale groupoid for which the homomorphism Eff is injective (on arrows). The image $\mathrm{Eff}(G)$ of Eff is an open subgroupoid of $\Gamma(G_0)$ and hence effective; it is referred to as the *effect* of G.

Remarks 2.6 (1) The class of effective groupoids is stable under weak equivalences among étale groupoids. In other words, if two étale groupoids are weakly equivalent, then one is effective if and only if the other is too.

(2) The construction of the effect of an étale groupoid can be used to compute the holonomy groupoid of foliations. Indeed, let $\mathcal{F}$ be a foliation of M, and let S be a complete transversal of $\mathcal{F}$. Recall that for any two points $x, y \in S$ on the same leaf L of $\mathcal{F}$, the arrows in $\mathrm{Mon}_S(M, \mathcal{F})$ from x to y are the homotopy classes of paths from x to y inside L. But the holonomy class of such a path α may be faithfully represented by the germ of a locally defined diffeomorphism on S, namely by $\mathrm{hol}^{S,S}(\alpha)$. It follows that the effect homomorphism of $\mathrm{Mon}_S(M, \mathcal{F})$ is given by the holonomy $\mathrm{Mon}_S(M, \mathcal{F}) \to \Gamma(S)$, and

$$\mathrm{Eff}(\mathrm{Mon}_S(M, \mathcal{F})) = \mathrm{Hol}_S(M, \mathcal{F})\,.$$

In particular, the étale holonomy groupoid $\mathrm{Hol}_S(M, \mathcal{F})$ is effective.

(3) Let $f\colon M \to N$ be a surjective submersion with connected fibers, and $\mathcal{F}$ the associated foliation of M. Then the étale holonomy groupoid is weakly equivalent to the manifold N (regarded as a unit groupoid).

(4) If an orbifold groupoid (Example 2.5 (6)) is weakly equivalent to an effective groupoid, it represents a classical reduced orbifold as introduced by Satake [55].

2.4 Some general constructions

In this section we will define transformations between homomorphisms of Lie groupoids, and discuss some ways of constructing new Lie groupoids out of given ones (cf. [29]).

Induced groupoids. Let G be a Lie groupoid and $\phi : M \to G_0$ a smooth map. Then one can define the *induced groupoid* $\phi^*(G)$ over M in which the arrows from x to y are the arrows in G from $\phi(x)$ to $\phi(y)$, i.e.

$$\phi^*(G)_1 = M \times_{G_0} G_1 \times_{G_0} M \ ,$$

and the multiplication is given by the multiplication in G. The space $\phi^*(G)_1$ can be constructed by two pull-backs as in the diagram

$$\begin{array}{ccccc} \phi^*(G)_1 & \longrightarrow & & & M \\ \downarrow & & & & \downarrow \phi \\ G_1 \times_{G_0} M & \xrightarrow{\mathrm{pr}_1} & G_1 & \xrightarrow{\mathrm{t}} & G_0 \\ \downarrow & & \downarrow \mathrm{s} & & \\ M & \xrightarrow{\phi} & G_0 & & \end{array}$$

The lower pull-back has a natural smooth structure because s is a submersion. If the composition $\mathrm{t} \circ \mathrm{pr}_1$ is also a submersion, the upper pull-back has a natural smooth structure as well. It follows that the diagram

$$\begin{array}{ccc} \phi^*(G)_1 & \longrightarrow & G_1 \\ {\scriptstyle (\mathrm{s},\mathrm{t})}\downarrow & & \downarrow{\scriptstyle (\mathrm{s},\mathrm{t})} \\ M \times M & \xrightarrow{\phi\times\phi} & G_0 \times G_0 \end{array}$$

is a pull-back. Therefore $\phi^*(G)$ is a Lie groupoid if the map

$$\mathrm{t} \circ \mathrm{pr}_1 \colon G_1 \times_{G_0} M \longrightarrow G_0$$

is a submersion.

Suppose that this map is a submersion. Then ϕ induces a homomorphism of Lie groupoids $\phi\colon \phi^*(G) \to G$, which is a weak equivalence if and only if the submersion $\mathrm{t} \circ \mathrm{pr}_1$ is surjective. If G is a foliation groupoid, then so is $\phi^*(G)$. If G is proper, then $\phi^*(G)$ is also proper.

Transformations. For two homomorphisms $\phi, \psi\colon G \to H$ of Lie groupoids, a *smooth natural transformation* (briefly transformation) from ϕ to

ψ is a smooth map

$$T\colon G_0 \longrightarrow H_1$$

such that for each $x \in G_0$, $T(x)$ is an arrow from $\phi(x)$ to $\psi(x)$ in H, and for each arrow $g\colon x \to y$ in G the square

$$\begin{array}{ccc} \phi(x) & \xrightarrow{T(x)} & \psi(x) \\ {\scriptstyle \phi(g)}\downarrow & & \downarrow{\scriptstyle \psi(g)} \\ \phi(y) & \xrightarrow{T(y)} & \psi(y) \end{array}$$

commutes. We write $T\colon \phi \to \psi$ to indicate that T is such a transformation from ϕ to ψ.

If $T : \phi \to \psi$ and $R : \psi \to \rho$ are two transformations, so is their product $RT : \phi \to \rho$ given by $RT(x) = R(x)T(x)$. In particular, the homomorphisms from G to H are themselves the objects of a groupoid with transformations as arrows. We will denote this groupoid by

$$\mathrm{Hom}(G, H)\ .$$

In fact, Lie groupoids, homomorphisms and transformations form a 2-category.

Sums and products. For two Lie groupoids G and H one can construct the product Lie groupoid

$$G \times H$$

in the obvious way, by taking the product manifolds $G_0 \times H_0$ and $G_1 \times H_1$. In a similar way one constructs the sum (disjoint union) Lie groupoid

$$G + H\ .$$

In fact, one can construct the sum Lie groupoid

$$\sum_i G_i$$

of a (countable) indexed family (G_i) of Lie groupoids. The sums and products have familiar universal property in the category Gpd (in fact, also in the 2-category) of Lie groupoids and homomorphisms.

If G and H are foliation (or étale, or proper) groupoids, then so are $G \times H$ and $G + H$.

Strong fibered products. For two homomorphisms $\phi\colon G \to K$ and

$\psi\colon H \to K$ one can construct the fibered products of the sets of objects and arrows:

$$(G \times_K H)_0 = G_0 \times_{K_0} H_0 = \{(x,y) \,|\, x \in G_0,\, y \in H_0,\, \phi(x) = \psi(y)\}$$

$$(G \times_K H)_1 = G_1 \times_{K_1} H_1 = \{(g,h) \,|\, g \in G_1,\, h \in H_1,\, \phi(g) = \psi(h)\}\ .$$

With multiplication defined component-wise, this defines a groupoid

$$G \times_K H\ .$$

However, in general this is not a Lie groupoid. It is if the fibered products $G_0 \times_{K_0} H_0$ and $G_1 \times_{K_1} H_1$ are transversal. For example, for $G_0 \times_{K_0} H_0$ this means that the map $\phi \times \psi\colon G_0 \times H_0 \to K_0 \times K_0$ is transversal to the diagonal $\Delta\colon K_0 \to K_0 \times K_0$, so that $G_0 \times_{K_0} H_0 = (\phi \times \psi)^{-1}(\Delta K_0)$ is indeed a manifold.

If the transversality condition is satisfied, this construction gives a fibered product (pull-back) with the familiar universal property in the category Gpd. Below we will consider an alternative, larger fibered product. To emphasize the distinction, we often refer to the present fibered product as the *strong* one.

Weak fibered products. Let $\phi\colon G \to K$ and $\psi\colon H \to K$ be homomorphisms of Lie groupoids. We define a new groupoid P as follows: Objects of P are triples (x,k,y), where $x \in G_0$, $y \in H_0$ and $k \in K(\phi(x), \psi(y))$. Arrows in P from (x,k,y) to (x',k',y') are pairs (g,h) of arrows $g \in G_1$ and $h \in H_1$ such that

$$k'\phi(g) = \psi(h)k\ .$$

The multiplication is given component-wise. Often (but not always) P has a structure of a Lie groupoid. Indeed, the set of objects may be considered as the fibered product

$$P_0 = G_0 \times_{K_0} K_1 \times_{K_0} H_0\ ,$$

and if this fibered product is transversal then P_0 inherits a natural structure of a submanifold of $G_0 \times K_1 \times H_0$. This is the case, for example, when either $\phi\colon G_0 \to K_0$ or $\psi\colon H_0 \to K_0$ is a submersion. If P_0 has a manifold structure as above, then

$$P_1 = G_1 \times_{K_0} K_1 \times_{K_0} H_1 = \{(g,k,h) \,|\, \phi(\mathrm{s}(g)) = \mathrm{s}(k),\, \psi(\mathrm{s}(h)) = \mathrm{t}(k)\}$$

is also a manifold. Indeed, in this case P_1 can be obtained from the two

fibered products

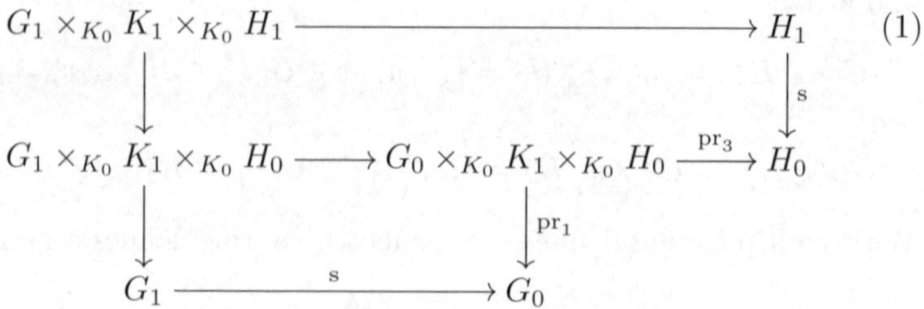

Thus, in this case P is a Lie groupoid, provided P_0 is Hausdorff (e.g. if K is Hausdorff). We refer to this groupoid P as the *weak pull-back* or the *weak fibered product*, and we denote it by

$$G \times_K^{(w)} H \, .$$

We will use weak fibered products more often than strong ones, and if not stated explicitly otherwise, 'fibered product' from now on will refer to the weak one, and will be simply denoted as $G \times_K H$.

We remark that if G and H are both foliation groupoids, respectively proper groupoids or étale groupoids, and if the weak fibered product $G \times_K H$ is a Lie groupoid, then $G \times_K H$ is also a foliation groupoid, proper groupoid or étale groupoid.

Comma groupoids. (i) Let G be a Lie groupoid and $x \in G_0$. We can view x as a homomorphism $1 \to G$ where 1 is the unit groupoid of a one-point space. The weak fibered product of $x\colon 1 \to G$ and $\mathrm{id}\colon G \to G$ can be described as the groupoid over $G(x,\text{-}) = \mathrm{s}^{-1}(x)$ whose arrows from $g \in G(x,\text{-})$ to $g' \in G(x,\text{-})$ are the arrows $h : y \to y'$ in G with $hg = g'$. This a Lie groupoid, denoted by x/G and called the *comma groupoid* of x over G.

(ii) Similarly, for a homomorphism $\phi\colon H \to G$ and $x \in G_0$ one can form the *comma groupoid*

$$x/\phi$$

as the weak fibered product of $x : 1 \to G$ and $\phi : H \to G$. It can be identified with the groupoid whose objects are pairs (g, y) where $y \in H_0$ and $g\colon x \to \phi(y)$ in G, and whose arrows $h\colon (g, y) \to (g', y')$ are arrows $h : y \to y'$ in H with $\phi(h)g = g'$. This comma groupoid x/ϕ is a Lie groupoid if ϕ is a submersion. One refers to x/ϕ as the *fiber of ϕ over x* (or sometimes as the weak fiber or homotopy fiber, to distinguished it from the 'strict' fiber $\phi^{-1}(x)$).

(iii) For $\phi\colon H \to G$ there is also a larger comma groupoid

$$H/\phi$$

whose objects are pairs (g, y) with $g \in G_1$ and $y \in H_0$ such that $\phi(y) = \mathrm{t}(g)$. Arrows $(g, y) \to (g', y')$ are pairs (h, f) where $f : \mathrm{s}(g) \to \mathrm{s}(g')$ in G and $h : y \to y'$ in H are such that $g'f = \phi(h)g$. This groupoid can be constructed as the weak fibered product of ϕ and the identity $\mathrm{id}\colon G \to G$. It is always a Lie groupoid.

Semi-direct products. Let G be a Lie groupoid.

(i) A left *action* of G on a manifold N *along* a smooth map $\epsilon\colon N \to G_0$ is given by a smooth map $\mu\colon G_1 \times_{G_0} N \to N$ (we write $\mu(g, y) = gy$), defined on the pull-back $G_1 \times_{G_0} N = \{(g, y) \,|\, \mathrm{s}(g) = \epsilon(y)\}$, which satisfies the following identities: $\epsilon(gy) = \mathrm{t}(g)$, $1_{\epsilon(y)}y = y$ and $g'(gy) = (g'g)y$, for any $g', g \in G_1$ and $y \in N$ with $\mathrm{s}(g') = \mathrm{t}(g)$ and $\mathrm{s}(g) = \epsilon(y)$. For such an action one can form the *action groupoid*

$$G \ltimes N$$

over N with $(G \ltimes N)_1 = G_1 \times_{G_0} N$, exactly as for Lie group actions (Example 2.1 (4)). This groupoid is a Lie groupoid, also referred to as *semi-direct product groupoid* of the G-action.

We define the quotient $G\backslash N$ as the space of orbits of the groupoid $G \ltimes N$. This space is in general not a manifold.

(ii) A right action of G on N is defined analogously, and such an action gives a semi-direct product $N \rtimes G$ and space of orbits N/G.

(iii) There is also a notion of a (left) action of a Lie groupoid G on another Lie groupoid H. It is given by two (left) actions of G on H_1 and on H_0, such that the groupoid structure maps of H are equivariant, i.e. compatible with the actions by G (note that the diagonal action of G on the domain of the multiplication map is well-defined). If we denote the action maps on H_i by $\epsilon_i\colon H_i \to G_0$ and $\mu_i\colon G_1 \times_{G_0} H_i \to H_i$, $i = 0, 1$, this implies in particular that $\epsilon_0 \circ \mathrm{s} = \epsilon_1 = \epsilon_0 \circ \mathrm{t}$. Thus the fibers $H_x = \epsilon_1^{-1}(x)$ are full subgroupoids of H over $\epsilon_0^{-1}(x)$, $x \in G_0$. These are Lie groupoids if ϵ_0 is a submersion, and for each arrow $g \in G_1(x', x)$ the action provides an isomorphism $H_{x'} \to H_x$ of Lie groupoids.

For such an action of G on H, one can form the *semi-direct product groupoid*

$$G \ltimes H$$

over H_0. For $y, y' \in H_0$, an arrow from y' to y in $G \ltimes H$ is a pair (g, h), where g is an arrow in $G(\epsilon_0(y'), \epsilon_0(y))$ and h is an arrow in $H(gy', y) \subset H_{\epsilon_0(y)}$. These arrows compose by the usual formula

$$(g, h)(g', h') = (gg', h(gh')) .$$

The groupoid $G \ltimes H$ has the natural structure of a Lie groupoid, as one sees, e.g. when the space of arrows is considered as the fibered product

$$H_1 \times_{G_0} G_1 = \{(h, g) \mid \epsilon_0(\mathrm{t}(h)) = \mathrm{t}(g)\} .$$

2.5 Principal bundles as morphisms between Lie groupoids

We begin by extending the notion of a principal G-bundle for Lie groups to the case where G is a Lie groupoid.

Let G be a Lie groupoid. A G-bundle over a manifold M is a manifold P equipped with a map $\pi\colon P \to M$ and a smooth right G-action μ on P (along $\epsilon\colon P \to G_0$) which is fiber-wise with respect to π, i.e. $\pi(pg) = \pi(g)$ for any $p \in P$ and any $g \in G_1$ with $\epsilon(p) = \mathrm{t}(g)$. Such a bundle P is said to be *principal* if

(i) π is a surjective submersion, and
(ii) the map $(\mathrm{pr}_1, \mu)\colon P \times_{G_0} G_1 \to P \times_M P$, sending (p, g) to (p, pg), is a diffeomorphism.

Note that in case G is a Lie group we recover the usual notion of a principal G-bundle.

For a principal G-bundle $\pi\colon P \to M$, we refer to the manifold P as the *total space* of the bundle, and we shall denote by $\delta\colon P \times_M P \to G_1$ the map $\mathrm{pr}_2 \circ (\mathrm{pr}_1, \mu)^{-1}$. This map is uniquely determined by the identity $p\delta(p, p') = p'$ and satisfies the equation $\delta(p, p')g = \delta(p, p'g)$.

An *equivariant* map between principal G-bundles $\pi : P \to M$ and $\pi'\colon P' \to M$ over M is a smooth map $f\colon P \to P'$ which commutes with all the structure maps, i.e. the identities $\pi'(f(p)) = \pi(p)$, $\epsilon'(f(p)) = \epsilon(p)$ and $f(pg) = f(p)g$ hold for any $p \in P$ and $g \in G_1$ with $\epsilon(p) = \mathrm{t}(g)$.

Remarks 2.7 (1) The space G_1 of arrows of a Lie groupoid G carries the structure of a principal G-bundle over G_0: for π one takes the target map and for ϵ the source map, while the right action is given by the multiplication in G. We call this bundle the *unit* bundle of G, and denote it by $U(G)$.

(2) If P is a principal G-bundle over M and $f\colon N \to M$ is a smooth

map, the pull-back $N \times_M P$ has the structure of a principal G-bundle over N. We denote this bundle by $f^*(P)$.

(3) Combining the previous two remarks, we see that for any map $\alpha\colon M \to G_0$ there is a principal G-bundle $\alpha^*(U(G))$ over M. Its total space is the space of pairs (m, g) where g is an arrow with target $\alpha(m)$. Bundles which are isomorphic to one of this form are called *trivial*.

(4) Let P be a principal G-bundle over M. Take any point $m \in M$, and choose a local section $\sigma\colon V \to P$ of π defined on an open neighbourhood V of m. Let $\alpha = \epsilon \circ \sigma\colon V \to G_0$. Then the map $\alpha^*(U(G)) \to P$, which sends (m, g) to $\sigma(m)g$, is an isomorphism from the trivial bundle $\alpha^*(U(G))$ to the restriction $P_V = \pi^{-1}(V)$. Its inverse sends p to $(\pi(p), \delta(\sigma(\pi(p)), p))$. Thus, any principal bundle is *locally trivial.*

(5) Every equivariant map $P \to P'$ between principal G-bundles over M is an isomorphism. In fact by (4) it is sufficient to check this for trivial bundles. But for $\alpha, \beta : M \to G_0$, a map $f : \alpha^*(U(G)) \to \beta^*(U(G))$ is completely determined by the map $\phi : M \to G_1$ sending m to $\mathrm{pr}_2(f(m, 1_{\alpha(m)}))$, since

$$f(m, g) = f(m, 1_{\alpha(m)})g = (m, \phi(m)g)\ .$$

Thus clearly f is an isomorphism, with inverse

$$f^{-1}(m, g) = (m, \phi(m)^{-1}g).$$

We recall the following lemma from [48, p. 146].

Lemma 2.8 *Let P be a principal G-bundle over M. Let Q be a manifold with a right G-action, and $f : Q \to P$ a submersion preserving the G-action. Then Q/G is a manifold and the quotient projection $Q \to Q/G$ is a principal G-bundle.*

Let G and H be Lie groupoids. A principal G-bundle over H is a principal G-bundle $\pi\colon P \to H_0$ over the manifold H_0,

$$\begin{array}{ccc} P & \xrightarrow{\epsilon} & G_0 \\ {\scriptstyle\pi}\big\downarrow & & \\ H_0 & & \end{array}$$

which is equipped with a left H-action on P along π, which commutes with the right G-action, i.e. $\epsilon(hp) = \epsilon(p)$ and

$$(hp)g = h(pg)$$

for any $h \in H_1$, $p \in P$ and $g \in G_1$ with $\mathrm{s}(h) = \pi(p)$ and $\epsilon(p) = \mathrm{t}(g)$.

We will think of such principal bundles as objects which represent abstract morphisms between Lie groupoids, and we shall write

$$P\colon H \longrightarrow G$$

to indicate that P is a principal G-bundle over H.

A map $P \to P'$ between principal G-bundles over H is a map of principal G-bundles over H_0 which also respects the H-action. As we have seen, any such map is an isomorphism.

With the idea of principal bundles as abstract morphisms in mind, we now discuss identity morphisms and composition of morphisms.

Let G be a Lie groupoid. Then the unit principal G-bundle $U(G)$ over G_0 has a natural left G-action, given by composition, and it is a principal G-bundle over G. (In the notation above we have $P = G_1$, $\pi = \mathrm{t}$ and $\epsilon = \mathrm{s}$.) We denote this bundle again by

$$U(G)\colon G \longrightarrow G\,.$$

Let G, H and K be Lie groupoids. Suppose $P\colon H \to G$ is a principal G-bundle over H and $Q\colon K \to H$ is a principal H-bundle over K. Then we construct a principal G-bundle over K

$$Q \otimes_H P\colon K \longrightarrow G$$

as follows: The fibered product $Q \times_{H_0} P$ carries a right H-action along $\epsilon \circ \mathrm{pr}_1 = \pi \circ \mathrm{pr}_2$ given by

$$(q,p)h = (qh, h^{-1}p)\,.$$

We denote its orbit space by $Q \otimes_H P$, and the orbit of (q,p) by $q \otimes p \in Q \otimes_H P$. Since Q is a principal H-bundle over K_0, Lemma 2.8 gives that $Q \otimes_H P$ is a smooth manifold and $Q \times_{H_0} P \to Q \otimes_H P$ is a principal H-bundle. Moreover, the fibered product $Q \times_{H_0} P$ carries a left K-action (along $\pi \circ \mathrm{pr}_1$ and on the Q-coordinate only) and a right G-action (along $\epsilon \circ \mathrm{pr}_2$ and on the P-coordinate only). These two actions respect the H-action, so they induce well-defined commuting actions on $Q \otimes_H P$, by K from the left and by G from the right,

$$k(q \otimes p)g = kq \otimes pg\,.$$

It remains to be proved that the right G-action is principal over K_0.

To see this, denote the inverse of the diffeomorphism $P \times_{G_0} G_1 \to P \times_{H_0} P$ by (pr_1, δ) and the inverse of the diffeomorphism $Q \times_{H_0} H_1 \to Q \times_{K_0} Q$ by (pr_1, δ'). Then the map $(Q \otimes_H P) \times_{G_0} G_1 \to (Q \otimes_H P) \times_{K_0}$

$(Q \otimes_H P)$ has an inverse which sends a point $(q \otimes p, q' \otimes p')$ into

$$(q \otimes p, \delta(p, \delta'(q, q')p')) .$$

We observe that this tensor product is associative up to isomorphism. Indeed, for $R: L \to K$, $Q: K \to H$ and $P: H \to G$, the evident map

$$R \otimes_K (Q \otimes_H P) \longrightarrow (R \otimes_K Q) \otimes_H P$$

is an isomorphism of principal G-bundles over L. Furthermore, the unit bundles act as units up to isomorphisms for this tensor product, by

$$U(H) \otimes_H P \longrightarrow P , \qquad P \otimes_G U(G) \longrightarrow P .$$

For example, the first maps $h \otimes p$ into hp, and its inverse maps p into $1_{\pi(p)} \otimes p$. Thus, Lie groupoids and *isomorphism classes* of principal bundles with this tensor product form a well-defined category, as we already suggested by the notation $P : H \to G$. We will denote this category GPD, and call the arrows in this category *generalized morphisms*, or simply just *morphisms* between Lie groupoids. (These morphisms are sometimes called Hilsum-Skandalis maps.) The category GPD is often referred to as the *Morita category* of Lie groupoids.

To conclude this section, we discuss isomorphisms in the category GPD. Following classical terminology from ring theory, we call a principal G-bundle over H a *Morita equivalence* if $\epsilon: P \to G_0$ is also (left) principal as an H-bundle over G_0. If such a P exists, we say that G and H are *Morita equivalent.* Later (Corollary 2.12) we shall see that two groupoids are Morita equivalent if and only if they are weakly equivalent. Here, we will show the following (see [50, 51]):

Proposition 2.9 *A principal G-bundle P over H represents an isomorphism in the category* GPD *if and only if P is a Morita equivalence. Hence two Lie groupoids are Morita equivalent if and only if they are isomorphic in the category* GPD.

Proof ($\Leftarrow$) Suppose that $P : H \to G$ is a Morita equivalence. Write $P^{-1} : G \to H$ for the 'opposite' bundle, namely the same manifold P with left G-action $\nu(g, p) = pg^{-1}$ and right H-action $\mu(p, h) = h^{-1}p$. Then there are isomorphisms

$$P \otimes_G P^{-1} \longrightarrow U(H) , \qquad P^{-1} \otimes_H P \longrightarrow U(G) .$$

For example, the second one maps $p \otimes p'$ into $\delta(p, p')$, where $\delta(p, p')$ is the unique arrow of G such that $p\delta(p, p') = p'$. One can easily check that this map is a map of principal bundles and hence an isomorphism.

($\Rightarrow$) Suppose that $P\colon H \to G$ is an isomorphism, so there is a principal bundle $Q\colon G \to H$ and isomorphisms

$$\alpha\colon P \otimes_G Q \longrightarrow U(H)\,, \qquad \beta\colon Q \otimes_H P \longrightarrow U(G)\,.$$

We again have the bundle P^{-1} as before, with a (principal) left G-action and a right H-action. To show that the H-action on P^{-1} is also principal, it is enough to show that there is an isomorphism

$$\tau\colon P^{-1} \longrightarrow Q$$

which preserves the actions. To define τ, note first that there is a canonical isomorphism (for any principal G-bundle P over H)

$$\theta\colon P \times_{H_0} (P \otimes_G Q) \longrightarrow P \times_{G_0} Q$$

given by $\theta(p, p' \otimes q) = (p, \delta(p,p')q)$, with inverse $\theta^{-1}(p,q) = (p, p \otimes q)$. Now define $\tau\colon P \to Q$ between manifolds as

$$\tau(p) = \mathrm{pr}_2(\theta(p, \alpha^{-1}(1_{\pi(p)})))\,.$$

This τ satisfies the identities

$$\tau(pg) = g^{-1}\tau(p)\,, \qquad \tau(hp) = \tau(p)h^{-1}\,.$$

In exactly the same way, we can define a map $\sigma\colon Q \to P$ as

$$\sigma(q) = \mathrm{pr}_2(\theta'(q, \beta^{-1}(1_{\pi(q)})))\,,$$

where $\theta' : Q \times_{G_0} (Q \otimes_H P) \to Q \times_{H_0} P$ is the canonical isomorphism. The map σ satisfies the identities

$$\sigma(qh) = h^{-1}\sigma(q)\,, \qquad \sigma(gq) = \sigma(q)g^{-1}\,.$$

Then $\tau \circ \sigma : Q \to Q$ and $\sigma \circ \tau : P \to P$ are maps of principal bundles, hence diffeomorphisms. But then τ and σ are diffeomorphisms too, and τ defines the required isomorphism $P^{-1} \to Q$ we were looking for. □

2.6 The principal bundles category as a universal solution

In this section we will prove that the category GPD of Lie groupoids and principal bundles is the universal solution to inverting the weak equivalences in the category Gpd of Lie groupoids and homomorphisms. This provides a good justification for considering principal bundles as morphisms between Lie groupoids. Indeed, in many cases it is natural to view two weakly equivalent Lie groupoids as representing the same geometric object. For example, for a foliation $\mathcal{F}$ of M and two complete transversals S and T, the étale groupoids $\mathrm{Hol}_S(M,\mathcal{F})$ and $\mathrm{Hol}_T(M\mathcal{F})$ are weakly equivalent, and both represent the 'leaf space' of the foliation.

Consider for a homomorphism $\phi\colon H \to G$ between Lie groupoids the principal G-bundle

$$\mathfrak{P}(\phi) = \phi^*(U(G))$$

over H_0. This bundle has in fact a natural structure of a principal G-bundle over H. Indeed, $\mathfrak{P}(\phi)$ is the space of pairs (y,g) where $y \in H_0$ and $g \in G_1$ with $\phi(y) = \mathrm{t}(g)$, and the actions of G and H are given by

$$h(y,g)g' = (\mathrm{t}(h), \phi(h)gg')$$

for any $h \in H_1$ with $\mathrm{s}(h) = y$ and $g' \in G_1$ with $\mathrm{s}(g) = \mathrm{t}(g')$. This construction has the following property [50, 51].

Proposition 2.10 *The construction of the principal bundle $\mathfrak{P}(\phi)$ out of a homomorphism ϕ of Lie groupoids defines a functor*

$$\mathfrak{P}\colon \mathsf{Gpd} \longrightarrow \mathsf{GPD}\,.$$

Moreover, a homomorphism in Gpd *is a weak equivalence if and only if this functor sends it to an isomorphism in* GPD.

Proof For the identity homomorphism $\mathrm{id} : G \to G$ we have $\mathfrak{P}(\mathrm{id}) = U(G)$, which is the identity on G in the category GPD. And for two homomorphisms $\phi\colon H \to G$ and $\psi\colon K \to H$ there is a canonical map

$$\tau\colon \mathfrak{P}(\psi) \otimes_H \mathfrak{P}(\phi) \longrightarrow \mathfrak{P}(\phi\circ\psi)\,,$$

which is given by

$$\tau((z,h)\otimes(y,g)) = (z, \phi(h)g)$$

for any $(z,h) \in \mathfrak{P}(\psi)$ and $(y,g) \in \mathfrak{P}(\phi)$. The map τ is well-defined on the tensor product and preserves the left K-action and the right G action, so it is an isomorphism. This proves that $\mathfrak{P}$ is a functor.

Finally, it is clear from the definitions that a homomorphism ϕ is a weak equivalence if and only if $\mathfrak{P}(\phi)$ is also H-principal, and by Proposition 2.9 this means precisely that $\mathfrak{P}(\phi)$ is an isomorphism in the category GPD. □

Theorem 2.11 *Let F : $\mathsf{Gpd} \to \mathcal{C}$ be a functor into any category $\mathcal{C}$ which sends weak equivalences into isomorphisms. Then F factors as*

$F = \tilde{F} \circ \mathfrak{P}$

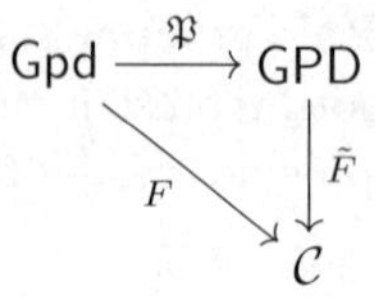

$$\begin{array}{ccc} \mathsf{Gpd} & \xrightarrow{\mathfrak{P}} & \mathsf{GPD} \\ & \searrow^{F} & \downarrow \tilde{F} \\ & & \mathcal{C} \end{array}$$

for a unique functor $\tilde{F} \colon \mathsf{GPD} \to \mathcal{C}$.

Proof We will present a proof based on general arguments, which can be applied in other contexts as well. However, some constructions in this proof lead us outside the scope of the Hausdorff conventions involved in the definition of the categories Gpd and GPD. Therefore, let us temporarily consider the larger category Gpd^* of all smooth groupoids G and homomorphisms, defined only by the conditions that G_0 and G_1 are manifolds (possibly non-Hausdorff) and s and t are submersions. All the constructions in Section 2.5 apply without any changes to this larger category, and in particular there is a corresponding category GPD^* of such (non-Hausdorff) groupoids and generalized morphisms. Now let Gpd' be the full subcategory of Gpd^* consisting of those groupoids which are weakly equivalent to (Lie) groupoids in Gpd, and let GPD' be the analogous category of generalized morphisms. We will prove the theorem for the categories Gpd' and GPD'. The theorem, as stated above, then follows formally, because the category GPD' is obviously equivalent to GPD, while the functor F extends to the category Gpd' because it sends weak equivalences to isomorphisms.

The main ingredient of the proof is the construction of the two-sided semi-direct product $H \ltimes P \rtimes G$ for any principal G-bundle P over H. This semi-direct product is a groupoid over P, in which arrows are triples (h, p, g), $h \in H$, $p \in P$ and $g \in G$, such that $\mathrm{s}(h) = \pi(p)$ and $\epsilon(p) = \mathrm{s}(g)$; in other words,

$$(H \ltimes P \rtimes G)_1 = H_1 \times_{H_0} P \times_{G_0} G_1 \, .$$

Furthermore, $\mathrm{s}(h, p, g) = p$, $\mathrm{t}(h, p, g) = hpg^{-1}$ and the multiplication is given by $(h, p, g)(h', p', g') = (hh', p', gg')$. Note that the groupoid $H \ltimes P \rtimes G$ is in the category Gpd' because G and H are (however, if G and H are non-Hausdorff Lie groupoids, this groupoid may have non-Hausdorff base).

The structure maps $\pi : P \to H_0$ and $\epsilon : P \to G_0$ of the principal bundle induce homomorphisms, which we again denote

$$H \xleftarrow{\pi} H \ltimes P \rtimes G \xrightarrow{\epsilon} G\,.$$

The condition that the G-action is principal implies that π is a weak equivalence. Thus, F maps it to an isomorphism, and we can define

$$\tilde{F}(P) = F(\epsilon) \circ F(\pi)^{-1}\,.$$

First of all, note that this is well-defined on isomorphism classes of principal bundles. Indeed, if $f : P \to P'$ is an isomorphism, it induces an isomorphism of groupoids $f\colon H \ltimes P \rtimes G \to H \ltimes P' \rtimes G$ for which the diagram

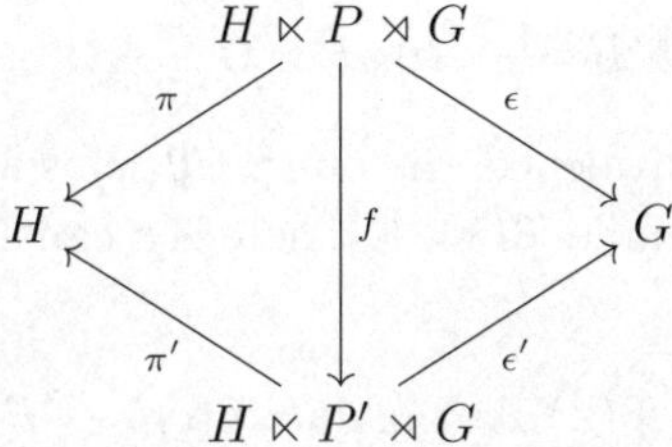

commutes. In particular,

$$\begin{aligned} F(\epsilon)F(\pi)^{-1} &= F(\epsilon)F(f)F(f^{-1})F(\pi)^{-1} \\ &= F(\epsilon \circ f)F(\pi \circ f)^{-1} \\ &= F(\epsilon')F(\pi')^{-1}\,. \end{aligned}$$

Before proving that $\tilde{F}$ thus defined is a functor, we show that it extends F, i.e. that $\tilde{F} \circ \mathfrak{P} = F$. To this end, first observe that if $\phi, \psi\colon H \to G$ are two homomorphisms and T a transformation from ϕ to ψ, then $F(\phi) = F(\psi)$. Indeed, let I be the groupoid with two objects 0 and 1 and one isomorphism between them, and consider for any groupoid H the (weak) equivalences

$$H \underset{i_1}{\overset{i_0}{\rightrightarrows}} H \times I \xrightarrow{\mathrm{pr}_1} H\,.$$

Here i_0 and i_1 are the evident inclusions and pr_1 is the projection, so $\pi \circ i_0 = \pi \circ i_1 = \mathrm{id}$. Since $F(\mathrm{pr}_1)$ is an isomorphism, we have $F(i_0) = F(i_1)$. The transformation T defines a homomorphism $T\colon H \times I \to G$ with $T \circ i_0 = \phi$ and $T \circ i_1 = \psi$. Therefore $F(\phi) = F(\psi)$.

From this basic property of F it follows immediately that $\tilde{F}$ extends

F. For if $\phi: H \to G$ is a homomorphism, the diagram

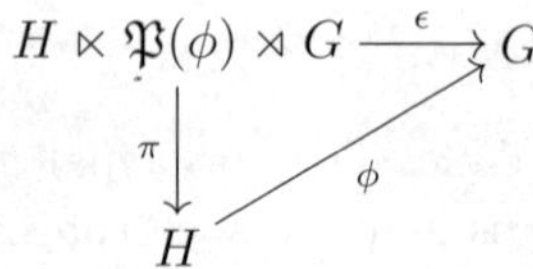

commutes up to a transformation $T: \phi \circ \pi \to \epsilon$. For a point $p = (y, g)$ in $\mathfrak{P}(\phi)$ we have $\pi(p) = y$ and $\epsilon(p) = \mathrm{s}(g)$, and T is given by

$$T(p) = g : \phi(\pi(p)) \longrightarrow \epsilon(p) \,.$$

To prove that $\tilde{F}$ is a functor and that it is unique, consider for $P: H \to G$ once again the two-sided semi-direct product

$$H \xleftarrow{\pi} H \ltimes P \rtimes G \xrightarrow{\epsilon} G \,.$$

Since π is a weak equivalence, the bundle $\mathfrak{P}(\pi)$ is invertible, with the inverse the opposite bundle $\mathfrak{P}(\pi)^{-1}$. There is a canonical map of principal bundles

$$\mathfrak{P}(\pi)^{-1} \otimes_{H \ltimes P \rtimes G} \mathfrak{P}(\epsilon) \longrightarrow P$$

which sends $(p, h) \otimes (x, g)$ into $h^{-1}pg$. Since this map is necessarily an isomorphism, we find that

$$P = \mathfrak{P}(\epsilon) \circ \mathfrak{P}(\pi)^{-1}$$

in the category GPD', so any functor $\mathsf{GPD}' \to \mathcal{C}$ which extends F has to satisfy the defining identity for $\tilde{F}$. This shows that $\tilde{F}$ is unique. Thus, it only remains to be proved that $\tilde{F}$ is a functor. Clearly $\tilde{F}$ preserves the identities. To see that $\tilde{F}$ preserves composition, consider morphisms $Q : K \to H$ and $P : H \to G$ in GPD', and the associated two-sided semi-direct products

$$H \xleftarrow{\pi} H \ltimes P \rtimes G \xrightarrow{\epsilon} G \,,$$

$$K \xleftarrow{\pi'} K \ltimes Q \rtimes H \xrightarrow{\epsilon'} H$$

and

$$K \xleftarrow{\pi''} K \ltimes (Q \otimes_H P) \rtimes G \xrightarrow{\epsilon''} G \,.$$

Construct the weak pull-back S of ϵ' and π, and observe that there is an evident projection $\tau: S \to K \ltimes (Q \otimes_H P) \rtimes G$ which makes the following diagram commutative:

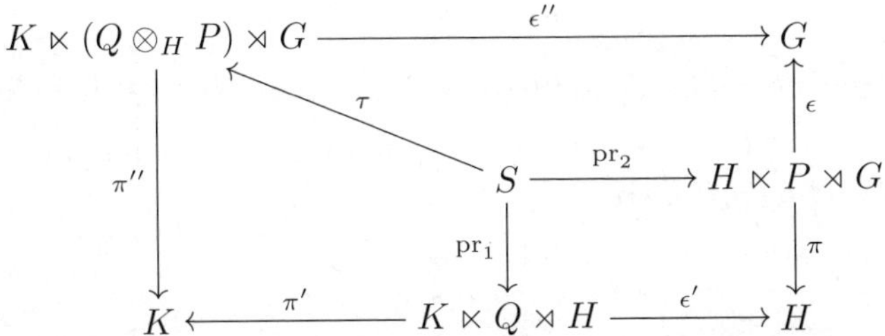

We know that in this diagram, π, π' and π'' are weak equivalences. Also pr_1 is a weak equivalence by [48, Proposition 5.12 (iv)]. Thus τ is also a weak equivalence, and $F(\tau)$ an isomorphism. It now follows by a diagram chase that $\tilde{F}$ preserves the composition:

$$\begin{aligned}\tilde{F}(Q \otimes_H P) &= F(\epsilon'') \circ F(\pi'')^{-1} \\ &= F(\epsilon'' \circ \tau) F(\pi'' \circ \tau)^{-1} \\ &= F(\epsilon \circ \mathrm{pr}_2) \circ F(\pi' \circ \mathrm{pr}_1)^{-1} \\ &= F(\epsilon) \circ F(\mathrm{pr}_2) \circ F(\mathrm{pr}_1)^{-1} \circ F(\pi')^{-1} \\ &= F(\epsilon) \circ F(\pi)^{-1} \circ F(\epsilon') \circ F(\pi')^{-1} \\ &= \tilde{F}(P) \circ \tilde{F}(Q) .\end{aligned}$$

□

Finally, let us observe the following two immediate consequences of the two-sided semi-direct product constructed in the proof [50, 51]:

Corollary 2.12 *Two Lie groupoids are weakly equivalent if and only if they are isomorphic in the category* GPD.

Proof ($\Rightarrow$) This direction is clear from the fact that the functor $\mathfrak{P}$: Gpd $\to$ GPD sends weak equivalences to isomorphisms.

($\Leftarrow$) If $P \colon H \to G$ is invertible, then P is also H-principal, and in the diagram of homomorphisms

$$H \xleftarrow{\pi} H \ltimes P \rtimes G \xrightarrow{\epsilon} G$$

not only π but also ϵ is a weak equivalence. Although $H \ltimes P \rtimes G$ may not be a Lie groupoid because its base P may be non-Hausdorff, it is weakly equivalent to a Lie groupoid, so the Lie groupoids G and H are weakly equivalent. □

Corollary 2.13 *For two Lie groupoids G and H, morphisms $H \to G$ in*

GPD *correspond to equivalence classes of diagrams of homomorphisms*

$$H \xleftarrow{w} K \xrightarrow{\phi} G$$

with w *a weak equivalence. Two such diagrams* $H \xleftarrow{w} K \xrightarrow{\phi} G$ *and* $H \xleftarrow{w'} K' \xrightarrow{\phi'} G$ *are equivalent (i.e. represent the same morphism in* GPD*) if and only if there is a diagram*

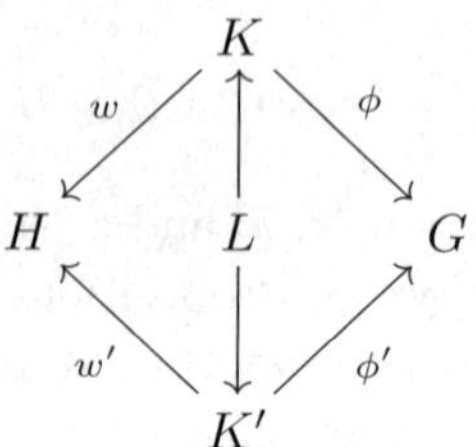

which commutes up to transformations and in which $L \to K$ *and* $L \to K'$ *are weak equivalences.*

3
Sheaves on Lie groupoids

In Chapter 2 we have introduced the notion of a Lie groupoid, and we have shown how singular spaces (such as the space of leaves of a foliation) can be represented by weak equivalence classes of Lie groupoids.

The purpose of the present chapter is two-fold. First, we will introduce the category of equivariant sheaves for a Lie groupoid G, as well as its derived category. As will become apparent, these categories are particularly relevant for étale groupoids, and more generally for foliation groupoids. Secondly, we will give an explicit construction of the fundamental groupoid $\pi_1(G)$ associated to a Lie groupoid G. These constructions are related in a Grothendieck style way, by the fact that the fundamental groupoid is determined by the locally constant sheaves.

We begin in Section 3.1 with an explicit description of equivariant sheaves for a groupoid, and give some examples. In Section 3.2 we study how the category of G-sheaves behaves with respect to homomorphisms of Lie groupoids. We show that the category of G-sheaves is in fact functorial with respect to the generalized morphisms, and that a weak (or Morita) equivalence between Lie groupoids induces an equivalence of the associated categories of sheaves.

In Section 3.3 we discuss the fundamental group of a Lie groupoid. More precisely, we construct for each Lie groupoid G a new Lie groupoid $\pi_1(G)$. This groupoid is weakly equivalent to a discrete groupoid. In particular, the isotropy groups of $\pi_1(G)$ are discrete. If G is connected, these isotropy groups of $\pi_1(G)$ are all isomorphic to each other, and will be referred to as 'the' fundamental group of G. In the special case where G is étale, we recover the fundamental group of G as defined earlier by [6, 50]. The construction is also closely related to the construction of loop spaces for orbifolds in [36].

We will prove that the fundamental groupoid $\pi_1(G)$ has a universal property. In fact, there is a canonical map $G \to \pi_1(G)$ which is universal among maps $G \to H$ into discrete groupoids, cf. Corollary 3.23 below. We will also prove that the construction of the fundamental groupoid is functorial for generalized homomorphisms. This will in particular imply that fundamental groups are invariant under weak equivalence.

To conclude this section, we use our construction of the fundamental group to give an explicit description of the kernel of the natural surjection

$$\pi_1(M) \longrightarrow \pi_1(M, \mathcal{F})$$

of the fundamental groupoid of a manifold M to the fundamental groupoid of the holonomy groupoid of a foliation $\mathcal{F}$ of M. From this description, we will be able to deduce a sharpening of Haefliger's original theorem on the existence of analytic codimension 1 foliations of compact manifolds [22] (see also [18, 33]).

In Section 3.4 we return to the general context of equivariant sheaves, and consider sheaves of R-modules over Lie groupoids. We show that these sheaves form an abelian category which has enough injectives, satisfies Grothendieck's axiom AB5, and has a small set of generators. These properties allow us to apply the standard machinery of homological algebra to this category. In particular, it allows us to construct the derived category, as we will show in Section 3.5. These general constructions and facts will later be used in Chapter 4, in our discussion of cohomology.

3.1 Sheaves on groupoids

From now on, we assume that the reader has some familiarity with the basic notions of the theory of sheaves on topological spaces. We will briefly recall some of the definitions, often only to fix the terminology and the notation. There are many good treatments of sheaf theory available, including [4, 5, 32].

We will continue to work in the context of smooth manifolds and smooth maps, although much of this section applies more generally.

One way of defining a sheaf (of sets) on a manifold M is as a local diffeomorphism $\pi: E \to M$; here we do not require E to be Hausdorff. We will often refer to π as an *étale map*, and to E as an *étale space over* M. Also, we will often just write E for $(\pi: E \to M)$. The category of sheaves of sets over M is the category of such étale spaces, where the arrows from $\pi : E \to M$ to $\pi' : E' \to M$ are the continuous maps $f\colon E \to E'$ with $\pi' \circ f = \pi$; such f are necessarily étale.

A seemingly different but equivalent way of defining the notion of a sheaf is via local sections. One then defines a sheaf $\mathcal{S}$ of sets on M to consists of sets $\mathcal{S}(U)$, one for each open subset U of M, together with 'restriction maps' $\rho_{V,U}\colon \mathcal{S}(U) \to \mathcal{S}(V)$ for $V \subset U$. The restriction maps are required to be functorial, in the sense that $\rho_{U,U} = \mathrm{id}$ and $\rho_{W,V} \circ \rho_{V,U} = \rho_{W,U}$ if $W \subset V \subset U$. Moreover, the following gluing condition should be satisfied for a union $U = \bigcup_{i \in I} U_i$ of open sets: for any family $\sigma_i \in \mathcal{S}(U_i)$, $i \in I$, compatible on overlaps in the sense that $\rho_{U_i \cap U_j, U_i}(\sigma_i) = \rho_{U_i \cap U_j, U_j}(\sigma_i)$ for all $i, j \in I$, there is a unique $\sigma \in \mathcal{S}(U)$ with $\rho_{U_i,U}(\sigma) = \sigma_i$ for all $i \in I$. For two such sheaves $\mathcal{S}$ and $\mathcal{S}'$, the maps $\phi : \mathcal{S} \to \mathcal{S}'$ are the natural transformations, given by functions $\phi_U \colon \mathcal{S}(U) \to \mathcal{S}'(U)$ commuting with the restrictions of $\mathcal{S}$ and $\mathcal{S}'$.

The equivalence between these two definitions is explained as follows: Given a local diffeomorphism $\pi\colon E \to M$, one defines a sheaf $\mathcal{S}_\pi$ by

$$\mathcal{S}_\pi(U) = \Gamma(U, E) = \{\sigma\colon U \to E \,|\, \pi \circ \sigma = \mathrm{id}\}\ ,$$

the set of smooth sections of π. Conversely, given a sheaf $\mathcal{S}$ as above, one first defines the *stalk* $\mathcal{S}_x$ of $\mathcal{S}$ at $x \in M$ as the colimit

$$\mathcal{S}_x = \varinjlim_{x \in U} \mathcal{S}(U)\ .$$

Thus, elements of $\mathcal{S}_x$ are equivalence classes of pairs (σ, U), where U is an open neighbourhood of x and $\sigma \in \mathcal{S}(U)$, and (σ, U) is equivalent to (σ', U') if and only if there is a neighbourhood $W \subset U \cap U'$ of x such that $\rho_{W,U}(\sigma) = \rho_{W,U'}(\sigma')$. The equivalence class of (σ, U) is denoted by σ_x and called the *germ* of σ at x. The union of the stalks $\mathcal{S}_x$, $x \in M$, carries a natural topology, called the *sheaf topology*. The basic open sets for this topology are the sets of the form

$$B_\sigma = \{\sigma_x \,|\, x \in U\}\ ,$$

where U is an open subset of M and $\sigma \in \mathcal{S}(U)$. In this way the union of the stalks form a topological space, denoted

$$\mathrm{Et}(\mathcal{S})$$

and referred to as the étale space of the sheaf $\mathcal{S}$. The evident projection

$$\pi\colon \mathrm{Et}(\mathcal{S}) \longrightarrow M$$

restricts to a homeomorphism on each basic open B_σ, i.e. π is a local homeomorphism. In particular, we can pull back the manifold structure on $\mathrm{Et}(\mathcal{S})$, making π into a local diffeomorphism.

These constructions establish in fact an equivalence of categories, between sheaves $\mathcal{S}$ of sets on M and étale spaces E over M. More precisely, the constructions of the étale space $\mathrm{Et}(\mathcal{S})$ from $\mathcal{S}$, and of the sheaf $\mathcal{S}_\pi$ from $\pi : E \to M$, are both functorial, and inverse to each other up to natural isomorphisms.

We will denote either of these equivalent categories by

$$\mathrm{Sh}(M) ,$$

and often change point of view, by tacitly moving from a sheaf $\mathcal{S}$ to the associated étale space and vice versa. We will also usually work with sheaves with more structure, such as sheaves of abelian groups or vector spaces.

Now let G be a Lie groupoid. We define a (right) G-sheaf to be a sheaf (étale space) $\pi : E \to G_0$ equipped with a smooth right G-action along π

$$E \times_{G_0} G_1 \longrightarrow E , \qquad (e, g) \longmapsto eg .$$

The category of such sheaves is denoted by

$$\mathrm{Sh}(G) ,$$

an arrow $E \to E'$ being a (smooth) map $f : E \to E'$ which respects the projection to G_0 as well as the action (i.e. $\pi'(f(e)) = \pi(e)$ and $f(eg) = f(e)g$, for any $e \in E$ and $g \in G_1$ with $\mathrm{t}(g) = \pi(e)$). We denote the set of such arrows by $\mathrm{Hom}(E, E')$, or sometimes by $\mathrm{Hom}_G(E, E')$ for emphasis.

Remark. Our convention to work with *right* G-sheaves is somewhat arbitrary, and we could as well have chosen to work with left G-sheaves. In fact, if E is a right G-sheaf, we will sometimes have occasion to use the associated left G-sheaf E^{op}, defined by the same étale space $E \to G_0$ and with the left action ge defined as eg^{-1}.

Example 3.1 Recall that for any set A the corresponding *constant G-sheaf* on G_0 is the étale space $\mathrm{pr}_2 : A \times G_0 \to G_0$, where A is given the discrete topology. It has the natural structure of a G-sheaf, where we take the trivial action

$$(a, y)g = (a, x)$$

for any $g : x \to y$ in G.

If G is a connected Lie group (G_0 just a point), every G-sheaf is constant, and $\mathrm{Sh}(G)$ is just the category of sets. In particular, this category contains no information about G. On the other hand, if G is a Lie groupoid weakly equivalent to an étale groupoid, there are many G-sheaves, as we will see, and in fact from the category $\mathrm{Sh}(G)$ one can essentially recover the groupoid G, at least up to weak equivalence.

Example 3.2 Let G be an étale groupoid. Consider the sheaf $\mathcal{A}^0$ of smooth functions on G_0, i.e.

$$\mathcal{A}^0(U) = C^\infty(U, \mathbb{R}) ,$$

for any open $U \subset G_0$. The stalk $\mathcal{A}^0_x$ at a point $x \in G_0$ is the vector space of germs of smooth functions. This sheaf has the natural structure of a G-sheaf. Indeed, if $f_y \in \mathcal{A}^0_y$ and $g\colon x \to y$ is an arrow in G, we can define a germ $f_y g$ at x as the composition of germs

$$f_y g = f_y \circ \mathrm{Eff}(g) .$$

More explicitly, we first choose a small neighbourhood V of y such that f_y is the germ of a smooth function $f\colon V \to \mathbb{R}$ at y, and next we choose a small neighbourhood U of x such that the source map $\mathrm{s} : G_1 \to G_0$ has a section $\sigma : U \to G_1$ through g. Choosing U small enough so that $\mathrm{t}\colon G_1 \to G_0$ maps $\sigma(U)$ into V, we then define $f_y g$ as the germ at x of $f \circ \mathrm{t} \circ \sigma\colon U \to \mathbb{R}$.

This example generalizes in fact to the general principle that for any étale groupoid G, any sheaf on G_0 constructed from the intrinsic smooth structure of G_0 has the structure of a G-sheaf. For example, the sheaf $\mathcal{A}^n$ of differential n-forms on G_0 is a G-sheaf, with action defined exactly as for $\mathcal{A}^0$: In the notation above, if $\omega_y \in \mathcal{A}^n_y$ is the germ at y of a differential n-form ω on an open V and $g\colon x \to y$ then

$$\omega_y g = \mathrm{Eff}(g)^*(\omega_y) = (\mathrm{t} \circ \sigma)^*(\omega)_x ,$$

where $\mathrm{t} \circ \sigma\colon U \to V$ as above.

Example 3.3 Let $(M, \mathcal{F})$ be a foliated manifold, and let $\mathrm{Hol}(M, \mathcal{F})$ be its holonomy groupoid. Let $\mathcal{A}^n_{\mathrm{bas}}$ be the sheaf on M of germs of basic n-forms. Recall that an n-form ω on an open $U \subset M$ is called *basic* if $i_X(\omega) = L_X(\omega) = 0$ for any vector field X on U which is tangent to the leaves of $\mathcal{F}$. If $\phi\colon U \to \mathbb{R}^p \times \mathbb{R}^q$ is a foliation chart for $\mathcal{F}$, the basic n-forms on U are exactly the forms obtained as pull-backs of n-forms along $\mathrm{pr}_2 \circ \phi\colon U \to \mathbb{R}^q$. From this description, it is easy to see that $\mathcal{A}^n_{\mathrm{bas}}$

carries a canonical action by the holonomy groupoid. Thus $\mathcal{A}^n_{\mathrm{bas}}$ is a $\mathrm{Hol}(M,\mathcal{F})$-sheaf.

Remark. Let G be a Lie groupoid. Let $\epsilon : E \to G_0$ be a manifold equipped with a right G-action, where ϵ is any smooth map. Then the sheaf of smooth sections of ϵ carries a natural action of the étale groupoid $\mathrm{Bis}(G)$ of germs of bisections of G [48, p. 115]. If G is itself étale, then $\mathrm{Bis}(G) = G$, and in fact all the G-sheaves in Example 3.2 can be obtained from a G-vector bundle E in this way.

Remark. Let $f : N \to M$ be a submersion. Then f can be factored uniquely (up to diffeomorphism) as $f = f^{(0)} \circ c$,

$$N \xrightarrow{c} N^{(0)} \xrightarrow{f^{(0)}} M ,$$

where $N^{(0)}$ is a (possibly non-Hausdorff) manifold, c is a submersion with connected fibers, and $f^{(0)}$ is an étale map. (Thus the fibers of $f^{(0)}$ are the connected components of the fibers of f, and we refer to $N^{(0)}$ as the *sheaf of connected components* of f.) Also, observe that this construction is stable under pull-back, in the sense that for any smooth map $M' \to M$ the canonical map $(M' \times_M N)^{(0)} \to M' \times_M N^{(0)}$ is a diffeomorphism.

Example 3.4 Let G be a Lie groupoid, and let $\epsilon\colon E \to G_0$ be a manifold with a right G-action. Assume that ϵ is a submersion. Then the sheaf $E^{(0)} \to G_0$ constructed in the previous remark has the structure of a G-sheaf. If $[e] \in E^{(0)}$ denotes the component of $\epsilon^{-1}(y)$ which contain the point $e \in \epsilon^{-1}(y)$, then the action is simply described by

$$[e]g = [eg]$$

for any arrow $g : x \to y$ in G. In fact, this construction provides a left adjoint to the inclusion of the étale G-spaces (G-sheaves) into the category of G-spaces $\epsilon\colon E \to G_0$ for which ϵ is a submersion. Explicitly, if $E \to F$ is any G-equivariant map from such an ϵ into an étale G-space, then it factors uniquely through $E \to E^{(0)}$ by a map of G-sheaves $E^{(0)} \to F$.

Example 3.5 Let G be a Lie groupoid. As a special case of the previous example, consider for each open $U \subset G_0$ the submersion

$$\mathrm{s}\colon \mathrm{t}^{-1}(U) \longrightarrow G_0$$

with the evident right G-action given by multiplication in G. Write

$$\tilde{U} = (\mathrm{s}^{(0)} : \mathrm{t}^{-1}(U)^{(0)} \to G_0)$$

for the associated G-sheaf of connected components. If $E \to G_0$ is any other G-sheaf, then there is an obvious bijection

$$\phi\colon \Gamma(U, E) \longrightarrow \mathrm{Hom}_G(\mathrm{t}^{-1}(U), E) ,$$

defined by $\phi(\sigma)(g) = \sigma(\mathrm{t}(g))g$. By the adjointness property of Example 3.4, we obtain an isomorphism

$$\Gamma(U, E) \longrightarrow \mathrm{Hom}_G(\tilde{U}, E) .$$

Thus the G-sheaf $\tilde{U}$ 'classifies' or 'represents' the sections over U. It follows that for every G-sheaf E there exists a surjection of the form

$$\coprod \tilde{U}_i \longrightarrow E$$

from a sum of copies of such G-sheaves $\tilde{U}_i$.

Example 3.6 The category $\mathrm{Sh}(G)$ has 'internal homs', and these are constructed as internal homs of ordinary sheaves on the space G_0. To be more explicit, recall first that if E and F are sheaves on the space G_0, then the sheaf $Hom(E, F)$ is defined, by setting for each open $U \subset G_0$,

$$Hom(E, F)(U) = \mathrm{Hom}(E|_U, F_U) ,$$

the set of sheaf maps $E|_U \to F|_U$ between the restricted sheaves over the subspace U. Now if E and F have the structure of G-sheaves, this sheaf $Hom(E, F)$ can also be equipped with a G-action, as follows. Suppose $g\colon x \to y$ is an arrow in G. Let $\tilde{g}\colon U \to G_1$ be a bisection through g, defined on a neighbourhood U of x, so that $\mathrm{t} \circ \tilde{g}\colon U \to V$ is a diffeomorphism onto a neighbourhood V of y. Now if $\alpha : E|_V \to F|_V$ represents an element of $\mathrm{Hom}(E, F)_y$, we define $\alpha g \in Hom(E, F)_x$ to be represented by the map $\alpha g\colon E|_U \to F|_U$, given for any $z \in U$ and $e \in E_z$ by

$$(\alpha g)(e) = \alpha(e\tilde{g}(z)^{-1})\tilde{g}(z) .$$

The bisection $\tilde{g}$ is not unique; but since $\mathrm{s}\colon G_1 \to G_0$ is a submersion, it holds that for any other such section $\tilde{g}'$ through g there is a neighbourhood $U_0 \subset U$ of x such that for any $z \in U_0$, $\tilde{g}(z)$ and $\tilde{g}'(z)$ lie in the same connected component of the fiber $\mathrm{s}^{-1}(z)$ of $\mathrm{s}\colon G_1 \to G_0$. Now consider for each $a \in F_z$ the open subset $W_a = \{h \in \mathrm{s}^{-1}(z) \mid \alpha(eh^{-1})h = a\}$ of $\mathrm{s}^{-1}(z)$. These sets form a partition of $\mathrm{s}^{-1}(z)$, i.e. $W_a \cap W_b = \emptyset$ or $W_a = W_b$ for any two $a, b \in F_z$. Thus $\tilde{g}(z)$ and $\tilde{g}'(z)$ lie in the same W_a, i.e. $\alpha(e\tilde{g}(z)^{-1})\tilde{g}(z) = \alpha(e\tilde{g}'(z)^{-1})\tilde{g}'(z)$. Since this holds for any $z \in U_0$, the definition of αg is independent of the choice of the admissible section $\tilde{g}$.

The G-sheaf $Hom(E, F)$ thus defined has the familiar adjointness property, given by the natural bijection

$$\mathrm{Hom}_G(D, Hom(E, F)) \cong \mathrm{Hom}_G(D \times E, F)$$

for any G-sheaf D (here Hom_G denotes the set of arrows in the category $\mathrm{Sh}(G)$, and $D \times E$ is the product of G-sheaves).

3.2 Functoriality and Morita equivalence

In this section we will discuss how the category $\mathrm{Sh}(G)$ of sheaves (of sets) on a Lie groupoid G behaves under morphisms between Lie groupoids. Later, we will consider the induced operations at the level of chain complexes of sheaves of modules.

Let $\phi\colon H \to G$ be a homomorphism between Lie groupoids. If $E \to G_0$ is a right G-space (i.e. a manifold with a right G-action), then clearly the pull-back $\phi^*(E) = H_0 \times_{G_0} E$ is a right H-space under the induced action

$$(y, e)h = (z, e\phi(h))$$

for $h\colon z \to y$ in H and $e \in E_{\phi(y)}$. Since the pull-back of an étale map is again étale, we obtain in this way a functor

$$\phi^*\colon \mathrm{Sh}(G) \longrightarrow \mathrm{Sh}(H)\ .$$

Proposition 3.7 *Let $\phi : H \to G$ be a homomorphism between Lie groupoids. Then there exists a functor*

$$\phi_*\colon \mathrm{Sh}(H) \longrightarrow \mathrm{Sh}(G)$$

which is right adjoint to ϕ^*.

Remark. The adjointness means that for any G-sheaf $\mathcal{S}$ and any H-sheaf $\mathcal{T}$ there is a natural isomorphism

$$\mathrm{Hom}_H(\phi^*\mathcal{S}, \mathcal{T}) \cong \mathrm{Hom}_G(\mathcal{S}, \phi_*\mathcal{T})\ , \qquad (1)$$

and this property determines ϕ_* uniquely up to natural isomorphism.

Proof Let $\mathcal{T}$ be any H-sheaf. If we assume for a moment that ϕ_* exists, then the adjointness property and the representability of local sections by sheaves of the form $\tilde{U}$ (Example 3.5) yield an explicit description of $\phi_*\mathcal{T}$. Indeed, for any open set $U \subset G_0$ we must have

$$\Gamma(U, \phi_*\mathcal{T}) = \mathrm{Hom}_G(\tilde{U}, \phi_*\mathcal{T}) = \mathrm{Hom}_H(\phi^*\tilde{U}, \mathcal{T}) .$$

Thus, we take this expression as the definition of $\phi_*\mathcal{T}$. Note that since the factorization $\mathrm{t}^{-1}(U) \to \mathrm{t}^{-1}(U)^{(0)} \to G_0$ defining $\tilde{U}$ is preserved by pull-back along $\phi\colon H_0 \to G_0$, we can also write

$$\phi_*\mathcal{T}(U) = \mathrm{Hom}_H(\phi^*\mathrm{t}^{-1}(U), \mathcal{T}) ,$$

where the latter Hom_H is that of manifolds with H-action. It is clear that $\phi_*\mathcal{T}$ thus defined is a sheaf on G_0. We have to show that it carries a G-action, and that it is indeed adjoint.

To see that G acts on $\phi_*\mathcal{T}$, consider an arrow $g : x \to y$ in G. Let $\tilde{g}\colon U \to G_1$ be a bisection through g, defined on a neighbourhood U of x. Then $\mathrm{t} \circ \tilde{g}$ defines a diffeomorphism $U \to V$ onto an open neighbourhood V of y. This section $\tilde{g}$ induces a map of G-spaces

$$\tilde{g}_* : \mathrm{t}^{-1}(U) \longrightarrow \mathrm{t}^{-1}(V) , \qquad k \longmapsto \tilde{g}(\mathrm{t}(k))k$$

and hence a map

$$\tilde{g}_* : \tilde{U} \longrightarrow \tilde{V}$$

between the sheaves of fiber-wise components. We claim that, for U small enough, the latter map does not depend on the choice of the bisection $\tilde{g}$ through g. Indeed, if $\tilde{g}'$ is another one, then $\tilde{g}'(x) = g = \tilde{g}(x)$, so for x' close to x the arrows $\tilde{g}(x')$ and $\tilde{g}'(x')$ will also lie in the same component if the fiber of $\mathrm{s}\colon \mathrm{t}^{-1}(V) \to G_0$ over x'. Hence if $k : y' \to x'$ then $\tilde{g}_*(k)$ and $\tilde{g}'_*(k)$ will lie in the same component of the fiber of $\mathrm{t}^{-1}(V) \to G_0$ over y'.

Now if $a_y \in (\phi_*\mathcal{T})_y$ is the germ of a map $a\colon \phi^*(\tilde{V}) \to \mathcal{T}$ defined for a neighbourhood V of y, define the action of $g\colon x \to y$ on a_y by

$$a_y g = (a \circ \phi^*(\tilde{g}_*))_x .$$

This is well-defined because for a neighbourhood U of x which is small enough the map $\tilde{g}_* : \tilde{U} \to \tilde{V}$ does not depend on the choice of $\tilde{g}$ as we have seen.

Finally, for the adjointness property, consider an H-sheaf $\mathcal{T}$ as above and a G-sheaf $\mathcal{S}$. Then the desired bijection (1) between G-maps $\lambda\colon \mathcal{S} \to$

$\phi_*\mathcal{T}$ and H-maps $\mu\colon \phi^*\mathcal{S} \to \mathcal{T}$ can explicitly be described as follows: Given μ, define $\lambda_U\colon \mathcal{S}(U) \to \phi_*\mathcal{T}(U)$ as the composition

$$\begin{array}{ccc}
\mathrm{Hom}_G(\tilde{U},\mathcal{S}) & \xrightarrow{\phi^*} & \mathrm{Hom}_H(\phi^*\tilde{U},\phi^*\mathcal{S}) \\
\uparrow \cong & & \downarrow \mu\circ - \\
\mathcal{S}(U) & \xrightarrow{\lambda_U} & \mathrm{Hom}_H(\phi^*\tilde{U},\mathcal{T}) = \phi_*\mathcal{T}(U)
\end{array}$$

Conversely, given λ, define μ stalk-wise for points $y \in H_0$ as the composite

$$\mu_y\colon (\phi^*\mathcal{S})_y = \mathcal{S}_{\phi(y)} \xrightarrow{\lambda_{\phi(y)}} (\phi_*\mathcal{T})_{\phi(y)} \xrightarrow{\mathrm{ev}} \mathcal{T}_y\ ,$$

where the evaluation map $\mathrm{ev}\colon (\phi_*\mathcal{T})_{\phi(y)} \to \mathcal{T}_y$ sends the germ at $\phi(y)$ of a map $a\colon \phi^*\tilde{U} \to \mathcal{T}$ defined on the neighbourhood U of $\phi(y)$ to its value at the component of $1_{\phi(y)} \in \mathrm{t}^{-1}(U)_{\phi(y)} = \phi^*(\mathrm{t}^{-1}(U))_y$.

We leave it to the reader to check that these constructions of λ from μ and vice versa are mutually inverse, and establish the required bijection (1). □

Examples 3.8 (1) Let $f\colon N \to M$ be a smooth map between manifolds. Then the functor $f_*\colon \mathrm{Sh}(N) \to \mathrm{Sh}(M)$ is given by the composition with f^{-1}, i.e.

$$f_*(\mathcal{S})(U) = \mathcal{S}(f^{-1}(U))\ .$$

(2) Let G be a Lie groupoid and $\phi\colon G \to \{pt\}$ the canonical map into one-point space. Then $\phi_*\colon \mathrm{Sh}(G) \to \mathrm{Sets}$ is given by

$$\phi_*(E) = \Gamma_{\mathrm{inv}}(G,E)\ .$$

Here $\Gamma_{\mathrm{inv}}(G,E)$ denotes the set of G-*invariant sections* of E, i.e. of global sections σ of E which are invariant in the sense that $\sigma(y)g = \sigma(x)$ for any $g\colon x \to y$.

(3) Let G be a Lie groupoid, and denote by $q\colon G \to |G|$ the quotient projection. This projection again induces a functor $q_*\colon \mathrm{Sh}(G) \to \mathrm{Sh}(|G|)$, although here $|G|$ in not necessarily a manifold and the sheaves $\mathrm{Sh}(|G|)$ should be understood in topological sense, without the smooth structure. In this case we have

$$q_*(E)(U) = \Gamma_{\mathrm{inv}}(G|_{q^{-1}(U)},E)\ .$$

Next, suppose $P : H \to G$ is a generalized morphism between Lie groupoids. Thus P is a right principal G-bundle over H_0 equipped with a left H-action, and we write

$$\begin{array}{ccc} P & \xrightarrow{\epsilon} & G_0 \\ {\scriptstyle \pi}\downarrow & & \\ H_0 & & \end{array}$$

for the structure maps, as in Section 2.5. The bundle P induces a functor $P \otimes_G$ - from manifolds equipped with a left G-action to manifolds with a left H-action. Explicitly, if $M \to G_0$ is such a G-manifold, then $P \otimes_G M$ is obtained from the pull-back $P \times_{G_0} M$ by identifying (pg, m) and (p, gm) for any $g : x \to y$ in G, $p \in P_y$ and $m \in M_x$. This quotient space is a manifold, because the action of G on $P \times_{G_0} M$ is principal (Lemma 2.8).

Proposition 3.9 *The functor $P \otimes_G$ - maps left G-sheaves to left H-sheaves.*

Proof Suppose that $f : E \to G_0$ is an étale space over G_0, equipped with a left G-action. We have to show that the projection

$$\rho : P \otimes_G E \longrightarrow H_0\,, \qquad p \otimes e \mapsto \pi(p)$$

is again étale. To this end, consider the diagram

$$\begin{array}{ccc} P \times_{G_0} E & \xrightarrow{\mathrm{pr}_2} & P \\ {\scriptstyle q}\downarrow & & \downarrow{\scriptstyle \pi} \\ P \otimes_G E = (P \times_{G_0} E)/G & \xrightarrow{\rho} & P/G = H_0 \end{array}$$

where q is the quotient map. Now $P \to H_0$ is a principal G-bundle, and hence (Lemma 2.8) $P \times_{G_0} E$ is a principal G-bundle over $P \otimes_G E$. Since every map of principal G-bundles over the same base is an isomorphism, the square above is necessarily a pull-back. Now pr_2 is the pull-back of the étale map $E \to G_0$, hence is itself étale. Since π is a surjective submersion, ρ must also be étale (see e.g. [48, Exercise 5.16]). □

Thus, using the equivalence $E \to E^{\mathrm{op}}$ between left and right G-sheaves, we obtain a functor of right sheaves

$$P^* : \mathrm{Sh}(G) \longrightarrow \mathrm{Sh}(H)$$

defined on étale spaces by

$$P^*(E) = (P \otimes_G E^{\mathrm{op}})^{\mathrm{op}}\,.$$

From the unit and associativity properties of the tensor product, we immediately obtain that this construction is functorial in P.

Proposition 3.10 *Let E be a G-sheaf.*

(i) For the unit bundle $U(G)\colon G \to G$, there is a canonical isomorphism

$$U(G)^*(E) \longrightarrow E$$

natural in E.

(ii) For generalized morphisms $Q\colon K \to H$ and $P\colon H \to G$, there is a canonical isomorphism

$$Q^*(P^*(E)) \longrightarrow (Q \otimes_H P)^*(E)$$

natural in E.

Corollary 3.11 (Morita invariance) *Let $P : H \to G$ be a Morita equivalence. Then $P^*\colon \mathrm{Sh}(G) \to \mathrm{Sh}(H)$ is an equivalence of categories.*

Next, we observe that this construction of the functor P^* for generalized morphisms extends that of the pull-back functor ϕ^* for ordinary homomorphisms:

Proposition 3.12 *Let $\phi : H \to G$ be a homomorphism between Lie groupoids, and let $\mathfrak{P}(\phi) : H \to G$ be the associated principal G-bundle over H. For any G-sheaf E there exists a canonical isomorphism*

$$\phi^* E \longrightarrow (\mathfrak{P}(\phi) \otimes_G E^{\mathrm{op}})^{\mathrm{op}} = \mathfrak{P}(\phi)^*(E)$$

natural in E.

Proof Recall that the points of $\mathfrak{P}(\phi)$ are pairs (y, g) where $y \in H_0$ and $g\colon x \to \phi(y)$ in G. The isomorphism sends $e \in (\phi^* E)_y$ to $(y, 1_{\phi(y)}) \otimes e$, and its inverse sends $(y, g) \otimes e$ to eg^{-1}. □

Corollary 3.13 *Let G and H be Lie groupoids.*

(i) If $\phi\colon H \to G$ is a weak equivalence, then $\phi^* : \mathrm{Sh}(G) \to \mathrm{Sh}(H)$ is an equivalence of categories.

(ii) Let $P\colon H \to G$ be a generalized morphism, with associated two-sided semi-direct product

$$H \xleftarrow{\pi} H \ltimes P \rtimes G \xrightarrow{\epsilon} G$$

as in Section 2.6. Then there is a natural isomorphism

$$P^* \cong \pi_* \circ \epsilon^*$$

of functors $\mathrm{Sh}(G) \to \mathrm{Sh}(H)$.

(iii) For any generalized morphism $P : H \to G$*, the functor* $P^* : \mathrm{Sh}(G) \to \mathrm{Sh}(H)$ *has a right adjoint*

$$P_* = \epsilon_* \circ \pi^* : \mathrm{Sh}(H) \longrightarrow \mathrm{Sh}(G) .$$

Proof Part (i) follows from Proposition 3.10 and Proposition 3.12. For part (ii), we have $P = \mathfrak{P}(\epsilon) \circ \mathfrak{P}(\pi)^{-1}$ as generalized morphisms (Section 2.6), hence $P \circ \mathfrak{P}(\pi) = \mathfrak{P}(\epsilon)$ and $\pi^* \circ P^* \cong \epsilon^*$ by Proposition 3.10. But π is a weak equivalence, so π^* is an equivalence of categories. Then so is its adjoint π_*, and the unit of the adjunction $\mathrm{id} \to \pi_* \circ \pi^*$ is an isomorphism. Thus from $\pi^* \circ P^* \cong \epsilon^*$ we obtain a natural isomorphism

$$P^* \cong \pi_* \circ \pi^* \circ P^* \cong \pi_* \circ \epsilon^* .$$

Part (iii) now follows from (ii), because π_* is an equivalence of categories with inverse π^* and ϵ^* has a right adjoint ϵ_*. □

3.3 The fundamental group and locally constant sheaves

Let G be a Lie groupoid. We say that G is *connected* if its space of orbits $|G|$ is (path-)connected. This motivates us to define a G*-path* (or *path in* G) as a sequence

$$\sigma_n g_n \sigma_{n-1} \cdots \sigma_1 g_1 \sigma_0 ,$$

where $\sigma_0, \ldots, \sigma_n : [0,1] \to G_0$ are paths in G_0 and $g_1, \ldots, g_n$ are arrows in G such that $g_i : \sigma_{i-1}(1) \to \sigma_i(0)$, $i = 1, \ldots, n$. We say that $\sigma_n g_n \cdots g_1 \sigma_0$ is a path from $\sigma_0(0)$ to $\sigma_n(1)$, and that it has *order* $n \geq 0$.

$$\bullet \xrightarrow{\sigma_0} \bullet \overset{g_1}{\dashrightarrow} \bullet \xrightarrow{\sigma_1} \bullet \quad \cdots \quad \bullet \xrightarrow{\sigma_{n-1}} \bullet \overset{g_n}{\dashrightarrow} \bullet \xrightarrow{\sigma_n} \bullet$$

With this definition, we can say that G is connected if and only if for any two points $x, y \in G_0$ there exists a G-path from x to y. If $\sigma'_m g'_m \cdots g'_1 \sigma'_0$ is another G-path with $\sigma'_0(0) = \sigma_n(1)$, we can *concatenate* these two into a new G-path

$$\sigma'_m g'_m \cdots g'_1 \sigma'_0 1_{\sigma_n(1)} \sigma_n g_n \cdots g_1 \sigma_0 .$$

For any Lie groupoid G, a connected component of G is a maximal (non-empty) connected subgroupoid of G. Any connected component of

G is an open full subgroupoid of G, and G decomposes into a disjoint union of its connected components,

$$G = \sum_i H_i \,.$$

A connected component H_i of G is just the restriction of G to the inverse image of a connected component of $|G|$ along the quotient projection $G_0 \to |G|$.

Let G be a Lie groupoid. We shall denote by $\mathcal{P}G$ the set of all G-paths. We shall now define when two G-paths are G-homotopic (with fixed end-points). This will later enable us to define the fundamental group of G. First, we define an equivalence relation on $\mathcal{P}G$, called simply *equivalence of paths*, to be generated by the following:

(i) (multiplication equivalence) the G-paths

$$\sigma_n g_n \cdots \sigma_{i+1} g_{i+1} \sigma_i g_i \sigma_{i-1} \cdots g_1 \sigma_0$$

and

$$\sigma_n g_n \cdots \sigma_{i+1} (g_{i+1} g_i) \sigma_{i-1} \cdots g_1 \sigma_0$$

are equivalent if σ_i is a constant path $(0 < i < n)$, and

(ii) (concatenation equivalence) the G-paths

$$\sigma_n g_n \cdots g_{i+1} \sigma_i g_i \sigma_{i-1} g_{i-1} \cdots g_1 \sigma_0$$

and

$$\sigma_n g_n \cdots g_{i+1} (\sigma_i \sigma_{i-1}) g_{i-1} \cdots g_1 \sigma_0$$

are equivalent if $g_i = 1_{\sigma_{i-1}(1)}$ $(1 \le i \le n)$. (Here $\sigma_i \sigma_{i-1}$ is the usual reparametrized concatenation of the paths σ_{i-1} and σ_i.)

A *deformation* between G-paths $\sigma_n g_n \cdots g_1 \sigma_0$ and $\sigma'_n g'_n \cdots g'_1 \sigma'_0$ of the same order from x to y consists of homotopies

$$D_i \colon [0,1]^2 \longrightarrow G_0 \,, \qquad i = 0, 1, \ldots, n \,,$$

from $D_i(0, \text{-}) = \sigma_i$ to $D_i(1, \text{-}) = \sigma'_i$, and paths

$$d_i \colon [0,1] \longrightarrow G_1 \,, \qquad i = 1, \ldots, n \,,$$

from g_i to g'_i, which satisfy

(a) $\mathrm{s} \circ d_i = D_{i-1}(\text{-}, 1)$ and $\mathrm{t} \circ d_i = D_i(\text{-}, 0)$ for all $i = 1, 2, \ldots, n$, and

(b) $D_0([0,1], 0) = \{x\}$ and $D_n([0,1], 1) = \{y\}$.

We may see such a deformation as a continuous family of G-paths of order n,

$$D_n(t, -)d_n(t)\dots d_n(t)D_0(t, -),$$

from x to y, $t \in [0,1]$.

Two G-paths in $\mathcal{P}G$ are *G-homotopic* (with fixed end-points) if one can pass from one to another by a sequence of deformations and equivalences. With the multiplication induced by concatenation, the G-homotopy classes of G-paths form a groupoid over G_0, which we call the *fundamental groupoid* of the Lie groupoid G, and denote by

$$\pi_1(G).$$

The *fundamental group* of G with respect to a base-point $x_0 \in G_0$ is the isotropy group

$$\pi_1(G, x_0) = \pi_1(G)_{x_0}.$$

It consists of G-homotopy classes of *(G, x_0)-loops* (or loops in the *pointed groupoid* (G, x_0)), which are by definition the G-homotopy classes of G-paths from x_0 to x_0. Note that this definition generalizes the definition of the fundamental group of an étale groupoid given in [6].

We now define a smooth structure on $(\pi_1(G))_1$ in the same way as one defines the smooth structure on the fundamental groupoid of a manifold: for any G-path $\sigma = \sigma_n g_n \dots g_1 \sigma_0$ from x to y, take a simply connected chart U around x and a simply-connected chart V around y in G_0. Now a basic open neighbourhood $B(U, V, \sigma)$ of the G-homotopy class $[\sigma]$ of σ in $(\pi_1(G))_1$ consists of the G-homotopy classes of G-paths from $x' \in U$ to $y' \in V$ of the form

$$\tau\sigma\gamma,$$

where γ is any path in U from x' to x and τ is any path in V from y to y'. It is easy to see that such subsets of $(\pi_1(G))_1$ form a basis for a topology. Furthermore, each $B(U, V, \sigma)$ is naturally homeomorphic to $U \times V$ by the map (s, t), and this gives a smooth structure on $(\pi_1(G))_1$. It is then clear that $\pi_1(G)$ becomes a Lie groupoid with the property that $(\mathrm{s}, \mathrm{t})\colon G_1 \to G_0 \times G_0$ is a local diffeomorphism. In particular, the fundamental group of G at any point x_0 is discrete.

If we decompose G into the sum of its connected components $G = \sum_i H_i$, then

$$\pi_1(G) = \sum_i \pi_1(H_i).$$

Each Lie groupoid $\pi_1(H_i)$ is transitive. In particular, if G is connected, the fundamental group of G does not depend (up to an isomorphism) on the choice of the base-point.

Let $\phi\colon G \to H$ be a homomorphism of Lie groupoids. Then ϕ induces a function

$$\mathcal{P}\phi\colon \mathcal{P}G \longrightarrow \mathcal{P}H$$

by

$$\mathcal{P}\phi(\sigma_n g_n \dots g_1 \sigma_0) = (\phi \circ \sigma_n)\phi(g_n) \dots \phi(g_1)(\phi \circ \sigma_0)\ .$$

This function maps G-homotopic G-paths to H-homotopic H-paths, and therefore it induces a map

$$\pi_1(\phi) = \phi_* \colon \pi_1(G) \longrightarrow \pi_1(H)\ ,$$

which is clearly a homomorphism of Lie groupoids. In particular, for any $x_0 \in G_0$ the homomorphism of Lie groupoids $\phi\colon G \to H$ induces a homomorphism of fundamental groups

$$\phi_* \colon \pi_1(G, x_0) \longrightarrow \pi_1(H, \phi(x_0))\ .$$

Proposition 3.14 *Let G be a Lie groupoid.*

(i) The natural map $G \to \pi_1(G)$ over G_0 is a homomorphism of Lie groupoids.

(ii) The fundamental groupoid $\pi_1(G)$ is weakly equivalent to a discrete groupoid.

(iii) If G is connected, the inclusion $\pi_1(G, x_0) \to \pi_1(G)$ is a weak equivalence for any $x_0 \in G_0$, and the map $(\mathrm{s}, \mathrm{t})\colon \pi_1(G) \to G_0 \times G_0$ is a covering projection.

Proof (i) The natural map $G \to \pi_1(G)$ sends an arrow $g\colon x \to y$ to the class of the G-path $\overline{y} g \overline{x}$, where $\overline{x}$ and $\overline{y}$ denote the constant paths. This map clearly preserves the groupoid structure, and it remains to check that it is smooth. To this end, take any arrow $g\colon x \to y$ in G, and choose a connected open neighbourhood W of g in G such that $U = \mathrm{s}(W)$ is a simply-connected chart around x. We may also assume that $\mathrm{t}(W)$ lies inside a simply-connected chart V around y. An arrow $g' \colon x' \to y'$ in W is mapped into the G-homotopy class of the G-path

$$\overline{g'} = \overline{y'} g' \overline{x'}\ .$$

Since W is connected, we may choose a path γ in W from g to g'. It follows that $\overline{g'}$ is G-homotopic to the G-path

$$(\mathrm{t}\circ\gamma)g(\mathrm{s}\circ\gamma)^{-1}.$$

Therefore the functor $G \to \pi_1(G)$ maps W into $B(U,V,\overline{g})$, and the composition of this functor with the diffeomorphism $B(U,V,\sigma) \to U\times V$ is exactly the map (s,t), which is smooth.

(iii) It is clear from the construction of the smooth structure on $\pi_1(G)$ that (s,t) is a covering projection if G is connected, therefore the inclusion $\pi_1(G,x_0) \to \pi_1(G)$ is a weak equivalence by [48, Proposition 5.15].

(ii) We apply result (iii) to any component of G. □

Let G be a connected Lie groupoid. A *covering space* over G is a covering space $\pi : E \to G_0$ over G_0 equipped with a right G-action $E\times_{G_0} G_1 \to E$ along π. Thus in particular, any covering space over G is a sheaf over G. A covering space over G is also referred to as a *locally constant G-sheaf.* Morphisms between two covering spaces E and F over G are equivariant maps $f : E \to F$; any such morphism is necessarily a covering projection. The category of covering spaces over G, which is a full subcategory of the category $\mathrm{Sh}(G)$ of sheaves over G, will be denoted by

$$\mathrm{Cs}(G).$$

Assume now that G is a connected Lie groupoid and that $\pi : E \to G_0$ is a covering space over G. The map π extends to a homomorphism of Lie groupoids

$$\pi : E \rtimes G \longrightarrow G$$

by $\pi(e,g) = g$. The G-paths have the following 'unique path lifting property', which generalizes the familiar property of covering spaces over manifolds: Let e_0 be a base-point in E, denote $x_0 = \pi(e_0)$, and suppose that $\sigma_n g_n \dots g_1\sigma_0$ is a G-path from x_0 to x. Then there are unique paths $\tilde\sigma_0,\dots,\tilde\sigma_n$ in E with $\pi\circ\tilde\sigma_i = \sigma_i$, $\tilde\sigma_0(0) = e_0$ and $\tilde\sigma_i(0)g_i = \tilde\sigma_{i-1}(1)$. In other words, if we write $\tilde\sigma_i(0) = e_i$ and $(e_i,g_i) : e_i g_i \to e_i$ for the arrows in $E\rtimes G$, then

$$\tilde\sigma_n(e_n,g_n)\dots(e_1,g_1)\tilde\sigma_0$$

is the unique $(E\rtimes G)$-path starting at e_0 which projects to $\sigma_n g_n \cdots g_1\sigma_0$ along $\mathcal{P}\pi$. Since G-homotopic paths are in this way lifted to $(E\rtimes G)$-homotopic paths, it follows that any covering space over G with a base-point e_0 as above has a natural fiber-wise action of $\pi_1(G,x_0)$.

Examples 3.15 (1) Let M be a connected manifold. If we consider M as the unit groupoid, then any path in M is equivalent to a path in the manifold M in the usual sense. Furthermore, $\pi_1(M)$ is the usual fundamental groupoid of M.

(2) Let G be a discrete group. If we consider G as a groupoid over a one-point space, the fundamental groupoid $\pi_1(G)$ is just the group G itself. In fact, the natural homomorphism $G \to \pi_1(G)$ is an isomorphism of groupoids for any discrete groupoid G.

(3) Let G be a connected Lie groupoid. Then the target-fiber $\pi_1(G)(\text{-}, x_0)$ over a base-point $x_0 \in G_0$ is a covering space over G_0 by the source map,

$$\mathrm{s}\colon \pi_1(G)(\text{-}, x_0) \longrightarrow G_0 \,.$$

In fact, this map is a left principal $\pi_1(G, x_0)$-bundle. Furthermore, there is a natural right action of G on $\pi_1(G)(\text{-}, x_0)$ along this map, given by the composition of the homomorphism $G \to \pi_1(G)$ and the multiplication in $\pi_1(G)$, which makes it into a covering space over G. (In fact, this is the universal covering space over G.)

As an example, we shall now compute the fundamental group of the action groupoid $M \rtimes G$ associated to an action of a discrete group G on M.

Proposition 3.16 *Let M be a connected manifold with a smooth right action of a discrete group G, and let $x_0 \in M$. Then there is an isomorphism of groups*

$$\pi_1(M \rtimes G, x_0) \cong \{(g, \varsigma) \,|\, g \in G,\, \varsigma \in \pi_1(M)(x_0, x_0 g)\} \,,$$

where the latter is a group with the multiplication given by

$$(g', [\sigma'])(g, [\sigma]) = (g'g, [R_g \circ \sigma'][\sigma]) \,.$$

In particular, there is a short exact sequence of groups

$$1 \longrightarrow \pi_1(M, x_0) \longrightarrow \pi_1(M \rtimes G, x_0) \longrightarrow G \longrightarrow 1 \,.$$

Remark. Here we denoted by $R_g : M \to M$ the right translation $x \mapsto xg$.

Proof Denote $P = \{(g, \varsigma) \,|\, g \in G,\, \varsigma \in \pi_1(M)(x_0, x_0 g)\}$. It is easy to check that P is indeed a group for the multiplication described in the proposition. Denote by $\mathcal{P}_{x_0}$ the set of loops in $(M \rtimes G, x_0)$, and let

$f\colon \mathcal{P}_{x_0} \to P$ be the function which maps $\sigma_n(x_n, g_n)\ldots\sigma_1(x_1, g_1)\sigma_0$ into

$$(g_n \ldots g_2 g_1, [R_{g_n \ldots g_2 g_1} \circ \sigma_n] \ldots [R_{g_1} \circ \sigma_1][\sigma_0]) .$$

It is easy to see that f identifies $(M \rtimes G)$-homotopic loops, hence induces a function

$$\pi_1(M \rtimes G, x_0) \longrightarrow P$$

which is clearly a surjective homomorphism. This homomorphism is in fact an isomorphism. To see this, it is enough to observe that there is an obvious deformation between the $(M \rtimes G)$-paths

$$\sigma_i(x_i, g_i)\sigma_{i-1}$$

and

$$\overline{\sigma_i(1)}(\sigma_i(1), g_i)((R_{g_i} \circ \sigma_i)\sigma_{i-1}) .$$

By induction, this implies that any $(M \rtimes G, x_0)$-loop is $(M \rtimes G)$-homotopic to a loop of the form $\overline{x_0}(x_0, g_1)\sigma_0$ in $(M \rtimes G, x_0)$, which maps to the unit of P only if $g_1 = 1$ and σ_0 is homotopic to the constant loop in M. □

Example 3.17 Let $\mathcal{F}$ be the Reeb foliation of S^3. We may choose a complete transversal section S diffeomorphic to $\mathbb{R}$ such that the associated étale holonomy groupoid $\mathrm{Hol}_S(T^2, \mathcal{F})$ is isomorphic to the action groupoid of an effective action of $\mathbb{Z} \oplus \mathbb{Z}$ on $\mathbb{R}$ (see Example 2.3 (2)): one of the generators acts as a contraction on $\mathbb{R}^+$ and as identity on $\mathbb{R}^-$, and the other as contraction on $\mathbb{R}^-$ and as identity on $\mathbb{R}^+$. The associated fundamental group is therefore again $\mathbb{Z} \oplus \mathbb{Z}$.

We shall now prove that the fundamental group is invariant under weak equivalence. This will show that the fundamental group of the Reeb foliation in the previous example does not depend on our choice of a transversal section.

Theorem 3.18 *Let G be a connected Lie groupoid with a base-point x_0. Then the functor*

$$\mathrm{Cs}(G) \longrightarrow \pi_1(G, x_0)\text{-}sets ,$$

which sends a covering space E over G to its fiber E_{x_0}, is an equivalence of categories.

Proof First, recall that the fiber E_{x_0} has a natural right action of

$\pi_1(G, x_0)$ given by the path-lifting. For any set A with a right $\pi_1(G, x_0)$-action define

$$E = A \times_{\pi_1(G,x_0)} \pi_1(G)(\text{-}, x_0) .$$

This is a covering space over G, because $\mathrm{s} : \pi_1(G)(\text{-}, x_0) \to G_0$ is a covering with a free properly discontinuous action of $\pi_1(G, x_0)$. It is now easy to check that this construction gives us the inverse (up to a natural isomorphism) of the functor above. □

Corollary 3.19 *Let G be a connected Lie groupoid with a base-point x_0, and let K be a discrete group. Then there is a natural bijection between the generalized maps of Lie groupoids $G \to K$ and conjugacy classes of homomorphisms of groups $\pi_1(G, x_0) \to K$,*

$$\mathsf{GPD}(G, K) \cong [\pi_1(G, x_0), K] .$$

In particular, for any abelian discrete group A, we have a natural bijection

$$\mathsf{GPD}(G, A) \cong \mathrm{Hom}(\pi_1(G, x_0), A) .$$

Proof A principal K-bundle over G is in particular a covering space over G, and its fiber over x_0 has a principal right K-action and a natural left $\pi_1(G, x_0)$-action. These two actions on the fiber commute with each other, so the fiber is in fact a principal K-bundle over $\pi_1(G, x_0)$. As in the proof of Theorem 3.18 we can then show that we have an equivalence between the category of principal K-bundles over G and principal K-bundles over $\pi_1(G, x_0)$. Finally, any principal K-bundle E over $\pi_1(G, x_0)$ comes from a homomorphism of groups $f: \pi_1(G, x_0) \to K$: choose a base point $e_0 \in E$, and define f by

$$f(\varsigma)e_0 = e_0\varsigma$$

for any $\varsigma \in \pi_1(G, x_0)$. If we choose a different base-point, we get a homomorphism conjugate to f. □

Proposition 3.20 *Let $P : H \to G$ be a principal G-bundle over H, where G and H are connected Lie groupoids. Then the functor $P^* : \mathrm{Sh}(G) \to \mathrm{Sh}(H)$ maps $\mathrm{Cs}(G)$ into $\mathrm{Cs}(H)$.*

Proof Let E be a covering space over G. We have to prove that the associated H-sheaf

$$\rho: P \otimes_G E \longrightarrow H_0 , \qquad p \otimes e \mapsto \pi(p)$$

is a covering projection. As in the proof of Proposition 3.9, consider the pull-back

$$\begin{array}{ccc} P \times_{G_0} E & \xrightarrow{\mathrm{pr}_2} & P \\ {\scriptstyle q}\downarrow & & \downarrow{\scriptstyle \pi} \\ P \otimes_G E = (P \times_{G_0} E)/G & \xrightarrow{\rho} & P/G = H_0 \end{array}$$

The map pr_2 is a covering projection because it is a pull-back of the covering projection $E \to G_0$. Since π is a surjective submersion, so is q, and hence ρ is a covering as well. □

Corollary 3.21 *If $\phi\colon H \to G$ is a weak equivalence between Lie groupoids, then so is $\pi_1(\phi)\colon \pi_1(H) \to \pi_1(G)$. In particular, the fundamental groupoids of Morita equivalent Lie groupoids are Morita equivalent. If G and H are Morita equivalent connected Lie groupoids, then their fundamental groups are isomorphic.*

Proof First notice that ϕ induces an equivalence of categories $\mathrm{Cs}(G) \to \mathrm{Cs}(H)$, by Proposition 3.20. Thus, if G and H are connected, the corollary follows from Theorem 3.18, Proposition 3.14 (iii) and the fact that a discrete group K can be recovered from the category of K-sets uniquely up to isomorphism; in fact it is the group of automorphisms of the forgetful functor from K-sets to sets. In the general case, we apply this result to the components of the Lie groupoids. □

Corollary 3.22 *For any Lie groupoid G we have a natural isomorphism of Lie groupoids*

$$\pi_1(\pi_1(G)) \cong \pi_1(G)\,.$$

If G is weakly equivalent to a discrete groupoid, then the natural homomorphism $G \to \pi_1(G)$ is an isomorphism of Lie groupoids.

Proof Let G be weakly equivalent to a discrete group. Then the inclusion $G_{x_0} \to G$ is a weak equivalence, hence $\pi_1(G_{x_0}) \to \pi_1(G)$ is a weak equivalence by Corollary 3.21. But the natural map $G_{x_0} \to \pi_1(G_{x_0})$ is also an isomorphism, so we have the diagram of homomorphisms of Lie groupoids

$$\begin{array}{ccc} G_{x_0} & \xrightarrow{w.e.} & G \\ {\scriptstyle \cong}\downarrow & & \downarrow \\ \pi_1(G_{x_0}) & \xrightarrow{w.e.} & \pi_1(G) \end{array}$$

It follows that $G \to \pi_1(G)$ must be a weak equivalence as well. Any weak equivalence which is isomorphism on objects (in fact it is identity in our case) is an isomorphism of Lie groupoids.

If G is weakly equivalent to a discrete groupoid, we use this argument for each of the components of G to show that $G \to \pi_1(G)$ is an isomorphism of Lie groupoids.

Now the isomorphism $\pi_1(\pi_1(G)) \cong \pi_1(G)$ follows because $\pi_1(G)$ is weakly equivalent to a discrete groupoid by Proposition 3.14 (ii). □

Denote by $\mathsf{GPD}_{\mathrm{dis}}$ the full subcategory of GPD of Lie groupoids weakly equivalent to discrete groupoids. By Proposition 3.14 (ii) and Corollary 3.21 it follows that π_1 induces a functor

$$\pi_1 \colon \mathsf{GPD} \longrightarrow \mathsf{GPD}_{\mathrm{dis}} .$$

Corollary 3.23 *The functor* $\pi_1 : \mathsf{GPD} \to \mathsf{GPD}_{\mathrm{dis}}$ *is left adjoint to the inclusion* $\mathsf{GPD}_{\mathrm{dis}} \to \mathsf{GPD}$.

Proof It is sufficient to show (cf. [39]) that for any Lie groupoid G and any discrete groupoid K, any generalized morphism $G \to K$ factors uniquely through the natural homomorphism $G \to \pi_1(G)$ as a generalized morphism. By passing to connected components, it is enough to show this for G connected and K a discrete group. In this way, the corollary follows from Corollary 3.19 because the conjugacy classes of homomorphisms between discrete groups are exactly the generalized morphisms between them. □

Remark. We can also consider categories of homomorphisms (instead of generalized maps): write $\mathsf{Gpd}_{\mathrm{dis}}$ for the full subcategory of Gpd of Lie groupoids weakly equivalent to discrete groupoids. Analogously to Corollary 3.23, the functor

$$\pi_1 \colon \mathsf{Gpd} \longrightarrow \mathsf{Gpd}_{\mathrm{dis}}$$

is left adjoint to the inclusion of $\mathsf{Gpd}_{\mathrm{dis}}$ into Gpd. Indeed, for any Lie groupoid G and any Lie groupoid H weakly equivalent to a discrete groupoid, any homomorphism $\phi \colon G \to H$ factors uniquely through the canonical homomorphism $G \to \pi_1(G)$. (To see this, one uses the fact that for G and H connected, $\pi_1(G)(x, \text{-})$ is the universal covering space over G while $H(\phi(x), \text{-})$ is a covering space over H, for any $x \in G_0$ (cf. [48, p. 133]). This gives us a unique equivariant map $\pi_1(G)(x, \text{-}) \to H(\phi(x), \text{-})$ whose composition with $G(x, \text{-}) \to \pi_1(G)(x, \text{-})$ agrees with ϕ.)

Example 3.24 Let $\mathcal{F}$ be a foliation of a manifold M. Consider the natural homomorphism of Lie groupoids

$$M \longrightarrow \mathrm{Hol}(M, \mathcal{F})$$

given by the inclusion. It induces a homomorphism of Lie groupoids

$$\pi_1(M) \longrightarrow \pi_1(\mathrm{Hol}(M, \mathcal{F})) .$$

Note that this homomorphism is surjective. Indeed, any $\mathrm{Hol}(M, \mathcal{F})$-path $\sigma_n g_n \dots g_1 \sigma_0$ is $\mathrm{Hol}(M, \mathcal{F})$-homotopic to the image of the path

$$\sigma_n \tau_n \dots \tau_1 \sigma_0$$

in M, where τ_i is any path inside a leaf of $\mathcal{F}$ representing the arrow g_i. The groupoid $\pi_1(\mathrm{Hol}(M, \mathcal{F}))$ will be referred to as the *fundamental groupoid of the foliated manifold* $(M, \mathcal{F})$, and denoted simply by $\pi_1(M, \mathcal{F})$.

By the same argument we can see that the inclusion $M \to \mathrm{Mon}(M, \mathcal{F})$ induces a surjective homomorphism

$$\pi_1(M) \longrightarrow \pi_1(\mathrm{Mon}(M, \mathcal{F})) .$$

In fact, this is an isomorphism of Lie groupoids. Indeed, for any $\mathrm{Mon}(M, \mathcal{F})$-path $\sigma_n g_n \dots g_1 \sigma_0$, the homotopy class of the associated path $\sigma_n \tau_n \dots \tau_1 \sigma_0$ as above is uniquely determined by the $\mathrm{Mon}(M, \mathcal{F})$-homotopy class of $\sigma_n g_n \dots g_1 \sigma_0$. Thus this construction defines an inverse of $\pi_1(M) \to \pi_1(\mathrm{Mon}(M, \mathcal{F}))$.

Alternatively, one can show this by using Theorem 3.18, and by the isomorphism $\mathrm{Cs}(M) \cong \mathrm{Cs}(\mathrm{Mon}(M, \mathcal{F}))$ (assuming M is connected). The later is true because any covering space over M admits a unique action of $\mathrm{Mon}(M, \mathcal{F})$ by path-lifting.

This, in particular, implies the following: if $\mathcal{F}$ is a foliation with injective holonomy homomorphisms (e.g. if any leaf of $\mathcal{F}$ is simply connected), then

$$\pi_1(M, \mathcal{F}) \cong \pi_1(M) .$$

Indeed, in this case we have $\mathrm{Hol}(M, \mathcal{F}) = \mathrm{Mon}(M, \mathcal{F})$.

Let $\mathcal{F}$ be a foliation of a manifold M. To describe the kernel of the natural homomorphism $\pi_1(M) \to \pi_1(\mathrm{Hol}(M, \mathcal{F}))$, let us introduce the following terminology.

Suppose that $\sigma = \sigma_n g_n \dots g_1 \sigma_0$ is a $\mathrm{Hol}(M, \mathcal{F})$-path. If τ_i is a path inside a leaf representing g_i, for any $i = 1, \dots, n$, we will say that the concatenation of paths in M

$$\tilde{\sigma} = \sigma_n(\tau_n \dots (\sigma_1(\tau_1\sigma_0))\dots)$$

is a *realization* of σ. Every $\mathrm{Hol}(M,\mathcal{F})$-path σ has a realization, and such a realization represents the same arrow in $\pi_1(\mathrm{Hol}(M,\mathcal{F}))$ as σ. However, not every deformation between two $\mathrm{Hol}(M,\mathcal{F})$-paths can be realized by a homotopy between their realizations. We will now show that it can be realized by a more general kind of '$\mathcal{F}$-homotopy'.

Let α and β be two paths from x to y in M. Then an *$\mathcal{F}$-homotopy* from α to β is a map

$$H\colon I^2(k) \to M\ ,$$

where $I^2(k)$ is the square $[0,1]^2$ with $k \geq 0$ holes in the interior (thus it is a compact manifold of dimension 2), such that $H(0,\text{-}) = \alpha$, $H(1,\text{-}) = \beta$, $H([0,1],0) = \{x\}$, $H([0,1],1) = \{y\}$, and the boundary of each hole is mapped by H to a path with trivial holonomy inside a leaf of $\mathcal{F}$. The paths α and β are *$\mathcal{F}$-homotopic* if there exists an $\mathcal{F}$-homotopy between them. Note that if α and β are homotopic, they are also $\mathcal{F}$-homotopic. In particular, 'being $\mathcal{F}$-homotopic' is a well-defined equivalence relation between the homotopy classes of paths in M.

Proposition 3.25 *Let $\mathcal{F}$ be a foliation of a manifold M. Then two arrows in $\pi_1(M)$ map to the same arrow along the natural homomorphism*

$$\pi_1(M) \longrightarrow \pi_1(\mathrm{Hol}(M,\mathcal{F}))$$

if and only if they are $\mathcal{F}$-homotopic.

Proof Note that any path in M is at the same time a $\mathrm{Hol}(M,\mathcal{F})$-path of order 0, and that it is the unique realization of itself. Therefore the proposition follows from the following lemma. □

Lemma 3.26 *Let $\mathcal{F}$ be a foliation of a manifold M. Suppose that σ and σ' are two* $\mathrm{Hol}(M,\mathcal{F})$*-homotopic* $\mathrm{Hol}(M,\mathcal{F})$*-paths from x to y, with realizations $\tilde{\sigma}$ and $\tilde{\sigma}'$. Then $\tilde{\sigma}$ and $\tilde{\sigma}'$ are $\mathcal{F}$-homotopic.*

Proof Since $\mathcal{F}$-homotopies, just like usual homotopies, can be concatenated, it is enough to show that if either

(i) there exists a deformation between σ and σ', or
(ii) σ and σ' differ by a multiplication equivalence, or
(iii) σ and σ' differ by a concatenation equivalence,

then there exists an $\mathcal{F}$-homotopy between $\tilde{\sigma}$ and $\tilde{\sigma}'$. Let $\sigma =$

$\sigma_n g_n \dots g_1 \sigma_0$ and $\sigma' = \sigma'_{n'} g'_{n'} \dots g'_1 \sigma'_0$ be the Hol$(M, \mathcal{F})$-paths, and write $\tilde{\sigma} = \sigma_n \tau_n \dots \tau_1 \sigma_0$ and $\tilde{\sigma}' = \sigma'_{n'} \tau'_{n'} \dots \tau'_1 \sigma'_0$ for their realizations.

(i) First note that in this case we have $n = n'$. Observe that any deformation $(D_n, d_n, \dots, d_1, D_0)$ between σ and σ' can be *locally realized* in the following sense: for any $t' \in [0,1]$ and any realization

$$\tilde{\sigma}^{t'} = D_n(t', \text{-})\rho_n \dots \rho_1 D_0(t', \text{-})$$

of $\sigma^{t'} = D_n(t', \text{-}) d_n(t') \dots d_n(t') D_0(t', \text{-})$, there exists a neighbourhood W of t' in $[0,1]$ such that the realization $\tilde{\sigma}^{t'}$ can be extended to a *realization* of the deformation $(D_n, d_n, \dots, d_1, D_0)$ *over* W, which is given by a collection of maps

$$\nu_i \colon W \times [0,1] \longrightarrow M \ , \qquad i = 1, \dots, n$$

such that $\nu_i(t', \text{-}) = \rho_i$ and

$$D_n(t, \text{-})\nu_n(t, \text{-}) \dots \nu_1(t, \text{-}) D_0(t, \text{-})$$

is a realization of σ^t for any $t \in W$.

It follows that we can find $0 = t_1 < t_2 < \dots < t_m < t_{m+1} = 1$ and realizations $\tilde{\sigma}^{t_j}$ of σ^{t_j}, $j = 1, \dots, m$, such that $\tilde{\sigma}^0 = \tilde{\sigma}$ and $\tilde{\sigma}^{t_j}$ can be extended to a realization (given by $(\nu^j_n, \dots, \nu^j_1)$) of the deformation $(D_n, d_n, \dots, d_1, D_0)$ over $[t_j, t_{j+1}]$ with $D_n(1, \text{-})\nu^m_n(1, \text{-}) \dots \nu^m_1(1, \text{-}) D_0(1, \text{-})$ equal to $\tilde{\sigma}'$. Note that for any $i = 1, \dots, n$ and any $j = 1, \dots, m$, the path $\omega^j_i = \nu^j_i(t_{j+1}, \text{-})$ represents the same arrow in Hol$(M, \mathcal{F})$ as $\alpha^{j+1}_i = \nu^{j+1}_i(t_{j+1}, \text{-})$.

By a reparametrization of the maps (ν^j_i) we can now define an $\mathcal{F}$-homotopy from $\tilde{\sigma}$ to $\tilde{\sigma}'$ defined on $I^2(n(m-1))$, as illustrated in Figure 3.1 for the case $m = 3$ and $n = 1$.

(ii) Note that in this case we can assume without loss of generality that $n = 2$, $n' = 1$ and σ_1 is a constant path. Note that $\sigma_0 = \sigma'_0$, $\sigma_2 = \sigma'_1$ and $\tau_2 \tau_1$ has the same holonomy as τ'_1. Now we can find an $\mathcal{F}$-homotopy between $\tilde{\sigma}$ and $\tilde{\sigma}'$ defined on the square with only one hole as indicated in Figure 3.2.

(iii) Here we can assume without loss of generality that $n = 1$, $n' = 0$ and g_1 is a unit arrow. Thus we have $\sigma_1 \sigma_0 = \sigma'_0$ and τ_1 has trivial holonomy. Now Figure 3.3 illustrates how to construct an $\mathcal{F}$-homotopy between $\tilde{\sigma}$ and $\tilde{\sigma}'$, defined on $I^2(1)$. □

As an application, we will now show that the fundamental group of an analytic foliation of codimension one is non-trivial. This statement is

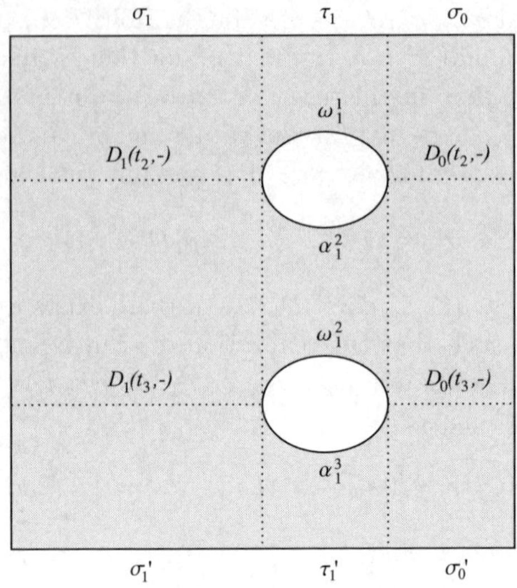

Figure 3.1 Case (i)

in fact a refinement of the Haefliger's theorem, with an analogous proof (see also [18, 33]).

Proposition 3.27 *Let $\mathcal{F}$ be an analytic codimension one foliation of a manifold M. Then any $\gamma : S^1 \to M$ transversal to $\mathcal{F}$ represents a non-trivial element of $\pi_1(M, \mathcal{F})$ (and hence also of $\pi_1(M)$).*

Proof Assume that γ represents a unit element of $\pi_1(M, \mathcal{F})$. By Lemma 3.26 it follows that there exists an $\mathcal{F}$-homotopy between γ (reparametrized by $[0,1]$)) and the constant loop (say at the base point $\gamma(1)$). We can reparametrize this $\mathcal{F}$-homotopy to a smooth map

$$H : B^2(m) \longrightarrow M ,$$

defined on a disk $B^2(m)$ with m holes (thus a smooth compact manifold) such that H extends $\gamma : S^1 \to M$ and maps the boundary of each hole into a path with trivial holonomy inside a leaf.

Now we can proceed in the same way as in the proof of Haefliger's theorem, cf. [48]. We can deform H a little so that the pull-back $H^*(\mathcal{F})$ of $\mathcal{F}$ along H is a foliation of $B^2(m)$ with finitely many Morse singularities lying on different leaves, transversal to the outer boundary (which maps as γ), and the boundaries of the holes are leaves of $H^*(\mathcal{F})$ with trivial holonomy. In particular, the foliation $H^*(\mathcal{F})$ has concentric circles around the holes. Now we can replace the holes with centres, and

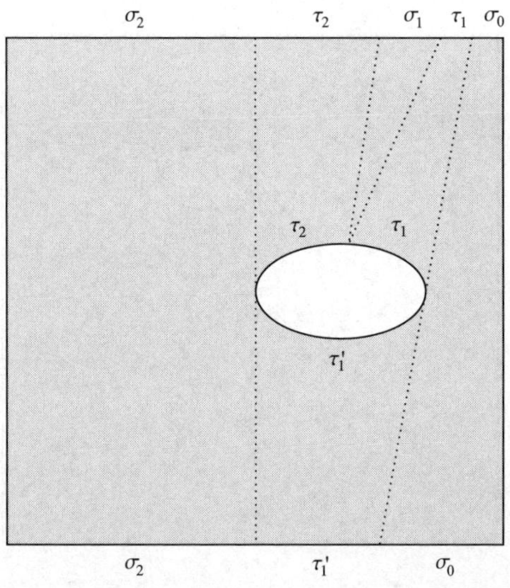

Figure 3.2 Case (ii)

exactly the same argument as in the proof of Haefliger's theorem shows that the foliation $\mathcal{F}$ can not be analytic. □

Corollary 3.28 *Let $\mathcal{F}$ be an analytic codimension one foliation of a compact manifold M with base-point x_0. Then the fundamental group $\pi_1(M, \mathcal{F})_{x_0}$ (and hence also $\pi_1(M, x_0)$) is non-trivial.*

Proof Since M is compact, there exists a transversal loop in $(M, \mathcal{F})$ (see e.g. [48, Lemma 2.28]). □

3.4 *G*-sheaves of *R*-modules

From now on we will assume that the reader is familiar with the basics of homological algebra. There are many good expositions, e.g. [11, 31, 38, 60], and we will sometimes use the latter as an explicit reference.

Let R be a commutative ring with unit. In practice, R will always be the field $\mathbb{R}$ of reals or the field $\mathbb{C}$ of complex numbers. If G is a Lie groupoid, we denote by

$$\mathrm{Sh}_R(G)$$

the category of G-sheaves of R-modules. These are G-sheaves A on G_0 for which each stalk A_x has the structure of an R-module. Moreover,

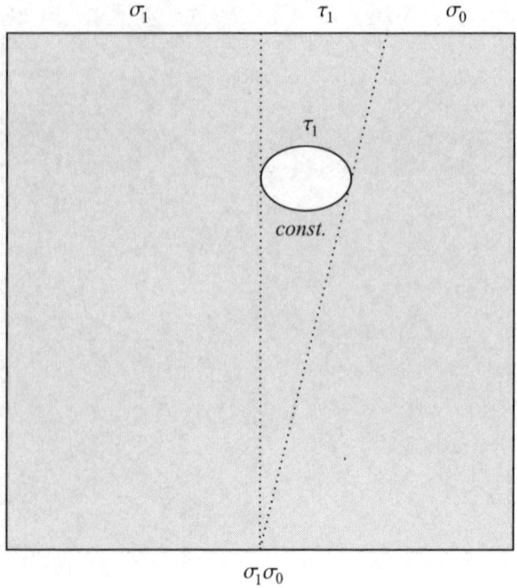

Figure 3.3 Case (iii)

this structure is required to vary continuously in x, and to be preserved by the action $A_y \to A_x$ of any arrow $g: x \to y$ in G.

If $\phi: H \to G$ is a homomorphism of Lie groupoids, the adjoint functors $\phi^* : \mathrm{Sh}(G) \to \mathrm{Sh}(H)$ and $\phi_* : \mathrm{Sh}(H) \to \mathrm{Sh}(G)$ of Section 3.1 clearly preserve the structure of such sheaves of R-modules, and define adjoint functors

$$\phi^* : \mathrm{Sh}_R(G) \rightleftarrows \mathrm{Sh}_R(H) : \phi_*$$

The same is true for the functor P^* induced by a general morphism $P : H \to G$, and in fact all the statements in Section 3.2 extend to the categories $\mathrm{Sh}_R(G)$ of sheaves of R-modules on Lie groupoids. In particular, $\mathrm{Sh}_R(G)$ and $\mathrm{Sh}_R(H)$ are equivalent categories whenever G and H are Morita equivalent groupoids.

Recall that for any topological space X, the category $\mathrm{Sh}_R(X)$ is an abelian category. Moreover, for any point $x \in X$, the functor $\mathrm{Sh}_R(X) \to \{R\text{-modules}\}$, which sends a sheaf A to its stalk A_x, is an exact functor. Thus, since a map $A \to B$ between sheaves of R-modules is an isomorphism if and only if each map of stalks $A_x \to B_x$ is an isomorphism of R-modules, the category $\mathrm{Sh}_R(X)$ inherits many of the exactness properties from the more familiar category of R-modules.

In exactly the same way, $\mathrm{Sh}_R(G)$ is an abelian category. For example, if A and B are two G-sheaves of R-modules, their sum $A \oplus B$ as sheaves of R-modules on G_0 inherits a natural G-action, and this defines the sum in $\mathrm{Sh}_R(G)$. The same remark applies to the construction of kernels and cokernels, and of infinite sums. Furthermore, for any homomorphism $\phi: H \to G$ between Lie groupoids, the functor $\phi^* : \mathrm{Sh}_R(G) \to \mathrm{Sh}_R(H)$ is exact, as is clear by comparing the stalks:

$$\phi^*(A)_y = A_{\phi(y)}$$

for any point $y \in H_0$. The functor ϕ^* also preserves infinite sums, because it has a right adjoint.

We consider some properties of the abelian category $\mathrm{Sh}_R(G)$ which are important for developing homological algebra in this category.

First of all, as we already indicated, $\mathrm{Sh}_R(G)$ has arbitrary (small) colimits,, which can be constructed from cokernels and infinite sums. Furthermore, the construction of *directed* (or *filtered*) colimits indexed by a category $\mathcal{I}$ is an exact functor $\mathrm{Sh}_R(G)^{\mathcal{I}} \to \mathrm{Sh}_R(G)$, again because it suffices to check this at the level of stalks. Thus, the abelian category $\mathrm{Sh}_R(G)$ satisfies the Grothendieck axiom (AB5) [60, p. 57].

Next, although $\mathrm{Sh}_R(G)$ is a large category, for a given sheaf of R-modules A there is only a set of subsheaves of R-modules $B \subset A$. Indeed, this is clear because such a B is a subspace of the étale space of A having some additional properties. A large category with the property that there is only a set of isomorphism classes of monomorphisms $B \to A$ into a given object A is called *well-powered* [60, p. 385].

Next we consider injective objects. Recall that an object I in an abelian category $\mathcal{A}$ is injective if for any monomorphism $A \to B$ in $\mathcal{A}$, any map $A \to I$ can be extended to a map $B \to I$. The category $\mathcal{A}$ is said to have *enough injectives* if for every object A there exists a monomorphism $A \to I$ into some injective object I. We recall, without proof, the following elementary facts concerning adjoint functors and injectives.

Lemma 3.29 Let $\phi^* : \mathcal{A} \leftrightarrows \mathcal{B} : \phi_*$ be functors between abelian categories, and assume that ϕ^* is a left adjoint to ϕ_*.

(i) The functor ϕ^* is faithful if and only if for each object A of $\mathcal{A}$ the unit $\eta_A : A \to \phi_*\phi^* A$ is a monomorphism [39].

(ii) If ϕ^* preserves monomorphisms then ϕ_* preserves injectives.

(iii) If ϕ^* is faithful and exact, and $\mathcal{B}$ has enough injectives, then so does $\mathcal{A}$.

This lemma can be used to construct explicit functorial injective resolutions in our categories $\mathrm{Sh}_R(G)$ of G-sheaves of R-modules. Recall that the usual category of R-modules has enough injectives. First, we deduce from this that our category $\mathrm{Sh}_R(X)$ of sheaves of R-modules on a topological space X has enough injectives. To this end, write X^δ for the space X with the discrete topology, and $p\colon X^\delta \to X$ for the identity map. Then a sheaf of R-modules B on X^δ is simply a family $\{B_x\}_{x\in X}$ of R-modules. If A is a sheaf of R-modules on X, and $A_x \to J_x$ is an embedding of the stalk A_x into an injective R-module J_x, then the family $\{A_x \to J_x\}$ can be viewed as an embedding $p^*(A) \to J$ in $\mathrm{Sh}_R(X^\delta)$, and the desired embedding of A into an injective object of $\mathrm{Sh}_R(X)$ can be constructed as the composite $A \to p_*p^*(A) \to p^*(J)$. Thus $\mathrm{Sh}_R(X)$ has enough injectives. Now consider a Lie groupoid G. The unit homomorphism $u\colon G_0 \to G$ induces adjoint functors

$$u^* : \mathrm{Sh}_R(G) \rightleftarrows \mathrm{Sh}_R(G_0) : u_* \ ,$$

and for an object B of $\mathrm{Sh}_R(G)$, an embedding into an injective can be constructed from such an embedding $u^*(B) \to I$ in $\mathrm{Sh}_R(G_0)$, just shown to exist, again as a composite of the form $B \to u_*u^*(B) \to u_*(I)$. This proves that $\mathrm{Sh}_R(G)$ has enough injectives.

Now we consider the 'free' sheaves of R-modules. The forgetful functor

$$\mathrm{Sh}_R(G) \longrightarrow \mathrm{Sh}(G)$$

from G-sheaves of R-modules to G-sheaves of sets has a left adjoint. We will denote this adjoint by

$$E \mapsto R[E] \ ,$$

where E is any G-sheaf of sets. As a sheaf, $R[E]$ is simply the direct sum $\oplus_{r\in R}E$, and its G-action is inherited from that of E. Thus for the stalks we have the identity

$$R[E]_x = R[E_x] \ ,$$

where the right hand side is the usual free R-module on the set E_x. The adjointness property means that for any object A of $\mathrm{Sh}_R(G)$, any map of G-sheaves $E \to A$ extends uniquely to a map $R[E] \to A$ in $\mathrm{Sh}_R(G)$.

If one combines this construction with the covering of any sheaf E by a family $\{\tilde{U}_i\}$ of sheaves as in Example 3.5, one can conclude that any object A in $\mathrm{Sh}_R(G)$ is covered by a family of objects of the form $R[\tilde{U}_i]$, or in other words, there exists an exact sequence of the form

$$\bigoplus_{i\in I} R[\tilde{U}_i] \longrightarrow A \longrightarrow 0 \ .$$

This means that the family of G-sheaves of R-modules of the form $R[\tilde{U}]$, for any open subset U of G_0, *generates* the category $\mathrm{Sh}_R(G)$ [39, p. 127].

We summarize the properties of the category $\mathrm{Sh}_R(G)$ discussed so far in the following proposition.

Proposition 3.30 *Let G be a Lie groupoid.*

(i) The category $\mathrm{Sh}_R(G)$ is an abelian category with enough injectives, it satisfies Grothendieck's axiom AB5 and has a small set of generators (indexed by open sets $U \subset G_0$).

(ii) Each homomorphism $\phi\colon H \to G$ between Lie groupoids induces adjoint functors

$$\phi^*\colon \mathrm{Sh}_R(G) \rightleftarrows \mathrm{Sh}_R(H) : \phi_* \ .$$

The left adjoint ϕ^* is an exact functor (hence ϕ_* preserves injectives).

(iii) If $\phi\colon H \to G$ is a weak equivalence, then $\phi^*\colon \mathrm{Sh}_R(G) \to \mathrm{Sh}_R(H)$ is an equivalence of abelian categories.

3.5 Derived categories

In this section we review the construction of the derived category associated to the abelian category $\mathrm{Sh}_R(G)$ of sheaves of R-modules on a Lie groupoid G. Given the properties of such abelian categories established in the previous section, the material in this section is simply a particular instance of the general theory as exposed e.g. in [60, Chapter 10] and many readers will be familiar with it.

We shall write

$$\mathrm{Ch}_R(G)$$

for the category of complexes in $\mathrm{Sh}_R(G)$. Unless explicitly stated otherwise, we will work with *cochain* complexes

$$\ldots \longrightarrow A^n \xrightarrow{\ d\ } A^{n+1} \longrightarrow \ldots \qquad\qquad n \in \mathbb{Z}\,,$$

and view chain complexes $C = C_\bullet$ as cochain complexes by reindexing $C^n = C_{-n}$. As usual, we call a complex $A = A^\bullet$ bounded below if there exists a $k \in \mathbb{Z}$ such that $A^n = 0$ for any $n < k$. The full subcategory of $\mathrm{Ch}_R(G)$ consisting of bounded bellow complexes is denoted by $\mathrm{Ch}_R^+(G)$. We also use the categories $\mathrm{Ch}_R^-(G)$ and $\mathrm{Ch}_R^b(G)$ of complexes which are bounded above, respectively bounded (from both sides).

Later, we will usually work over the field $\mathbb{R}$ of reals, and we adopt the convention that $\mathrm{Ch}(G)$ stands for $\mathrm{Ch}_{\mathbb{R}}(G)$, and similarly for $\mathrm{Ch}^+(G)$, $\mathrm{Ch}^-(G)$ and $\mathrm{Ch}^b(G)$.

The standard notion of cochain homotopy [60, p. 15] defines an equivalence relation on the maps $A \to B$ between any two complexes A and B. This gives rise to a quotient of the category $\mathrm{Ch}_R(G)$, viz. the category

$$K_R(G)\,,$$

consisting of complexes in $\mathrm{Sh}_R(G)$ and homotopy classes of maps. There are similar quotient categories $K_R^+(G)$, $K_R^-(G)$ and $K_R^b(G)$, of $\mathrm{Ch}_R^+(G)$, $\mathrm{Ch}_R^-(G)$ and $\mathrm{Ch}_R^b(G)$.

A map $f\colon A \to B$ in $\mathrm{Ch}_R(G)$ is called a *quasi-isomorphism (q.i.)* if f induces isomorphisms

$$H^n(A) \longrightarrow H^n(B)$$

for each $n \in \mathbb{Z}$. Here $H^n(A)$ and $H^n(B)$ are the cohomology groups constructed as objects of $\mathrm{Sh}_R(G)$. By the remark in the previous section, a map $f : A \to B$ is a quasi-isomorphism if and only if for each point $x \in G_0$, the map $f_x : A_x \to B_x$ is a quasi-isomorphism between complexes of ordinary R-modules (instead of sheaves). Of course if f is chain homotopic to another map $f'\colon A \to B$, then f is a quasi-isomorphism if and only if f' is. So 'quasi-isomorphism' is also a well defined property of maps in $K_R(G)$. The *derived category* $D_R(G)$ is the category obtained from $K_R(G)$ by formally inverting all quasi-isomorphisms. It requires some care to prove that this category $D_R(G)$ actually exists, but the properties of $\mathrm{Sh}_R(G)$ established in Proposition 3.30 suffice for this (cf. [60, Remark 10.4.5]). A more explicit description of $D_R(G)$ is as follows: It has the same objects as $\mathrm{Ch}_R(G)$. Furthermore, a morphism $A \to B$ in $D_R(G)$ is an equivalence class of diagrams of the form

$$A \overset{q}{\longleftarrow} C \overset{f}{\longrightarrow} B$$

where q is a quasi-isomorphism. Two such diagrams $A \leftarrow C \to B$ and $A \leftarrow C' \to B$ are equivalent if there is a diagram of the form

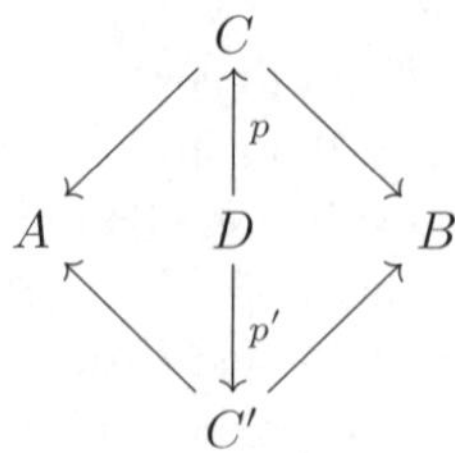

commuting in $K_R(G)$, i.e. up to homotopy, and where p and p' are quasi-isomorphisms. They then represent the same morphism $A \to B$ in $D_R(G)$.

There is an evident functor $\pi : \mathrm{Ch}_R(G) \to D_R(G)$, which sends an arrow $A \to B$ to the equivalence class of the diagram $A = A \to B$. This functor enables one to formulate the universal property of the category $D_R(G)$, similar to the property of Theorem 2.11. Namely, any functor $F : \mathrm{Ch}_R(G) \to \mathcal{C}$ into any category $\mathcal{C}$ which turns quasi-isomorphisms into isomorphisms can be factored as $F = \bar{F} \circ \pi$ for a unique functor $\bar{F}$ [60, Proposition 10.1.2].

In a similar way, one can construct derived categories

$$D_R^+(G)\,, \quad D_R^-(G)\,, \quad D_R^b(G)$$

by formally turning quasi-isomorphisms into isomorphisms.

In the case of bounded below complexes (i.e. $\mathrm{Sh}_R^+(G)$ and $\mathrm{Ch}_R^b(G)$) there is a even more concrete description of the derived category [60, Theorem 10.4.8]: The derived category $D_R^+(G)$ is equivalent to the full subcategory of $K_R^+(G)$ whose objects are bounded below complexes I with the property that each I^n is an injective object of $\mathrm{Sh}_R(G)$. Similarly, $D_R^b(G)$ is equivalent to the full subcategory of $K_R^+(G)$ whose objects are bounded below complexes I with each I^n injective, and with $H^n(I) = 0$ for n sufficiently large.

Let $\phi : H \to G$ be a homomorphism between Lie groupoids. The adjoint functors ϕ^* and ϕ_* between $\mathrm{Sh}_R(G)$ and $\mathrm{Sh}_R(H)$ induce adjoint functors

$$\phi^* : K_R(G) \rightleftarrows K_R(H) : \phi_*\,.$$

Since ϕ^* is exact, it maps quasi-isomorphisms to quasi-isomorphisms, and hence it immediately induces a functor

$$\phi^* : D_R(G) \longrightarrow D_R(H)\,.$$

If ϕ is a weak equivalence, then $\phi^* : \mathrm{Sh}_R(G) \to \mathrm{Sh}_R(H)$ is an equivalence of categories, and hence so are $\phi^* : K_R(G) \to K_R(H)$ and $\phi^* : D_R(G) \to D_R(H)$. In particular:

Proposition 3.31 *If G and H are Morita equivalent groupoids, then $D_R(G)$ and $D_R(H)$ are equivalent categories.*

Furthermore, if ϕ is a weak equivalence, then $\phi_* : \mathrm{Sh}_R(G) \to \mathrm{Sh}_R(H)$ is also an equivalence, inverse to ϕ^* up to a natural isomorphism. In particular, $\phi_* : \mathrm{Ch}_R(H) \to \mathrm{Ch}_R(G)$ preserves quasi-isomorphisms, and induces a functor $D_R(H) \to D_R(G)$, inverse to ϕ^*.

In general, however, ϕ_* does not preserve the quasi-isomorphisms. On the other hand, ϕ_* does preserve injectives, and one can define a *right derived functor*

$$R\phi_* \colon D_R^+(H) \longrightarrow D_R^+(H)$$

by $R\phi_*(A) = \phi_*(I)$, where $A \to I$ is any quasi-isomorphism of the complex A into a complex I of injectives. (This does not depend on the choice of I because I is unique up to homotopy equivalence, i.e. up to isomorphism in $K_R^+(H)$.) This functor $R\phi_*$ is right adjoint to $\phi^* \colon D_R^+(G) \to D_R^+(H)$.

These constructions extend to generalized morphisms $P \colon H \to G$, to give adjoint functors

$$P^* \colon D_R^+(G) \rightleftarrows D_R^+(H) : RP_* \ .$$

This follows by the universal property. Explicitly, we can factor P as $\phi \circ \epsilon^{-1}$ where $\phi \colon K \to G$ is a homomorphism and $\epsilon \colon K \to H$ a weak equivalence, and then we get

$$P^* = \epsilon_* \circ \phi^*$$

and

$$RP_* = R\phi_* \circ \epsilon^* \ .$$

We shall also make use of the *internal tensor* and *Hom*. If A and B are objects of $\mathrm{Ch}_R(G)$, one defines complexes $A \otimes B$ and $Hom(A, B)$ by

$$(A \otimes B)^n = \bigoplus_{p+q=n} A^p \otimes_R B^q$$

and

$$Hom(A, B)^n = \prod_{q-p=n} Hom(A^p, B^q)$$

as usual. (Observe that with this grading, there is indeed a bijective correspondence between maps $A \otimes B \to C$ of degree d and maps $A \to Hom(B, C)$ of degree d.)

In general for a fixed A, the functor $A \otimes$ - : $\mathrm{Ch}_R(G) \to \mathrm{Ch}_R(H)$ does not preserve quasi-isomorphisms. However, we will mainly work over a field (e.g. $R = \mathbb{R}$), in which case it does, so that $A \otimes B$ is well defined as a functor $\otimes \colon D(G) \times D(G) \to D(G)$. (Remember our convention that omitting the subscript R means that we take $R = \mathbb{R}$.) In the general case it is possible to obtain a well defined functor $\otimes_L : D^-_R(G) \times D^-_R(G) \to D^-_R(G)$, by replacing a bounded above complex B by a quasi-isomorphic complex F consisting of free R-modules.

For a fixed A, the functor $\mathit{Hom}(A, \text{-})\colon \mathrm{Ch}_R(G) \to \mathrm{Ch}_R(G)$ does not preserve quasi-isomorphisms either. However, just as for $R\phi_*$, one can obtain a well defined functor

$$R\mathit{Hom}(A, \text{-})\colon D^+_R(G) \longrightarrow D_R(G)$$

by $R\mathit{Hom}(A,B) = \mathit{Hom}(A,I)$ where $B \to I$ is a quasi-isomorphism into a bounded below complex of injectives. Moreover, if I is such a complex then $\mathit{Hom}(\text{-}\,, I)$ is exact, so it maps a quasi-isomorphism $A \to A'$ between bounded above complexes to a quasi-isomorphism $\mathit{Hom}(A', I) \to \mathit{Hom}(A, I)$. (Indeed, by a mapping cone argument it suffices to prove that $\mathit{Hom}(A,I)$ is again acyclic (i.e. quasi-isomorphic to zero) whenever A itself is acyclic. But $\mathit{Hom}(A,I)$ is the product total complex of the double complex $\mathrm{Hom}(A^{-p}, I^q)$ which has acyclic rows (for q fixed). So $\mathit{Hom}(A,I)$ is acyclic by [60, Acyclic Assembly Lemma 2.7.3]. Thus, one obtains a well defined 'derived internal hom' functor

$$R\mathit{Hom}\colon D^-_R \times D^+_R(G) \longrightarrow D^+_R(G)\,.$$

We will also use the notation $\mathrm{Hom}(A,B)$ for the 'external hom': for A and B in $\mathrm{Sh}_R(G)$ this is an R-module, not a sheaf. At the level of derived categories this gives a functor

$$R\mathrm{Hom}\colon D^-_R(G) \times D^+_R(G) \longrightarrow D^+_R(\mathrm{pt}) = D^+_R$$

into the derived category of bounded below complexes of R-modules.

4

Sheaf cohomology

In this chapter we have attempted to give a systematic treatment of sheaf cohomology for étale groupoids. This cohomology applies in particular to Lie groupoids for which there are 'enough' equivariant sheaves. It is possible to give a more precise formulation of this property (in the style of Tannaka duality), but we will refrain from doing that here. We only observe that the class of Lie groupoids having enough equivariant sheaves is closed under weak equivalence (cf. Section 3.2), and includes étale (and foliation) groupoids (cf. Section 2.3).

In the context of foliations, the sheaf cohomology of étale groupoids was described in concrete terms by Haefliger [24]. His approach was based on the bar resolution, to be explained in Section 4.2 of this chapter. It is possible to approach the cohomology from a Grothendieck style point of view, using derived categories and derived functors. The latter approach has the advantage that many general properties such as functoriality, Morita invariance, duality, etc., will be easily derivable. Early references for the treatment of cohomology of étale groupoids along these lines and for the relation to the bar complex include [43, 58].

In Section 4.5 we will show how to describe the cohomology of such étale groupoids in terms of the much easier cohomology of small categories. More precisely, one can associate to each étale groupoid G a small discrete category $\mathrm{Emb}(G)$, and prove an isomorphism of the form $H^*(G, A) = H^*(\mathrm{Emb}(G), A)$, under suitable conditions (Theorem 4.21). This result was first proved in [44] and later in somewhat more general form in [16]. This approach to the sheaf cohomology of étale groupoids has already turned out to be very useful in the construction of characteristic classes of foliations [16], and in the description of extensions of groupoids [47], and we expect it to have many other applications.

Much of this chapter is an exposition of known constructions and results, which have appeared in scattered places. The only thing which is new is a detailed comparison, for a foliated manifold $(M, \mathcal{F})$, between the cohomology of the ambient manifold M and that of the holonomy groupoid $\mathrm{Hol}(M, \mathcal{F})$. This section was prompted by a question of Haefliger, as to what extent the manifold M itself behaves as a classifying space for the holonomy groupoid (see Boulder meeting).

There are many aspects of the cohomology of foliations and their groupoids which should have been included, but for which there was no space or time. In particular, we would have liked to discuss the relation to other cohomologies, such as the basic cohomology and the leaf-wise cohomology of foliations, and the connections to cyclic cohomology, Bredon cohomology and K-theory, and to the cohomology of the classifying space. The reader will find some of these points discussed in [7, 13, 21, 46, 56].

4.1 Sheaf cohomology of foliation groupoids

In this section we will discuss the general framework of sheaf cohomology for Lie groupoids. Although the definitions and some of the general statements hold for Lie groupoids in general, they are mainly of interest for foliation groupoids, i.e. groupoids which are Morita equivalent to étale ones. We recall that these include groupoids arising from foliations, such as the holonomy and monodromy groupoids. In later sections, we will restrict our attention to étale groupoids. We will work with sheaves over a ring R, as before, and leave R unspecified. Thus, in this section 'G-sheaf' means G-sheaf of R-modules.

Let G be a Lie groupoid, and let A be a G-sheaf. The sheaf cohomology of G with coefficients in A is defined by

$$H^n(G, A) = R^n p_*(A) \qquad n = 0, 1, 2, \ldots \tag{1}$$

Here $p\colon G \to 1$ is the unique homomorphism into the one-point groupoid, and $R^n p_*(A) = H^n(Rp_*(A))$ is the cohomology in degree n of the complex $Rp_*(A)$. Recall that p_* is the functor $\Gamma_{\mathrm{inv}}(G, \text{-})$, so that $H^n(G, A)$ can be described explicitly as the cohomology of the complex

$$\Gamma_{\mathrm{inv}}(G, I^0) \longrightarrow \Gamma_{\mathrm{inv}}(G, I^1) \longrightarrow \Gamma_{\mathrm{inv}}(G, I^2) \longrightarrow \ldots \tag{2}$$

where $0 \to A \to I^0 \to I^1 \to I^2 \to \ldots$ is any resolution of A by injective

G-sheaves. Recall also that for any G-sheaf B,

$$\Gamma_{\text{inv}}(G, B) = \text{Hom}_G(R, B)$$

is the R-module of morphisms of G-sheaves from the constant sheaf R into B.

Remark. We point out that in the following two cases, our definition agrees with the usual one:

(a) If M is a manifold, we can view M as the unit (étale) groupoid. An M-sheaf A is just a sheaf on A, and $H^n(M, A)$ is the usual sheaf cohomology of spaces [19].

(b) If G is a group, we can view G as a discrete groupoid over a one-point space, and $H^n(G, A)$ is the usual cohomology of the group G [60].

The definition (1) and the description (2) obviously apply to any bounded below complex A of G-sheaves. In this case one sometimes refers to $H^n(G, A)$ as the *hypercohomology*. If A is such a complex, one can form the cohomology G-sheaves $H^q(A)$ (as objects of the abelian category $\text{Sh}_R(G)$), and the hypercohomology can be expressed in terms of 'ordinary' cohomology by the so-called hypercohomology spectral sequence:

Proposition 4.1 *For any bounded below complex A of G-sheaves there is a spectral sequence*

$$E_2^{p,q} = H^p(G, H^q(A)) \Rightarrow H^{p+q}(G, A) .$$

Proof The proof is standard: There exists a double complex $A \to I^0 \to I^1 \to \ldots$ where each I^p is a bounded below complex of injective G-sheaves $I^{p,q}$, with the property that for each q, not only $0 \to A^q \to I^{0q} \to I^{1q} \to \ldots$ is an injective resolution of A^q, but the same is true for the boundaries $0 \to B^q(A) \to B^q(I^{0,\bullet}) \to B^q(I^{1,\bullet}) \to \ldots$, the cycles $0 \to Z^q(A) \to Z^q(I^{0,\bullet}) \to Z^q(I^{1,\bullet}) \to \ldots$ and the cohomology G-sheaves $0 \to H^q(A) \to H^q(I^{0,\bullet}) \to H^q(I^{1,\bullet}) \to \ldots$ (cf. e.g. [60, p. 149] or [31, p. 301]).

The spectral sequence of the double complex $\Gamma_{\text{inv}}(G, I^{p,q})$ is the spectral sequence of the proposition. Indeed, the total complex $\oplus_{p+q=n} I^{p,q}$ is a complex of injectives which is quasi-isomorphic to A, hence the total complex of $\Gamma_{\text{inv}}(G, I^{p,q})$ computes $H^{p+q}(G, A)$.

On the other hand, for a fixed $p \geq 0$, one has

$$H^q(\Gamma_{\text{inv}}(G, I^{p,\bullet})) = \Gamma_{\text{inv}}(G, H^q(I^{p,\bullet}))$$

because $B^q(I^{p,\bullet})$, $Z^q(I^{p,\bullet})$ and $H^q(I^{p,\bullet})$ are all injective. Since

$$H^q(I^{0,\bullet}) \longrightarrow H^q(I^{1,\bullet}) \longrightarrow \dots$$

is an injective resolution of $H^q(A)$, one has

$$H^p H^q \Gamma_{\mathrm{inv}}(G, I) = H^p \Gamma_{\mathrm{inv}}(G, H^q(A)) ,$$

giving the desired description of the E_2-term. □

Remark. It follows by the same double complex that if A is a bounded below complex with $H^p(G, A^q) = 0$ for each q and each $p > 0$, then

$$H^p(G, A) = H^p(\Gamma_{\mathrm{inv}}(G, A)) .$$

Indeed, in this case $\Gamma_{\mathrm{inv}}(G, I^{\bullet,q})$ is an injective resolution of $\Gamma_{\mathrm{inv}}(G, A^q)$, and $\Gamma_{\mathrm{inv}}(G, I)$ is a double complex with acyclic columns (p fixed). So the complex $\Gamma_{\mathrm{inv}}(G, A)$ computes the cohomology of the total complex which we know to be $H^*(G, A)$.

The cohomology groups $H^q(G, A)$ are covariant in the coefficients A and contravariant in the groupoid G. The first property is obvious from the fact that Rp_* is a functor. For the second, consider a homomorphism $\phi \colon G' \to G$ between Lie groupoids. Then for any G-sheaf (or any bounded below complex of G-sheaves) A, one can construct a canonical map

$$\phi^* \colon H^n(G, A) \longrightarrow H^n(G', \phi^*(A)) , \tag{3}$$

explicitly in term of injective resolutions, as follows: If $A \to I$ is an injective resolution of G-sheaves (i.e. a quasi-isomorphism into a complex consisting of injectives), then $\phi^* A \to \phi^* I$ is again a quasi-isomorphism. Let $\phi^* I \to J$ be an injective resolution of G'-sheaves. Then the composite $\phi^* A \to \phi^* I \to J$ is an injective resolution of $\phi^* A$. The pull-back of invariant sections

$$\Gamma_{\mathrm{inv}}(G, I) \longrightarrow \Gamma_{\mathrm{inv}}(G', \phi^* I) \tag{4}$$

when composed with $\Gamma_{\mathrm{inv}}(G', \phi^* I) \to \Gamma_{\mathrm{inv}}(G', J)$ gives a map $\Gamma_{\mathrm{inv}}(G, I) \to \Gamma_{\mathrm{inv}}(G', J)$, and hence a morphism (3) after taking cohomology.

In the special case where ϕ is a weak equivalence, the functor ϕ^* is an equivalence of categories, hence it preserves injectives, and (4) is an isomorphism. Thus (3) is an isomorphism as well. For the record, we state this explicitly:

Proposition 4.2 (Morita invariance of cohomology) *A weak equivalence $\phi\colon G' \to G$ induces isomorphisms*

$$H^n(G, A) \longrightarrow H^n(G', \phi^* A) \qquad n = 0, 1, 2, \ldots$$

for any complex A of G-sheaves which is bounded below.

Remark 4.3 (Leray spectral sequence) The cohomology groups $H^n(G, A)$ depend covariantly on A through the *Leray spectral sequence*

$$E_2^{p,q} = H^p(G, R^q\phi_*(B)) \Rightarrow H^{p+q}(G', B)$$

associated to a homomorphism $\phi\colon G' \to G$ and a bounded below complex B of G'-sheaves. Here $R^q\phi_*(B)$ is the q-th cohomology G-sheaf $H^q(R\phi_*(B))$ of the complex of G-sheaves $R\phi_*(B)$. This spectral sequence can be viewed as a special case of the hypercohomology spectral sequence, as follows: Let $B \to I$ be an injective resolution, and consider the complex $\phi_* I$ and its spectral sequence $H^p(G, H^q(\phi_* I)) \Rightarrow H^{p+q}(G, \phi_* I)$ as in Proposition 4.1. Here we have $H^q(\phi_* I) = R^q\phi_*(B)$ by definition, while $H^n(G, \phi_* I) = H^n(G', I) = H^n(G', B)$ because $\Gamma_{\mathrm{inv}}(G, \phi_* I) = \Gamma_{\mathrm{inv}}(G', I)$ and I is a resolution of B.

4.2 The bar resolution for étale groupoids

In this section we continue the discussion of sheaf cohomology of groupoids, but we focus on étale groupoids. The results in this section do not apply to Lie groupoids generally, although by the Morita invariance of cohomology they do extend (in a suitable form) to foliation groupoids.

Fix a Lie groupoid G, and consider for each $n \geq 0$ the space

$$G_n = \{(g_1, \ldots, g_n) \,|\, \mathrm{s}(g_i) = \mathrm{t}(g_{i+1})\} \,.$$

Thus G_n is the space of strings

$$\bullet \xleftarrow{g_1} \bullet \xleftarrow{g_2} \cdots \xleftarrow{g_n} \bullet$$

of composable strings in the groupoid G. One can construct G_n as an iterated fibered product $G_1 \times_{G_0} G_1 \times_{G_0} \cdots \times_{G_0} G_1$ along submersions (the maps s and t and their pull-backs), so G_n is a smooth manifold for each $n \geq 0$. For $n = 1, 2$, we recover the spaces G_0 and G_1 of objects and arrows in G. These spaces G_n together form a *simplicial manifold* $G_\bullet$, called the *nerve* of G, denoted also Nerve(G). The simplicial structure maps

$$d_i \colon G_n \longrightarrow G_{n-1} \qquad i = 0, \ldots, n$$

are defined by

$$d_i(g_1, \ldots, g_n) = \begin{cases} (g_2, \ldots, g_n) & ; \quad i = 0 \\ (g_1, \ldots, g_i g_{i+1}, \ldots, g_n) & ; \quad 0 < i < n \\ (g_1, \ldots, g_{n-1}) & ; \quad i = n, \end{cases}$$

while the degeneracy maps

$$s_j \colon G_{n-1} \longrightarrow G_n \qquad j = 0, \ldots, n-1$$

map a string $(g_1, \ldots, g_{n-1})$ to the string obtained by inserting the identity arrow $1_{s(g_j)}$ between g_j and g_{j+1}.

Let $\lambda_n \colon G_n \longrightarrow G_0$ be the 'last vertex' map,

$$\lambda_n(g_1, \ldots, g_n) = s(g_n) \; .$$

Any G-sheaf A gives a sheaf

$$A_{(n)} = \lambda_n^*(A)$$

on G_n, with stalk $A_{s(g_n)}$ at $(g_1, \ldots, g_n)$. Furthermore, any simplicial structure map $d_i \colon G_n \to G_{n-1}$ induces an isomorphism

$$d_i^*(A_{(n-1)}) \longrightarrow A_{(n)} \tag{5}$$

obtained canonically from the given action of G on A. Indeed, the stalk of $d_i^*(A_{(n-1)})$ at $(g_1, \ldots, g_n)$ is either $A_{s(g_n)}$ (for $i < n$) or $A_{s(g_{n-1})}$ (for $i = n$), and the stalk of the isomorphism (5) is either the identity on $A_{s(g_n)}$ or the action by $g_n \colon A_{s(g_{n-1})} \to A_{s(g_n)}$. So by pulling back sections along $d_i \colon G_n \to G_{n-1}$ and composing with these isomorphisms (5), one obtains maps

$$\delta^i \colon \Gamma(G_{n-1}, A_{(n-1)}) \longrightarrow \Gamma(G_n, d_i^*(A_{(n-1)})) \cong \Gamma(G_n, A_{(n)}) \; .$$

These give the $\Gamma(G_n, A_{(n)})$ the structure of a cosimplicial R-module,

$$\Gamma(G_0, A_{(0)}) \rightrightarrows \Gamma(G_1, A_{(1)}) \Rrightarrow \Gamma(G_2, A_{(2)}) \;\; \ldots \tag{6}$$

By taking alternating sums of the δ^i in the usual way, we thus obtain a cochain complex $\Gamma(G_\bullet, A_{(\bullet)})$. (The cohomology of this complex can be thought of as some sort of Čech cohomology for the 'hypercover' $G_\bullet$ of the groupoid G.) The same constructions apply of course to any complex of G-sheaves A, to give a double complex $\Gamma(G_p, A^q_{(p)})$.

An important property of this construction for any *étale* groupoid G is that if A is an injective G-sheaf, each of the sheaves $A_{(n)}$ on G_n is again injective. For latter use, we state this in somewhat more general form, as follows:

Lemma 4.4 *Let G be a Lie groupoid, and let E be a right G-sheaf. Let $\mathrm{pr}\colon E \rtimes G \to G$ be the projection from the semi-direct product groupoid $E \rtimes G$.*

(i) The functor $\mathrm{pr}^*\colon \mathrm{Sh}_R(G) \to \mathrm{Sh}_R(E \rtimes G)$ has a left adjoint $\mathrm{pr}_!$.

(ii) This left adjoint $\mathrm{pr}_!$ preserves monomorphisms, so pr^* preserves injectives.

(iii) For a homomorphism $\phi\colon G \to G'$ of Lie groupoids, the associated fibered product

$$\begin{array}{ccc} \phi^*(E) \rtimes G' & \xrightarrow{\tilde{\phi}} & E \rtimes G \\ \Big\downarrow{\scriptstyle \mathrm{pr}'} & & \Big\downarrow{\scriptstyle \mathrm{pr}} \\ G' & \xrightarrow{\phi} & G \end{array}$$

satisfies the base change formula

$$\mathrm{pr}'_!\tilde{\phi}^*(B) \cong \phi^*\mathrm{pr}_!(B)$$

for any $(E \rtimes G)$-sheaf B, naturally in B.

Analogous assertions of course hold for any left G-sheaf E.

Proof Recall that the groupoid $E \rtimes G$ has the manifold E as space of objects, and arrows $g\colon eg \to e$ for any $g\colon y \to x$ in G and $e \in E_x$. The projection pr maps this arrow $g\colon eg \to e$ to $g\colon y \to x$. For a $(E \rtimes G)$-sheaf B of R-modules, one constructs the left adjoint $\mathrm{pr}_!(B)$ as the sheaf on G_0 with stalk

$$\mathrm{pr}_!(B)_x = \bigoplus_{e \in E_x} B_e$$

at any point $x \in G_0$. This sheaf has a natural G-action, because any arrow $g : y \to x$ induces an isomorphism $\mathrm{pr}_!(B)_x \to \mathrm{pr}_!(B)_y$ which maps the summand B_e for $e \in E_x$ to the summand B_{eg} via the action $B_e \to B_{eg}$ of the arrow $g : eg \to e$ in $E \rtimes G$. It is not difficult to verify that this indeed defines a left adjoint to pr^*. Furthermore, it is clear from this stalk-wise description that $\mathrm{pr}_!$ preserves monomorphisms and satisfies the base-change formula (iii). Finally, the assertion about injectives follows by Lemma 3.29. □

Let G be an étale groupoid. The simplicial manifold $G_\bullet$ described above is closely related to a simplicial right G-sheaf $B_\bullet(G)$,

$$B_0(G) \leftleftarrows B_1(G) \leftleftarrows B_2(G) \quad \dots$$

which we call the *bar resolution* of G. It is defined as follows: Let $B_n(G) = G_{n+1}$ be the space of $(n+1)$-strings $(g_1, \dots, g_n, h) \in G_{n+1}$.

$$\bullet \xleftarrow{g_1} \dots \xleftarrow{g_n} \bullet \xleftarrow{h} \bullet$$

The groupoid G acts on the right on $B_n(G)$. This action is along the last vertex map

$$\lambda = \lambda_n \colon B_n(G) \longrightarrow G_0$$

sending $(g_1, \dots, g_n, h)$ to $\mathrm{s}(h)$. This is an étale map because G is an étale groupoid. An arrow $k \colon z \to \mathrm{s}(h)$ in G acts on this string by

$$(g_1, \dots, g_n, h)k = (g_1, \dots, g_n, hk) .$$

The simplicial face maps

$$d_i \colon B_n(G) \longrightarrow B_{n-1}(G) \qquad i = 0, 1, \dots, n$$

are defined by

$$d_i(g_1, \dots, g_n, h) = \begin{cases} (g_2, \dots, g_n, h) & ; \ i = 0 \\ (g_1, \dots, g_i g_{i+1}, \dots, g_n, h) & ; \ 0 < i < n \\ (g_1, \dots, g_n h) & ; \ i = n, \end{cases}$$

These are obviously maps of G-sheaves.

Let $R[B_n(G)]$ be the free G-sheaf of R-modules on the G-sheaf $B_n(G)$ of sets. By taking alternating sum of the d_i, one obtains a cochain complex of G-sheaves of R-modules, naturally augmented by the constant sheaf R,

$$0 \longleftarrow R \longleftarrow R[B_0(G)] \longleftarrow R[B_1(G)] \longleftarrow \dots \tag{7}$$

Lemma 4.5 *Let G be an étale groupoid.*

(i) For each $n \geq 0$, the projection $p_n : B_n(G) \to G_n$, sending $(g_1, \dots, g_n, h)$ to $(g_1, \dots, g_n)$, induces a weak equivalence of étale groupoids

$$p_n \colon B_n(G) \rtimes G \longrightarrow G_n$$

compatible with the simplicial structures (here G_n is viewed as the unit groupoid).

(ii) Under this equivalence, the projection $\mathrm{pr}_n : B_n \rtimes G \to G$ *corresponds to* λ_n*, in the sense that for any G-sheaf A there is a natural isomorphism*

$$p_n^* \lambda_n^*(A) = \mathrm{pr}_n^*(A) \, .$$

(iii) For each G-sheaf A there is a natural isomorphism

$$\mathrm{Hom}_G(B_n(G), A) = \Gamma(G_n, A_{(n)})$$

(recall that $A_{(n)}$ *denotes the pull-back of A along* λ_n*).*

(iv) The augmented chain complex of G-sheaves

$$0 \longleftarrow R \longleftarrow R[B_0(G)] \longleftarrow R[B_1(G)] \longleftarrow \dots$$

is exact.

Proof We leave the easy proofs of (i) and (ii) to the reader. The isomorphism in (iii) is the map

$$\theta\colon \mathrm{Hom}_G(B_n(G), A) \longrightarrow \Gamma(G_n, A_{(n)})$$

sending a morphism $\alpha : B_n(G) \to A$ of G-sheaves to the section $\theta(\alpha)$, defined for any point $(g_1, \dots, g_n)$ in G_n by

$$\theta(\alpha)(g_1, \dots, g_n) = \alpha(g_1, \dots, g_n, 1_{s(g_n)}) \, .$$

Finally, for (iv), observe that the stalk at a point $y \in G_0$ of the simplicial G-sheaf $B_\bullet(G)$ is the simplicial set $(y/G)_\bullet$, the nerve of the comma category y/G. Like the nerve of any category with an initial object, the simplicial set $(y/G)_\bullet$ is contractible. It follows that the complex $R[B_\bullet(G)]$ is acyclic. Indeed, its stalk at y is the complex $R[B_\bullet(G)_y] = R[(y/G)_\bullet]$ which computes the homology $H_*(y/G, R)$. (An explicit chain homotopy $\xi : R[B_n(G)_y] \to R[B_{n+1}(G)_y]$ is defined on generators by $\xi(g_1, \dots, g_n, h) = (g_1, \dots, g_n, h, 1_{s(h)})$.) □

Proposition 4.6 *Let G be an étale groupoid. For any bounded below complex of G-sheaves A there is a spectral sequence*

$$E_1^{p,q} = H^q(G_p, A_{(p)}) \Rightarrow H^{p+q}(G, A)$$

natural in A.

Proof Let $A \to I$ be a quasi-isomorphism into a complex of injectives, and consider the double complex

$$\Gamma(G_p, I^q_{(p)}) \, .$$

For a fixed q, this is a complex of the form (6), and there are canonical isomorphisms

$$\Gamma(G_p, I^q_{(p)}) = \mathrm{Hom}_G(B_p(G), I^q) = \mathrm{Hom}_{\mathrm{Sh}_R(G)}(R[B_p(G)], I^q) ,$$

the first by Lemma 4.5 (iii) and the second by the adjunction (Section 3.4). Since $R[B_\bullet(G)]$ is acyclic (Lemma 4.5 (iv)) and I^q is injective, we find that $H^p(\Gamma(G_\bullet, I^q_{(\bullet)})) = 0$ for $p > 0$ and

$$H^0(\Gamma(G_\bullet, I^q_{(\bullet)})) = \mathrm{Hom}_{\mathrm{Sh}_R(G)}(R, I^q) = \Gamma_{\mathrm{inv}}(G, I^q) .$$

This shows that the total complex of $\Gamma(G_p, I^q_{(p)})$ computes $H^*(G, A)$. On the other hand, for fixed p, the map of complexes $A_{(p)} \to I_{(p)}$ is an injective resolution of the complex $A_{(p)}$, by Lemma 4.4 (ii) and Lemma 4.5 (ii). So

$$H^q(\Gamma(G_p, I_{(p)})) = H^q(G_p, A_{(p)}) ,$$

giving the E_1-term as described in the proposition. □

Corollary 4.7 *Let G be an étale groupoid, and let B be a G-sheaf. Let*

$$0 \longrightarrow B \longrightarrow A^0 \longrightarrow A^1 \longrightarrow \dots$$

be a resolution by sheaves A^q with the property that $H^q(G_p, A_{(p)}) = 0$ for each fixed p, and each $q > 0$. Then $H^*(G, B)$ is isomorphic to the cohomology of the double complex $\Gamma(G_p, A^q_{(p)})$.

Proof This follows from Proposition 4.6. Alternatively, one inspects the double complex directly. □

Examples 4.8 (1) For a G-sheaf B, any injective resolution $0 \to B \to I^0 \to I^1 \to \dots$ in $\mathrm{Sh}_R(G)$ satisfies the conditions of Corollary 4.7. Indeed, as observed in the proof of Proposition 4.6, each $I^q_{(p)}$ is an injective sheaf on G_p.

(2) (Godement resolution, see [60, p. 285]) Let X be a space, let X^δ be the same set with discrete topology, and let $p : X^\delta \to X$ be the evident (continuous) identity map. The Godement sheaf associated to a sheaf B on X is the sheaf

$$\mathcal{G}(B) = p_* p^*(B) .$$

A section of $\mathcal{G}(B)$ over $U \subset X$ is an arbitrary (not necessarily continuous) section from U into the étale space of B. The sheaf $\mathcal{G}(B)$ is flabby, in particular $H^n(X, \mathcal{G}(B)) = 0$ for $n > 0$. The Godement resolution is of the form

$$B \longrightarrow \mathcal{G}^0(B) \longrightarrow \mathcal{G}^1(B) \longrightarrow \ldots$$

where $\mathcal{G}^0(B) = \mathcal{G}(B)$ and $\mathcal{G}^{n+1}(B) = \mathcal{G}(\mathcal{G}^n(B))$. The construction of the Godement sheaf is stable under pull-back along local homeomorphism. More precisely, if $f\colon Y \to X$ is a local homeomorphism then $f^*\mathcal{G}(B) = \mathcal{G}(f^*B)$. In particular, if G is an étale groupoid and B is a G-sheaf, the Godement resolution $\mathcal{G}^\bullet(B)$ of B as a sheaf on G_0 carries a natural G-action, and for the pull-back $B_{(n)}$ to G_n we have $\mathcal{G}^\bullet(B_{(n)}) = \mathcal{G}^\bullet(B)_{(n)}$. Thus, as a special case of Lemma 4.5, the double complex

$$\Gamma(G_p, \mathcal{G}^q(B)_{(p)})$$

computes the cohomology $H^*(G, B)$ of G.

(3) Suppose G is a Hausdorff étale groupoid, and $R = \mathbb{R}$ is the field of reals. For any G-sheaf B there is a canonical *de Rham resolution*

$$0 \longrightarrow B \longrightarrow \Omega^0 \otimes B \longrightarrow \Omega^1 \otimes B \longrightarrow \ldots$$

by differential forms on G_0. For each of the G-sheaves $\Omega^q \otimes B$, the pull-back $(\Omega^q \otimes B)_{(p)} = \Omega^q \otimes B_{(p)}$ is the sheaf of differential forms on the space $G_{(p)}$ with coefficients in $B_{(p)}$. This is a fine sheaf, hence $H^n(G_p, \Omega^q \otimes B_{(p)}) = 0$ for $n > 0$. So the double complex

$$\Gamma(G_p, \Omega^q \otimes B_{(p)})$$

computes $H^*(G, B)$.

Remark. The assumption that G is Hausdorff is essential here. In fact $H^n(X, \Omega^q)$ need not be zero for $n > 0$ in case of a non-Hausdorff manifold X.

We conclude this section with another consequence of Lemmas 4.4 and 4.5.

Proposition 4.9 (Etale base change) *Let $\phi : H \to G$ be a homomorphism between Lie groupoids, and assume G is étale. Then for the fibered product square (Section 2.4)*

$$\begin{array}{ccc} G_0/\phi & \xrightarrow{v} & H \\ \downarrow{\scriptstyle\psi} & & \downarrow{\scriptstyle\phi} \\ G_0 & \xrightarrow{u} & G \end{array}$$

the following hold:

(i) For any H-sheaf E (of sets) the canonical map

$$u^*\phi_*(E) \longrightarrow \psi_* v^*(E)$$

is an isomorphism.

(ii)For any bounded below complex A of H-sheaves, the canonical map

$$u^* R\phi_*(A) \longrightarrow R\psi_*(v^*(A))$$

is an isomorphism.

Proof Under the Morita equivalence (Lemma 4.5 (i) and (ii)), the diagram corresponds to

$$\begin{array}{ccc} \phi^*(B_0(G)) \rtimes H & \xrightarrow{\mathrm{pr}'} & H \\ \Big\downarrow{\scriptstyle \tilde{\phi}} & & \Big\downarrow{\scriptstyle \phi} \\ B_0(G) \rtimes G & \xrightarrow{\mathrm{pr}} & G \end{array}$$

So (i) is equivalent to the assertion that there is a natural isomorphism $\mathrm{pr}^*\phi_* \to \tilde{\phi}_*\mathrm{pr}'^*$ for this square. By taking left adjoints of all functors involved, this is equivalent to the canonical map

$$\mathrm{pr}'_!\tilde{\phi}^* \longrightarrow \phi^*\mathrm{pr}_!$$

being an isomorphism, and this is indeed the case by Lemma 4.4 (iii). This proves (i). The property (ii) now follows because, again by Lemma 4.4, the exact functor v^* (or equivalently pr'^*) preserves injectives. □

4.3 Proper maps and orbifolds

We begin by reviewing some well-known facts about sheaves on paracompact Hausdorff spaces (cf. [19, 32]). First of all, recall that a sheaf B of abelian groups on a paracompact Hausdorff space X is called *soft* if for any inclusion $i: F \to X$ of a closed subspace, the restriction map $\Gamma(X, B) \to \Gamma(F, i^*B)$ is surjective. In other words, B is soft if any section over a closed subspace extends to the whole space. Soft sheaves are acyclic, i.e. $H^n(X, B) = 0$ for any $n > 0$ if B is soft. In particular, for an arbitrary sheaf A, one can compute $H^*(X, A)$ as the cohomology of $\Gamma(X, B)$, where $0 \to A \to B^0 \to B^1 \to \dots$ is a soft resolution. Related to this, any quasi-isomorphism $B \to C$ between bounded below complexes of soft sheaves induces a quasi-isomorphism $\Gamma(X, B) \to \Gamma(X, C)$.

Any injective sheaf is soft. If X is a smooth manifold, the sheaf Ω^n of differential n-forms on X is soft.

Next, recall that a map $f : Y \to X$ between Hausdorff spaces is proper if f is closed and each fiber of f is compact. For locally compact spaces, this is equivalent to the requirement that the inverse image of any compact subset of X is compact. Propriety is a local property, in the sense that $f: Y \to X$ is proper as soon as each $x \in X$ possesses an open neighbourhood U for which f restricts to a proper map $f^{-1}(U) \to U$. Also, propriety is stable under pull-backs, i.e. in a pull-back square

$$\begin{array}{ccc} Y' & \xrightarrow{b} & Y \\ {\scriptstyle f'}\downarrow & & \downarrow{\scriptstyle f} \\ X' & \xrightarrow{a} & X \end{array} \tag{8}$$

the map f' is proper whenever f is. We also recall the following 'proper base-change formulas':

Proposition 4.10 *Consider a pull-back square (8) with f proper.*

(i) For any sheaf of sets E on Y, the canonical map

$$a^* f_*(E) \longrightarrow f'_* b^*(E)$$

is an isomorphism.

(ii) For any bounded below complex A of sheaves of R-modules on Y, the canonical map

$$a^* Rf_*(A) \longrightarrow Rf'_*(b^* A)$$

is a quasi-isomorphism.

Proof See [19, 32]. □

Now consider a homomorphism between étale groupoids $\phi: H \to G$, and the following (weak) fibered products:

$$\begin{array}{ccccc} x/\phi & \xrightarrow{\tilde{x}} & G_0/\phi & \xrightarrow{v} & H \\ \downarrow & & {\scriptstyle \tilde{\phi}}\downarrow & & \downarrow{\scriptstyle \phi} \\ 1 & \xrightarrow{x} & G_0 & \xrightarrow{u} & G \end{array} \tag{9}$$

The comma groupoids G_0/ϕ and x/ϕ are again étale. Recall that the n-th space $(G_0/\phi)_n$ in the nerve of G_0/ϕ has as points the strings

$$(h_1, \ldots, h_n, g)$$

where $(h_1, \ldots, h_n) \in H_n$ and $g : x \to \phi(\mathrm{s}(h_n))$. Let us write $\tilde{\phi}_n : (G_0/\phi)_n \to G_0$ for the map which sends such a string to x. We observe the following consequence of 'proper base-change':

Proposition 4.11 *Let $\phi\colon H \to G$ be a homomorphism of étale groupoids. Suppose that H and G are Hausdorff, and that the map $\tilde{\phi}_n : (G_0/\phi)_n \to G_0$ is proper for each $n \geq 0$. Then for any bounded below complex A of H-sheaves, and any point $x \in G_0$,*

$$(R^n\phi_*(A))_x = H^n(x/\phi, A)$$

(here on the right A stands for its restriction to x/ϕ, i.e. for $\tilde{x}^*v^*(A)$ with $\tilde{x}$ and v as in (8)).

Proof Firstly, by étale base change (as in Section 4.2), $R\phi_*(A)_x = R\tilde{\phi}_*(v^*(A))_x$. Let $v^*(A) \to I$ be an injective resolution of (G_0/ϕ)-sheaves. Then, as in the proof of Corollary 4.7, $(R\tilde{\phi}_*(v^*(A)))_x$ is quasi-isomorphic to the total complex of the double complex

$$(\tilde{\phi}_n)_*(I^m_{(n)})_x\,,$$

where $I^m_{(n)}$ denotes the pull-back of I^m to $(G_0/\phi)_n$. (Use that $R[B_\bullet(H)]$ is an exact complex of H-sheaves (Lemma 4.5 (iv)) and $\psi_*(Hom(v^*R[B_n(H)], I^m)) = (\tilde{\phi}_n)_*(I^m_{(n)})$.) Since $\tilde{\phi}_n$ is proper, we have $(\tilde{\phi}_n)_*(I^m_{(n)})_x = \Gamma((x/\phi)_n, \tilde{x}^*(I^m_{(n)}))$. Since H is Hausdorff, each space $(x/\phi)_n$ is also Hausdorff, and I^m restricts to a soft sheaf $\tilde{x}^*(I^m_{(n)})$ on $(x/\phi)_n$. Thus, by Corollary 4.7 again, the double complex $\Gamma((x/\phi)_n, \tilde{x}^*(I^m_{(n)}))$ computes the cohomology $H^*(x/\phi_0, A)$. □

This proposition does not make use of any smooth structure, and applies to topological étale groupoids generally, provided the relevant spaces are paracompact Hausdorff. In particular, the following assertion is essentially a special case of the previous proposition:

Corollary 4.12 *Let G be a proper étale groupoid, let $|G|$ be its orbit space, and write $q\colon G \to |G|$ for the associated projection. Then for any bounded below complex A of G-sheaves, there is a canonical isomorphism*

$$(R^n q_*(A))_{q(x)} = H^n(G_x, A_x)$$

for any point $x \in G_0$ with isotropy group G_x.

Proof The statement is local in $|G|$. Write $q_i\colon G_i \to |G|$ (for $i = 0, 1, \ldots$), and consider a point $\xi \in |G|$. Choose $x \in G_0$ with $q_0(x) = \xi$. By [48, p. 142], there is a neighbourhood U of ξ such that the groupoid $q^{-1}(U) =$

$G|_{q_0^{-1}(U)}$ is Morita equivalent to the semi-direct product

$$U_x \ltimes G_x$$

where U_x is a neighbourhood of x and the isotropy group G_x acts on U_x, in such a way that q induces a homeomorphism $U_x/G_x \to U$. Thus, it suffices to prove the corollary in the special case where G is $U \rtimes G_x$. But now the statement follows from the previous proposition (by taking $U \rtimes G_x$ for H and the unit groupoid of the space U/G_x for G). Indeed, in the diagram

$$\dots \; U \times G_x^2 \rightrightarrows U \times G_x \rightrightarrows U \xrightarrow{q} U/G_x$$

all maps are proper, and $\xi/q = q^{-1}(\xi)$ is a finite groupoid equivalent to the isotropy group G_x. $\square$

Corollary 4.13 *Let G be a proper étale groupoid, and suppose the order $\#G_x$ of each isotropy group G_x is a unit in the ring R. Then for any G-sheaf A of R-modules we have*

$$H^*(G, A) = H^*(|G|, q_*(A)) .$$

Let Q be a paracompact Hausdorff space. Recall that an orbifold structure on Q (see Remark 2.6 (4)) is given in one of several equivalent ways by a proper effective groupoid G such that $|G| = Q$. We also refer to the map $q\colon G \to Q$ as an orbifold structure on Q. For any G-sheaf A, the cohomology $H^*(G, A)$ is called the *orbifold cohomology* of Q with coefficients in A. By the Leray spectral sequence for the map q, this cohomology is built up out of the cohomology of the underlying space Q and that of the isotropy groups G_x of the orbifold:

$$E_2^{p,s} = H^p(Q, R^s q_*(A)) \Rightarrow H^{p+s}(G, A) ,$$

where $R^s q_*$ is described explicitly by Corollary 4.12. Suppose now that we work with sheaves of modules over the field $\mathbb{R}$ of reals. Recall that any G-sheaf B of $\mathbb{R}$-modules has a resolution $0 \to B \to \Omega^0 \otimes B \to \Omega^1 \otimes B \dots$ by the G-sheaves of differential forms. The following result is a version of the de Rham theorem for orbifolds:

Corollary 4.14 *Let $q\colon G \to Q$ be an orbifold structure on Q. For any sheaf B of $\mathbb{R}$-modules on Q, there is a canonical isomorphism*

$$H^n(Q, B) = H^n(\Gamma_{\mathrm{inv}}(G, \Omega \otimes q^*(B))) \qquad i = 0, 1, \dots$$

Proof First observe that $q_*(\Omega^n \otimes q^*(B)) = q_*(\Omega^n) \otimes B$, and hence

$$\Gamma_{\text{inv}}(G, \Omega^n \times q^*(B)) = \Gamma(Q, q_*(\Omega^n) \otimes B) .$$

Next, q_* is exact by Corollary 4.12 (since we work over the reals), so $q_*(\Omega) \otimes B$ is a resolution of B. It thus suffices to show for each n that $q_*(\Omega^n) \otimes B$ is a soft sheaf on Q. This is a local property, so it suffices to prove this where $Q = U/G_x$ and $G = U \rtimes G_x$ as in the proof of Corollary 4.12. The details for this are standard: If $F \subset U/G_x$ is closed and $\sigma \in \Gamma(F, q_*(\Omega^n) \otimes B)$, then σ pulls back to a G_x-equivariant section of $\Omega^n \otimes q^*(B)$ on $q^{-1}(F) \subset U$. Since Ω^n is a fine sheaf, $\Omega^n \otimes q^*(B)$ is soft [19], so this section extends to a section $\tilde{\sigma}$ on all of U. By averaging over G_x, we can modify $\tilde{\sigma}$ into a G_x-equivariant section $\bar{\sigma}$. Then

$$\bar{\sigma} \in \Gamma_{\text{inv}}(U \rtimes G_x, \Omega^n \otimes q^*(B)) = \Gamma(U/G_x, q_*(\Omega^n) \otimes B)$$

is the required extension of σ. □

To conclude this section, we prove the analogue of the change of base formula (Proposition 4.10) for proper étale groupoids (and hence, in particular, for orbifolds).

Let $\phi : H \to G$ be a homomorphism from a proper Lie groupoid H into a Hausdorff étale groupoid G. Then we say that ϕ is *proper* if for each compact subset K of G_0, the orbit space $|K/\phi|$ of the comma groupoid K/ϕ is compact. Recall from the diagram (9) that the square

$$\begin{array}{ccc} G_0/\phi & \xrightarrow{v} & H \\ {\scriptstyle \tilde{\phi}}\downarrow & & \downarrow{\scriptstyle \phi} \\ G_0 & \xrightarrow{u} & G \end{array}$$

is a weak fibered product. Then since H is proper and G is Hausdorff, G_0/ϕ is a locally compact Hausdorff space, and ϕ is a proper map if and only if the map $|G_0/\phi| \to G_0$ induces by $\tilde{\phi}$ is a proper map of spaces in the usual sense. We note also that this definition of proper homomorphism is invariant under weak equivalences of groupoids.

Lemma 4.15 *In a weak pull-back square*

$$\begin{array}{ccc} P & \longrightarrow & H \\ {\scriptstyle \psi}\downarrow & & \downarrow{\scriptstyle \phi} \\ K & \longrightarrow & G \end{array}$$

the map ψ is proper whenever ϕ is.

Proof The induced square

$$\begin{array}{ccc} K_0/\psi & \longrightarrow & G_0/\phi \\ \downarrow & & \downarrow \\ K_0 & \longrightarrow & G_0 \end{array}$$

is again a fibered product (up to weak equivalence). This gives an ordinary fibered product of spaces

$$\begin{array}{ccc} |K_0/\psi| & \longrightarrow & |G_0/\phi| \\ {\scriptstyle |\psi|}\downarrow & & \downarrow{\scriptstyle |\phi|} \\ K_0 & \longrightarrow & G_0 \end{array}$$

so $|\psi|$ is proper whenever $|\phi|$ is. □

Proposition 4.16 (Proper base change) *For a weak pull-back square*

$$\begin{array}{ccc} P & \xrightarrow{b} & H \\ {\scriptstyle \psi}\downarrow & & \downarrow{\scriptstyle \phi} \\ K & \xrightarrow{a} & G \end{array}$$

of proper étale groupoids, and for any bounded below complex A of H-sheaves, the canonical map

$$a^* R\phi_*(A) \longrightarrow R\psi_*(b^*(A)) \tag{10}$$

is a quasi-isomorphism.

Proof Consider the rectangle

$$\begin{array}{ccccc} K_0/\psi & \xrightarrow{v} & P & \xrightarrow{b} & H \\ {\scriptstyle \tilde{\psi}}\downarrow & & \downarrow{\scriptstyle \psi} & & \downarrow{\scriptstyle \phi} \\ K_0 & \xrightarrow{u} & K & \xrightarrow{a} & G \end{array}$$

It is enough to prove that u^* maps (10) to a quasi-isomorphism. Thus, by étale base change for the left-hand square, it is enough to prove the base change formula for the composed rectangle, i.e. that

$$(au)^* R\phi_*(A) \longrightarrow R\tilde{\psi}_*((bv)^*(A))$$

is a quasi-isomorphism. This composed rectangle is the same as the composed rectangle of the fibered products

$$\begin{array}{ccccc} K_0/\psi & \longrightarrow & G_0/\phi & \longrightarrow & H \\ {\scriptstyle\tilde{\psi}}\downarrow & & {\scriptstyle\tilde{\phi}}\downarrow & & \downarrow{\scriptstyle\phi} \\ K_0 & \longrightarrow & G_0 & \xrightarrow{u} & G \end{array} \tag{11}$$

so by étale base change for the right-hand square, it remains to show the base change formula for the left-hand square of (11). This square decomposes into two,

$$\begin{array}{ccc} K_0/\psi & \longrightarrow & G_0/\phi \\ \downarrow & & \downarrow \\ |K_0/\psi| & \longrightarrow & |G_0/\phi| \end{array}$$

and

$$\begin{array}{ccc} |K_0/\psi| & \longrightarrow & |G_0/\phi| \\ \downarrow & & \downarrow \\ K_0 & \longrightarrow & G_0 \end{array}$$

the first of which satisfies base change by Corollary 4.12 and the second of which does by Proposition 4.10. □

4.4 A comparison theorem for foliations

As an application of étale base change, we will present a theorem which compares the cohomology of the leaf space of a foliation with that of the underlying manifold. Before stating this theorem, we recall that a topological space F is called *n-acyclic* if it is connected and $H^i(F, A) = 0$ for any abelian group A and any $0 < i \leq n$. (This is closely related to the stronger notion of n-connectedness: a space F is called *n-connected* if $\pi_i(F, x) = 0$ for $0 \leq i \leq n$ and any base point $x \in F$. For $n > 1$, a simply connected space F is n-connected if and only if it is n-acyclic.) A map $f\colon Y \to X$ is said to be *n-acyclic* if each of its fibers is an n-acyclic space.

Now consider a foliation $(M, \mathcal{F})$ and its holonomy groupoid $\mathrm{Hol}(M, \mathcal{F})$. For any $\mathrm{Hol}(M, \mathcal{F})$-sheaf A, the inclusion of the units u :

$M \to \mathrm{Hol}(M, \mathcal{F})$ induces a map

$$u^* : H^i(\mathrm{Hol}(M, \mathcal{F}), A) \longrightarrow H^i(M, u^*(A))$$

which should be thought as the pull-back along the quotient map $M \to M/\mathcal{F}$ from the manifold to the leaf space.

Theorem 4.17 *Let $(M, \mathcal{F})$ be a foliation, and suppose that for each leaf L the holonomy cover $\tilde{L}$ is an n-acyclic space. Then for each* $\mathrm{Hol}(M, \mathcal{F})$*-sheaf A the map*

$$H^i(\mathrm{Hol}(M, \mathcal{F}), A) \longrightarrow H^i(M, u^*(A))$$

is an isomorphism for $0 < i \leq n$. Similarly, the map

$$H^i(\mathrm{Mon}(M, \mathcal{F}), A) \longrightarrow H^i(M, u^*(A))$$

is an isomorphism for $0 < i \leq n$ if each leaf L has an n-acyclic universal cover.

In fact, this result is a special case of the following theorem:

Theorem 4.18 *Let G be an étale groupoid and let $p \colon M \to G_0$ be a left G-space. Suppose that the map p is an n-acyclic submersion. Then the homomorphism $p \colon G \ltimes M \to G$ of étale groupoids has the property that for any G-sheaf A,*

$$R^0 p_*(p^* A) = A$$

and

$$R^i p_*(p^* A) = 0 \qquad 0 < i \leq n\,.$$

In particular, for any such A,

$$p^* : H^i(G, A) \longrightarrow H^i(G \ltimes M, p^* A)$$

is an isomorphism for $0 \leq i \leq n$.

Remark. In this theorem, we do not need M to be Hausdorff, but we need to assume that the fibers of p are Hausdorff.

Proof (Proof of Theorem 4.17 from Theorem 4.18) We only discuss the case of the holonomy groupoid. The statement for the monodromy groupoid is proved in the same way. Let $j : T \to M$ be a complete transversal, with its associated weak equivalence $j : \mathrm{Hol}_T(M, \mathcal{F}) \to$

$\mathrm{Hol}(M,\mathcal{F})$, and construct the weak fibered product

$$\begin{array}{ccc} P & \xrightarrow{k} & M \\ \downarrow & & \downarrow u \\ \mathrm{Hol}_T(M,\mathcal{F}) & \xrightarrow{j} & \mathrm{Hol}(M,\mathcal{F}) \end{array}$$

Thus the manifold P_0 of objects of P consists of triples (t,α,m) where $t \in T$, $m \in M$ and α is an arrow from m to t in $\mathrm{Hol}(M,\mathcal{F})$. Arrows $\beta\colon (t,\alpha,m) \to (t',\alpha',m')$ in P are arrows $\beta\colon t \to t'$ such that $\beta\alpha = \alpha'$ (hence in particular $m = m'$). Thus P is the semi-direct product

$$P = \mathrm{Hol}_T(M,\mathcal{F}) \ltimes P_0 \,.$$

Moreover since j is a weak equivalence, so is k. So in the following commutative square the horizontal maps are isomorphisms:

$$\begin{array}{ccc} H^i(P,k^*u^*A) & \longleftarrow & H^i(M,u^*A) \\ \uparrow & & \uparrow \\ H^i(\mathrm{Hol}_T(M,\mathcal{F}),j^*A) & \longleftarrow & H^i(\mathrm{Hol}(M,\mathcal{F}),A) \end{array}$$

Since the fiber of $P_0 \to T$ over t is precisely the holonomy cover of the leaf through t, Theorem 4.17 follows by applying Theorem 4.18 to the $\mathrm{Hol}_T(M,\mathcal{F})$-space P_0. □

Proof (Proof of Theorem 4.18) Let us first observe that the last 'in particular' part follows from the first by general homological algebra. Indeed, if $p^*A \to I^0 \to I^1 \to \ldots$ is an injective resolution of $G \ltimes M$-sheaves, then the first part states that

$$0 \longrightarrow A = p_*p^*A \longrightarrow p_*I^0 \longrightarrow p_*I^1 \longrightarrow \ldots$$

is exact up to p_*I^n. Thus, if we extend this sequence on the right by an injective resolution of $\mathrm{Ker}(p_*(I^n \to I^{n+1})) = p_*\mathrm{Ker}(I^n \to I^{n+1})$, we obtain an injective resolution of A by G-sheaves of the form

$$0 \longrightarrow A \longrightarrow p_*I^0 \longrightarrow \ldots \longrightarrow p_*I^{n-1} \longrightarrow J^n \longrightarrow J^{n+1} \longrightarrow \ldots$$

and with $\mathrm{Ker}(J^n \to J^{n+1}) = \mathrm{Ker}(p_*(I^n \to I^{n+1}))$. Thus for $0 \le i \le n$ we find that $H^i(G,A) = H^i(\Gamma_{\mathrm{inv}}(G,p_*I)) = H^i(\Gamma_{\mathrm{inv}}(G \ltimes M,I)) = H^i(G \ltimes M, p^*A)$.

It thus suffices to prove the first assertion of the theorem. To this end, consider the square

$$\begin{array}{ccc} M & \xrightarrow{v} & G \ltimes M \\ {\scriptstyle p_0}\downarrow & & \downarrow{\scriptstyle p} \\ G_0 & \xrightarrow{u} & G \end{array}$$

where we have written $p_0\colon M \to G_0$ to distinguish it from the groupoid homomorphism $p\colon G \ltimes M \to G$, and $v\colon M \to G \ltimes M$ for the inclusion of the units. This square is a weak fibered product up to Morita equivalence, because $G_0/(G \ltimes M)$ is weakly equivalent to the manifold M (viewed as unit groupoid). Thus, by étale base change, there is a canonical quasi-isomorphism $u^* Rp_* = R(p_0)_* v^*$. Since a map of complexes of G-sheaves is a quasi-isomorphism if and only if it is mapped by u^* to a quasi-isomorphism of complexes of sheaves on G_0, we reduced the proof to the case of ordinary spaces, where it is known, and which we state in the following lemma. □

Lemma 4.19 *Let $p : M \to G_0$ be an n-acyclic submersion between smooth manifolds. Then for any sheaf A on G_0 we have*

$$R^0 p_*(p^* A) = A$$

and

$$R^i p_*(p^* A) = 0 \qquad 0 < i \leq n\,.$$

Remark. Recall that we do not assume that M is Hausdorff, although G_0 and the fibers of p are assumed to be Hausdorff.

Proof (cf. [1, 4]) Consider the Godement resolution

$$A \longrightarrow \mathcal{G}^0(A) \longrightarrow \mathcal{G}^1(A) \longrightarrow \dots$$

of A. This gives a resolution

$$p^* A \longrightarrow p^* \mathcal{G}^0(A) \longrightarrow p^* \mathcal{G}^1(A) \longrightarrow \dots$$

and by an argument exactly analogous to the proof of the first part of Theorem 4.18, it suffices to prove that $R^i p_*(p^*(\mathcal{G}^k(A)) = 0$ for $0 < i \leq n$ and for each k. Since $\mathcal{G}^k(A) = \mathcal{G}(\mathcal{G}^{k-1}(A))$ and A is arbitrary, we are thus left with proving that for any 'Godement sheaf' $\mathcal{G}(A)$ on G_0,

$$R^i p_*(p^*(\mathcal{G}(A)) = 0 \qquad 0 < i \leq n\,.$$

To this end, notice that $\mathcal{G}(A)$ is an infinite product of sheaves,

$$\mathcal{G}(A) = \prod_{x \in G_0} (i_x)_*(A_x)$$

where $i_x : 1 \to G_0$ is the inclusion of the point $x \in G_0$. The functor $p^* : \mathrm{Sh}(G_0) \to \mathrm{Sh}(M)$ preserves such products because it has a left adjoint (the functor sending a sheaf $E \to M$ to the sheaf of fiber-wise connected components of the composition $E \to M \to G_0$), so

$$p^*\mathcal{G}(A) = \prod_{x \in G_0} p^*(i_x)_*(A_x)\ . \tag{12}$$

Since the closed inclusions in the pull-back square

$$\begin{array}{ccc} f^{-1}(x) & \xrightarrow{p_x} & 1 \\ {\scriptstyle j_x}\big\downarrow & & \big\downarrow{\scriptstyle i_x} \\ M & \xrightarrow{p} & G_0 \end{array}$$

are proper maps, proper base change (Proposition 4.10) gives that $p^*(i_x)_* = (j_x)_* p_x^*$. So we can rewrite (12) as

$$p^*\mathcal{G}(A) = \prod_{x \in G_0} (j_x)_* p_x^*(A_x)\ .$$

Now choose for any $x \in G_0$ a resolution

$$0 \longrightarrow p_x^*(A_x) \longrightarrow I_x^0 \longrightarrow I_x^1 \longrightarrow \dots$$

of the constant sheaf $p_x^*(A_x)$ by injective sheaves on the fiber $f^{-1}(x)$. Since j_x is the inclusion of a closed subspace $f^{-1}(x) \subset M$, the functor $(j_x)_*$ is exact and preserves injectives, so

$$0 \longrightarrow (j_x)_* p_x^*(A_x) \longrightarrow (j_x)_* I_x^0 \longrightarrow (j_x)_* I_x^1 \longrightarrow \dots$$

is still an injective resolution. Taking the product over all $x \in G_0$ and using the isomorphism (12), we obtain an injective resolution

$$0 \longrightarrow p^*\mathcal{G}(A) \longrightarrow \prod_{x \in G_0} (j_x)_* I_x^0 \longrightarrow \prod_{x \in G_0} (j_x)_* I_x^1 \longrightarrow \dots$$

Now apply p_* to this, and notice that

$$p_*\Big(\prod_{x \in G_0} (j_x)_* I_x^k\Big) = \prod_{x \in G_0} p_*(j_x)_* I_x^k = \prod_{x \in G_0} (i_x)_*(p_x)_* I_x^k\ .$$

The stalk of this complex $\prod_{x \in G_0} (i_x)_*(p_x)_* I_x$ at a point $x \in G_0$ is $(p_x)_* I_x$, which computes $H^*(f^{-1}(x), A_x)$ hence is acyclic up to n. This shows that p_* maps the resolution of $p^*\mathcal{G}(A)$ above to a complex which is exact up to n, so $R^i p_*(p^*\mathcal{G}(A)) = 0$ for $0 < i \leq n$. □

4.5 The embedding category of an étale groupoid

We have seen that the cohomology of (étale) Lie groupoids generalizes the sheaf cohomology of manifolds as well as the cohomology of discrete groups. There is another generalization of group cohomology which involves no topology or smooth structure, and is purely combinatorial in nature, the cohomology of small categories. The aim of this section is, first to recall this cohomology, and then to show how the cohomology of an étale groupoid can be interpreted in terms of an associated small category. This result gives rise, among other things, to a Čech-de Rham model of the cohomology of étale groupoids, and in particular of leaf spaces of foliations.

Let $\mathcal{C}$ be a small category. This means that the objects and arrows of $\mathcal{C}$ form (small) sets $\mathcal{C}_0$ and $\mathcal{C}_1$; these sets are not assumed to have any further structure. Exactly as for groupoids, we can form an associated simplicial set

$$\mathcal{C}_0 \leftleftarrows \mathcal{C}_1 \leftleftarrows\!\!\!\!\leftarrow \mathcal{C}_2 \quad \cdots$$

called the *nerve* of $\mathcal{C}$. The set $\mathcal{C}_n$ is the set of strings of composable arrows

$$c_0 \xleftarrow{f_1} c_1 \xleftarrow{f_2} \cdots \xleftarrow{f_n} c_n$$

in the category $\mathcal{C}$, and the face maps $d_i : \mathcal{C}_n \to \mathcal{C}_{n-1}$ are defined as in Section 4.2, i.e. 'd_i omits c_i'. In particular, $d_0, d_1 : \mathcal{C}_1 \to \mathcal{C}_0$ are the source s and the target map t of $\mathcal{C}$.

For a ring R, a *presheaf* of R-modules on $\mathcal{C}$ is a contravariant functor A from $\mathcal{C}$ into the category of R-modules; we shall use any of the notations $A(f)(a) = a \cdot f = f^*(a)$ for an arrow f in $\mathcal{C}$ and $a \in A(\mathrm{t}(f))$. For such a functor A, one can form a cosimplicial abelian group $C^\bullet(\mathcal{C}, A)$, by defining

$$C^n(\mathcal{C}, A) = \prod_{(f_1,\ldots,f_n)\in\mathcal{C}_n} A(\mathrm{s}(f_n)) \,,$$

where the product ranges over all strings in $\mathcal{C}_n$. For $i = 0, 1, \ldots, n$, the face map $d_i : \mathcal{C}_n \to \mathcal{C}_{n-1}$ induces a map $\delta^i : C^{n-1}(\mathcal{C}, A) \to C^n(\mathcal{C}, A)$

defined for $f = (f_1, \ldots, f_n)$ by

$$\delta^i(a)(f) = \begin{cases} a(d_i(f)) & ; \ 0 \le i < n \\ a(d_n(f)) \cdot f_n & ; \ i = n \end{cases}$$

One obtains a cochain complex (also denoted) $C^\bullet(\mathcal{C}, A)$ by defining the differential $\delta \colon C^{n-1}(\mathcal{C}, A) \to C^n(\mathcal{C}, A)$ by $\delta = \sum_i (-1)^i \delta^i$, as usual. The cohomology of this complex is called the cohomology of the category $\mathcal{C}$ with coefficients in the presheaf A, and denoted

$$H^*(\mathcal{C}, A) .$$

Similarly, if A is a bounded below complex of presheaves, one defines a double complex $C^{p,q} = C^p(\mathcal{C}, A^q)$, and the cohomology of the associated total complex is the (hyper) cohomology of $\mathcal{C}$ with coefficients in the complex A, denoted again by $H^*(\mathcal{C}, A)$.

This cohomology of small categories has good general properties which are well-known and which we will not discuss. We observe only that, by the standard properties of double complexes, a quasi-isomorphism $A \to B$ of bounded below complexes of presheaves (that is, a map with the property that $A(c) \to B(c)$ is a quasi-isomorphism for any object c) induces an isomorphism $H^*(\mathcal{C}, A) \to H^*(\mathcal{C}, B)$.

Now let G be an étale groupoid, and let $\mathcal{U}$ be a basis of open sets in G_0. Typically, $\mathcal{U}$ will consist of small contractible 'balls'. With this, we can construct a small category

$$\mathrm{Emb}_{\mathcal{U}}(G) ,$$

called the *embedding category* of G (with respect to the basis $\mathcal{U}$). The objects of $\mathrm{Emb}_{\mathcal{U}}(G)$ are the members of $\mathcal{U}$. For $U, V \in \mathcal{U}$, an arrow $U \to V$ in $\mathrm{Emb}_{\mathcal{U}}(G)$ is a bisection (cf. [48, p. 115]) $\sigma : U \to G_1$ of G for which $\mathrm{t}(\sigma(U)) \subset V$. One can also think of such an arrow $\sigma : U \to V$ in $\mathrm{Emb}_{\mathcal{U}}(G)$ as a particular smooth family $\{\sigma(x)\}$ of arrows in G parametrized by $x \in U$. With this in mind, it is clear how the multiplication in G induces a composition of arrows in $\mathrm{Emb}_{\mathcal{U}}(G)$. Explicitly, for arrows $\sigma : U \to V$ and $\rho \colon V \to W$ in $\mathrm{Emb}_{\mathcal{U}}(G)$, the bisection $\rho\sigma \colon U \to W$ is given by

$$\rho\sigma(x) = \rho(\mathrm{t}(\sigma(x)))\sigma(x) .$$

Example 4.20 Let $\mathcal{F}$ be a foliation on a manifold M, and $\mathrm{Hol}_T(M, \mathcal{F})$ the étale holonomy groupoid of $(M, \mathcal{F})$ associated to a complete transversal T. Let $\mathcal{U}$ be a basis for T. An arrow $\sigma : U \to V$ in $\mathrm{Emb}_{\mathcal{U}}(\mathrm{Hol}_T(M, \mathcal{F}))$ is locally represented by a smooth family of paths

$\{\alpha_x\}$ parametrized by $x \in U$, such that each α_x is a path from x to $f(x) \in V$ inside a leaf of $\mathcal{F}$, and such that the map $f: U \to V$ given in this way is an embedding.

Let us return to the general case of an étale groupoid G and a basis $\mathcal{U}$ for G_0. Any G-sheaf A induces a presheaf $\Gamma(A)$ on the category $\mathrm{Emb}_{\mathcal{U}}(G)$, defined for $U \in \mathcal{U}$ by $\Gamma(A)(U) = \Gamma(U, A)$, the set of sections of A (as a sheaf on G_0) over U. This is indeed a contravariant functor, because each arrow $\sigma: U \to V$ in $\mathrm{Emb}_{\mathcal{U}}(G)$ induces a map

$$\sigma^*: \Gamma(V, A) \longrightarrow \Gamma(U, A)$$

defined for any $\alpha \in \Gamma(V, A)$ by

$$\sigma^*(\alpha)(x) = \alpha(\mathrm{t}(\sigma(x))) \cdot \sigma(x) .$$

Thus we obtain a functor

$$\Gamma: \mathrm{Sh}_R(G) \longrightarrow \{\text{presheaves of } R\text{-modules on } \mathrm{Emb}_{\mathcal{U}}(G)\} ,$$

for any ring R.

Before stating the main theorem of this section, we need one more definition. A G-sheaf A is said to be $\mathcal{U}$-*acyclic* if $H^n(U, A) = 0$ for any $n > 0$ and any open $U \in \mathcal{U}$. For example, any injective G-sheaf A is $\mathcal{U}$-acyclic, because A is still injective when viewed as a sheaf on G_0 by Lemma 4.4 (ii) and Lemma 4.5 (i). Also, the G-sheaf Ω^m of differential m-forms is $\mathcal{U}$-acyclic for each $m \geq 0$.

Theorem 4.21 *Let G be an étale groupoid and let $\mathcal{U}$ be a basis of open sets on G_0. Then for any bounded below complex A of $\mathcal{U}$-acyclic G-sheaves, there is a natural isomorphism*

$$H^*(G, A) = H^*(\mathrm{Emb}_{\mathcal{U}}(G), \Gamma(A)) .$$

Before we prove the theorem, we mention a few special cases.

Corollary 4.22 *Let G be an étale groupoid, and let $\mathcal{U}$ be a basis of contractible open sets on G_0. Then for any locally constant sheaf A we have*

$$H^*(G, A) = H^*(\mathrm{Emb}_{\mathcal{U}}(G), \Gamma(A)) .$$

Remark. Observe that in this case the presheaf $\Gamma(A)$ is also locally constant, in the sense that $\sigma^*: \Gamma(V, A) \to \Gamma(U, A)$ is an isomorphism for each arrow $\sigma: U \to V$ in $\mathrm{Emb}_{\mathcal{U}}(G)$.

Corollary 4.23 *Let G be an étale groupoid, and let $\mathcal{U}$ be a basis of open sets on G_0. Let A be a sheaf of $\mathbb{R}$-modules. Then*

$$H^*(G, A) = H^*(\mathrm{Emb}_{\mathcal{U}}(G), \Gamma(\Omega \otimes A)) .$$

Remark. This corollary states that, when working over the reals, $H^*(G, A)$ can be computed by some sort of 'Čech-de Rham double complex'. In bidegree p, q this complex is

$$\prod_{U_0 \leftarrow U_1 \leftarrow \ldots \leftarrow U_p} \Omega^q(U_p, A) \tag{13}$$

where the product is over strings of arrows in the embedding category and $\Omega^q(U_p, A)$ is the vector space of q-forms on U_p with coefficients in A. Therefore we refer to this last corollary as the *Čech-de Rham theorem for étale groupoids.*

Example 4.24 Let $(M, \mathcal{F})$ be a foliated manifold, and let $\mathcal{U}$ be a basis of open sets on a complete transversal T, as in Example 4.20 above. Then for any $\mathrm{Hol}(M, \mathcal{F})$-sheaf A, the Čech-de Rham double complex (13) computes $H^*(\mathrm{Hol}(M, \mathcal{F}), A)$. This follows by the invariance of cohomology under the Morita equivalence $\mathrm{Hol}(M, \mathcal{F}) \sim \mathrm{Hol}_T(M, \mathcal{F})$ and the corollary applied to $\mathrm{Hol}_T(M, \mathcal{F})$.

Proof (Proof of Theorem 4.21) Let $A \to I$ be a resolution of A by a bounded below complex of injective G-sheaves. Then, by Corollary 4.7, $H^*(G, A)$ is computed by the double complex

$$\Gamma(G_p, I^r) .$$

Also, since each A^r is $\mathcal{U}$-acyclic, $\Gamma(U, A) \to \Gamma(U, I)$ is a quasi-isomorphism for each $U \in \mathcal{U}$, so $H^*(\mathrm{Emb}_{\mathcal{U}}(G), \Gamma(A))$ is computed by the double complex

$$C^q(\mathrm{Emb}_{\mathcal{U}}(G), \Gamma(I^r)) = \prod_{U_0 \leftarrow U_1 \leftarrow \ldots \leftarrow U_p} \Gamma(U_q, I^r) .$$

We will prove the theorem by constructing a suitable complex $C(I)$ and explicit quasi-isomorphisms of total complexes

$$\Gamma(G_\bullet, I^\bullet) \longrightarrow C(I) \longleftarrow C^\bullet(\mathrm{Emb}_{\mathcal{U}}(G), \Gamma(I^\bullet)) .$$

For this consider the bisimplicial space $S_{p,q}$ whose p, q-simplices are of the form

$$U_0 \xleftarrow{\sigma_1} U_1 \xleftarrow{\sigma_2} \ldots \xleftarrow{\sigma_q} U_q \xleftarrow{g} x_0 \xleftarrow{g_1} x_1 \xleftarrow{g_2} \ldots \xleftarrow{g_p} x_p \tag{14}$$

where $\sigma_1, \ldots, \sigma_q$ are arrows in $\mathrm{Emb}_{\mathcal{U}}(G)$ and $g, g_1, \ldots, g_p$ are arrows in the groupoid G, the notation $U_q \stackrel{g}{\leftarrow} x_0$ indicating that the target of g lies in U_q. The topology of $S_{p,q}$ is that of the disjoint sum over all strings $(\sigma_1, \ldots, \sigma_q)$ of the fibered products $U_q \times_{G_0} G_p$. Each G-sheaf B induces a sheaf $B^{p,q}$ on $S_{p,q}$, by pulling back along the map $S_{p,q} \to G_0$ sending a point (14) to x_p. Write

$$C^{p,q}(B) = \Gamma(S_{p,q}, B^{p,q}) .$$

Then the simplicial face maps of $S_{p,q}$ make $C^{p,q}(B)$ into a double complex. There are natural augmentations

$$\Gamma(G_p, \lambda_p^*(B)) \longrightarrow C^{p,q}(B) \longleftarrow C^q(\mathrm{Emb}_{\mathcal{U}}(G), \Gamma(B)) \tag{15}$$

(recall that $\lambda_p^*(B)$ is the sheaf on G_p with stalk B_{x_p} at $x_0 \leftarrow \ldots \leftarrow x_p$). It is now enough to prove that for any injective G-sheaf B, each of the maps in (15) is a quasi-isomorphism into the total complex of $C^{p,q}(B)$. Indeed, when applied to $B = I^r$ for each r, we then obtain quasi-isomorphisms

$$\Gamma(G_s, I^r) \longrightarrow \bigoplus_{p+q=s} C^{p,q}(I^r) \longleftarrow C^s(\mathrm{Emb}_{\mathcal{U}}(G), \Gamma(I^r))$$

for each fixed r, and hence quasi-isomorphisms of total complexes, which proves the theorem.

So consider the maps in (15) for an injective G-sheaf B. For a fixed q, the complex $C^{\bullet,q}(B)$ is a product of bar complexes of the kind considered in Section 4.2, viz. for each string $U_0 \leftarrow \ldots \leftarrow U_q$ the bar complex of the étale comma groupoid G/U_q, with coefficients in the pull-back of the G-sheaf B along the evident homomorphism $G/U_q \to G$. Since this groupoid G/U_q is Morita equivalent to the manifold U_q, this bar complex computes the cohomology $H^*(U_q, B)$, and this vanishes in the positive degrees because B is injective. Thus, for each q the map

$$C^q(\mathrm{Emb}_{\mathcal{U}}(G), \Gamma(B)) \longrightarrow C^{\bullet,q}(B)$$

is a quasi-isomorphism. It follows that the right-hand map in (15) is a quasi-isomorphism into the total complex of $C^{p,q}(B)$.

To conclude the proof, we consider the left-hand map in (15), and show that

$$\Gamma(G_p, \lambda_p^*(B)) \longrightarrow C^{p,\bullet}(B)$$

is a quasi-isomorphism for each fixed p. To this end, write $\mathrm{pr}_{p,q} \colon S_{p,q} \to G_p$ for the projection of a point (14) to the string $x_0 \leftarrow \ldots \leftarrow x_p$. Then

$$C^{p,q}(B) = \Gamma(G_p, (\mathrm{pr}_{p,q})_*(B^{p,q})) .$$

Since $\lambda^*(B)$ and $(\mathrm{pr}_{p,q})_*(B^{p,q})$ are all injective sheaves on G_p, we need to show that

$$0 \longrightarrow \lambda_p^*(B) \longrightarrow (\mathrm{pr}_{p,0})_*(B^{p,q}) \longrightarrow (\mathrm{pr}_{p,1})_*(B^{p,q}) \longrightarrow \dots \qquad (16)$$

is a resolution of sheaves on G_p. At this point, it is convenient to change the notation. Let

$$\phi_p \colon G_p \longrightarrow G_0 \qquad\qquad \rho_{p,q} \colon S_{p,q} \longrightarrow G_0$$

be maps sending $x_0 \leftarrow \dots \leftarrow x_p$ to x_0, and (14) to x_0, respectively. Then by the G-action on B, the complex (16) is isomorphic to

$$0 \longrightarrow \phi_p^*(B) \longrightarrow (\mathrm{pr}_{p,0})_*\rho_{p,0}^*(B) \longrightarrow (\mathrm{pr}_{p,1})_*\rho_{p,1}^*(B) \longrightarrow \dots \qquad (17)$$

which we now show to be exact. The stalk of $(\mathrm{pr}_{p,q})_*\rho_{p,q}^*(B)$ at $x_0 \leftarrow \dots \leftarrow x_p$ is a colimit over all neighbourhoods U of x_0,

$$\varinjlim_{x_0 \in U} \prod_{U_0 \leftarrow \dots \leftarrow U_q \leftarrow U} \Gamma(U, B) \, .$$

For a fixed U, the complex inside this colimit computes the cohomology of the comma category $U/\mathrm{Emb}_{\mathcal{U}}(G)$ with constant coefficients. Since this comma category is contractible, its cohomology is trivial, and the map

$$\Gamma(U, B) \longrightarrow C^\bullet(U/\mathrm{Emb}_{\mathcal{U}}(G), \Gamma(U, B))$$

is a quasi-isomorphism. Taking the colimit over all neighbourhoods $U \in \mathcal{U}$ of x_0, we obtain a quasi-isomorphism

$$B_{x_0} \longrightarrow ((\mathrm{pr}_{p,\bullet})_*\rho_{p,\bullet}^*(B))_{x_0}$$

This shows that (17) is exact, and completes the proof of the theorem. □

Remark. Let $\mathcal{C}$ be a subcategory of $\mathrm{Emb}_{\mathcal{U}}(G)$. We say that $\mathcal{C}$ *generates* $\mathrm{Emb}_{\mathcal{U}}(G)$ if, for any arrow $\sigma \colon U \to V$ in $\mathrm{Emb}_{\mathcal{U}}(G)$ and any point $x \in U$ there exists an open neighbourhood $U_x \subset U$ of x and an open set $W \subset V$ with $\iota(\upsilon(U_x) \subset W$ such that the restriction $\sigma \colon U_x \to W$ belongs to $\mathcal{C}$.

For example, let $G = \mathrm{Hol}_T(M, \mathcal{F})$ and $\mathcal{U}$ a basis of open sets on T as in Example 4.20, and let $\mathcal{C}$ be the subcategory with the same objects as $\mathrm{Emb}_{\mathcal{U}}(G)$; arrows $\sigma : U \to V$ of $\mathcal{C}$ are arrows which are represented by a family of paths $\{\alpha_x\}$ parametrized by $x \in U$, such that each α_x is a path from x to $f(x) \in V$ inside a leaf of $\mathcal{F}$, and such that the map $f : U \to V$ given in this way is an embedding (see Example 4.20). In other words, arrows in $\mathcal{C}$ are globally represented by paths, where an arrow in $\mathrm{Emb}_{\mathcal{U}}(G)$ is only locally so represented. This $\mathcal{C}$ is a generating subcategory.

The proof of Theorem 4.21 applies to any such generating subcategory, and shows that

$$H^*(G, A) = H^*(\mathcal{C}, \Gamma(A))$$

for any A, as in Theorem 4.21.

4.6 Degree one cohomology and the fundamental group

In the previous section, we have shown that the cohomology of an étale groupoid G can be described in terms of a small category $\mathrm{Emb}_{\mathcal{U}}(G)$. In this section we will relate the covering spaces and the fundamental group of G to the small category $\mathrm{Emb}_{\mathcal{U}}(G)$.

Consider, for the moment, an arbitrary small category $\mathcal{C}$. A presheaf of sets on $\mathcal{C}$ is a contravariant set-valued functor on $\mathcal{C}$. Such a presheaf P is said to be *locally constant* if, for any arrow $f: c \to c'$ in $\mathcal{C}$, the map $P(f): P(c') \to P(c)$ is a bijection. We denote by

$$\mathrm{Lc}(\mathcal{C})$$

the full subcategory of the category of presheaves on $\mathcal{C}$ consisting of all locally constant presheaves.

For a discrete group K, a *(right) principal K-bundle* over $\mathcal{C}$ is a contravariant functor from $\mathcal{C}$ into the category of sets equipped with a right free and transitive action of K. The morphisms between such sets are functions which preserve the action. Since any such function is a bijection, any principal K-bundle over $\mathcal{C}$ is a locally constant presheaf.

Proposition 4.25 *Let G be a connected étale groupoid, and let $\mathcal{U}$ be a basis of simply connected open sets on G_0. The the functor*

$$\Gamma\colon \mathrm{Sh}(G) \longrightarrow \{\textit{presheaves of sets on } \mathrm{Emb}_{\mathcal{U}}(G)\}\,, \tag{18}$$

defined as in Section 4.5, restricts to an equivalence of categories between covering spaces and locally constant presheaves

$$\Gamma\colon \mathrm{Cs}(G) \longrightarrow \mathrm{Lc}(\mathrm{Emb}_{\mathcal{U}}(G))\,. \tag{19}$$

Furthermore, for each discrete group K, this equivalence restricts to an *equivalence between the category of principal K-bundles over G and that of principal K-bundles over* $\mathrm{Emb}_{\mathcal{U}}(G)$.

Proof Consider a G-sheaf E of sets. If the étale space $E \to G_0$ is a covering projection, then clearly for each arrow $\sigma: U \to V$ in $\mathrm{Emb}_{\mathcal{U}}(G)$ the map $\sigma^*: \Gamma(V, E) \to \Gamma(U, E)$ is and isomorphism. So the functor (18) maps covering spaces to locally constant presheaves.

To see that it is an equivalence, observe that there is also the following inverse construction. If P is a locally constant presheaf on $\mathrm{Emb}_{\mathcal{U}}(G)$, one can make a covering space $\hat{P}$ of G_0 as follows: points of $\hat{P}$ are equivalence classes of triples (x, U, p) where $x \in U \in \mathcal{U}$ and $p \in P(U)$. If $x \in U \subset V$ and $i: U \to V$ denotes the inclusion, then the equivalence relation identifies (x, V, p) with $(x, U, P(i)(p))$. For each $U \in \mathcal{U}$ and $p \in P(U)$, there is an obvious map $U \to \hat{P}$ sending $x \in U$ to the equivalence class of (x, U, p), and the topology on $\hat{P}$ is defined by requiring that each of these maps be an open embedding. It is routine to show that this defines a covering space over G_0. The groupoid G acts naturally on the space $\hat{P}$. Indeed, if (y, V, p) represents a point in $\hat{P}$ and if $g: x \to y$ is an arrow in G, we can find a small neighbourhood $U \in \mathcal{U}$ of x and a bisection $\sigma: U \to G_1$ through g such that σ defines an arrow $U \to V$ in $\mathrm{Emb}_{\mathcal{U}}(G)$. Then g acts on the equivalence class of (y, V, p) by mapping it to the equivalence class of $(x, U, P(\sigma)(p))$.

We leave it to the reader to check that this construction is inverse up to isomorphism to Γ, establishing the equivalence (19). It will then be clear that this equivalence maps principal K-bundles to principal K-bundles. □

Principal K-bundles over a small category $\mathcal{C}$ can be described by a 'non-abelian' cohomology in degree one. Explicitly, a *1-cocycle* on $\mathcal{C}$ with values in K is a map $\gamma: \mathcal{C}_1 \to K$ with the property that, for any composable pair $f: c \to c'$, $g: c' \to c''$ in $\mathcal{C}$,

$$\gamma(gf) = \gamma(g)\gamma(f) .$$

In other words, γ is just a functor from $\mathcal{C}$ into the group K, viewed as a category with one object only. Two such 1-cocycles γ and δ are called *cohomologous* it there exists a map $\alpha: \mathcal{C}_0 \to K$ (a *0-cochain*) such that for any arrow $f: c \to c'$ in $\mathcal{C}$

$$\delta(f) = \alpha(c')\gamma(f)\alpha(c)^{-1} .$$

This defines an equivalence relation on 1-cocycles. The set of equivalence classes is denoted by

$$H^1(\mathcal{C}, K)$$

and is the non-abelian cohomology of $\mathcal{C}$ with coefficients in K. If K is abelian then $H^1(\mathcal{C}, K)$ is an abelian group, and the definition agrees with the one of Section 4.5.

Proposition 4.26 *Let $\mathcal{C}$ be a small category and K a discrete group. There is a natural bijective correspondence between $H^1(\mathcal{C}, K)$ and the collection of isomorphism classes of principal K-bundles over $\mathcal{C}$.*

Proof Suppose P is a principal K-bundle over $\mathcal{C}$. Pick for each object c of $\mathcal{C}$ a point $p_c \in P(c)$. Such a point p_c defines an isomorphism of right K-sets $\phi_c \colon K \to P(c)$ by $\phi_c(k) = p_c k$. For any arrow $f : c \to c'$ in $\mathcal{C}$ we thus have a map of right K-sets

$$\phi_c^{-1} \circ P(f) \circ \phi_{c'} : K \longrightarrow K$$

which must be given by a left multiplication with a unique element in K, the inverse of which we denote by $\gamma_P(f)$. Thus the defining equation for $\gamma_P(f)$ is

$$\phi_c(\gamma_P(f)^{-1}k) = P(f)(\phi_{c'}(k)) .$$

It follows that γ_P is a 1-cocycle, because for another arrow $g : c' \to c''$ in $\mathcal{C}$ we have

$$\begin{aligned} \phi_c((\gamma_P(g)\gamma_P(f))^{-1}k) &= \phi_c(\gamma_P(f)^{-1}\gamma_P(g)^{-1}k) \\ &= P(f)\phi_{c'}(\gamma_P(g)^{-1}k) \\ &= P(f)P(g)\phi_{c''}(k) \\ &= P(gf)\phi_{c''}(k) , \end{aligned}$$

so $\gamma_P(g)\gamma_P(f)$ satisfies the defining identity for $\gamma_P(gf)$.

Suppose that $I : P \to Q$ is a map (necessarily an isomorphism) of principal K-bundles over $\mathcal{C}$. Let $\phi_c : K \to P(c)$ and $\psi_c : K \to Q(c)$ be isomorphisms as above. Then for each c, the composition

$$\psi_c^{-1} \circ I_c \circ \phi_c \colon K \longrightarrow K$$

is a map of right K-sets, necessarily given by left multiplication by an element $\alpha_I(c) \in K$. So the defining equation for $\alpha_I(c)$ is

$$\psi_c(\alpha_I(c)k) = I_c(\phi_c(k)) .$$

Then

$$\begin{aligned}\psi_c(\alpha_I(c)\gamma_P(f)^{-1}\alpha_I(c')^{-1}k) &= I_c\phi_c(\gamma_P(f)^{-1}\alpha_I(c')^{-1}k)\\ &= I_cP(f)\phi_{c'}(\alpha_I(c')^{-1}k)\\ &= P(f)I_{c'}\phi_{c'}(\alpha_I(c')^{-1}k)\\ &= P(f)\psi_{c'}(\alpha_I(c')\alpha_I(c')^{-1}k)\\ &= P(f)\psi_{c'}(k)\,,\end{aligned}$$

so $\alpha_I(c)\gamma_P(f)^{-1}\alpha_I(c')^{-1}$ satisfies the defining equation for $\gamma_Q(f)$. In particular, γ_Q and γ_P are cohomologous. This shows that the construction $P \to [\gamma_P]$, from principal K-bundles over $\mathcal{C}$ to $H^1(\mathcal{C}, K)$, is well defined on isomorphism classes, and in particular does not depend on the choice of the isomorphisms ϕ_c.

In the reverse direction, let $[\gamma] \in H^1(\mathcal{C}, K)$ be the cohomology class of a cocycle γ. Define a presheaf $P = P(\gamma)$ by $P(c) = K$ for any object c of $\mathcal{C}$, and by

$$P(f)(k) = \gamma(f)^{-1}k$$

for any arrow $f\colon c \to c'$ in $\mathcal{C}$. Then P is a functor because γ is a cocycle. So P is a principal K-bundle over $\mathcal{C}$.

We leave to the reader the easy verification that these two constructions are inverse to each other. □

Corollary 4.27 *Let G be a connected étale groupoid with a base-point $x_0 \in G_0$, and let $\mathcal{U}$ be a basis of simply connected open sets on G_0.*

(i) For any discrete group K, there is a natural isomorphism

$$H^1(\mathrm{Emb}_{\mathcal{U}}(G), K) \cong [\pi_1(G, x_0), K]\,,$$

where the right-hand side denotes the set of conjugacy classes of homomorphisms from $\pi_1(G, x_0)$ to K.

(ii) For each abelian group A there is a natural isomorphism

$$H^1(G, A) \cong \mathrm{Hom}(\pi_1(G, x_0), A)\,.$$

Proof (i) follows form the previous proposition and Corollary 3.19. If we choose the basis $\mathcal{U}$ to consists of contractible open sets, then $H^1(\mathrm{Emb}_{\mathcal{U}}(G), A) = H^1(G, A)$ by Corollary 4.22. Given this, (ii) follows from (i). □

5
Compactly supported cohomology

In this chapter we will discuss a cohomology with compact supports for étale groupoids, first introduced in [14], and further developed in [16, 45]. Our presentation is based on these sources. This compactly supported theory will again be invariant under Morita equivalence, hence is well defined for the wider class of foliation groupoids.

Given the fact that many of the étale groupoids arising in the context of foliations are non-Hausdorff, the notion of 'compact support' needs to be applied in the context of non-Hausdorff manifolds. The appropriate definition for which the usual properties known in the Hausdorff case extend to this wider context is somewhat subtle, and we have decided to devote the first section of this chapter to a detailed discussion of this matter. Subsequently, we introduce the cohomology theory with compact supports, and develop its main general properties. In particular, we discuss the covariant operation $\phi_!$, we derive a Hochschild-Serre type spectral sequence, and we prove the Morita invariance already referred to above. The theory is in some sense dual to the cohomology theory developed in the previous chapter. We will make this more precise by proving that the cohomology groups with compact supports are isomorphic to suitable homology groups of the embedding category. This result is parallel to the result for cohomology proved in Chapter 4, and leads to an easy proof of Poincaré duality for étale groupoids.

The compactly supported cohomology discussed here is related by natural maps to the basic cohomology with supports [25], and to foliated cohomology of [27, 49]. By the same maps, the Poincaré duality discussed here can be compared to the one of [56]. We refer to [14, 16] where some of these relations are made explicit.

5.1 Compact supports for sheaves over non-Hausdorff manifolds

In the previous chapter, we have used basic facts for sheaves on topological spaces, as available from any of the standard sources (e.g. [32, 4]). In this chapter, we will be concerned with compact supports in the context of étale groupoids such as holonomy groupoids of foliations. Here one is faced with the following difficulty: On the one hand, in order to develop the theory of compactly supported cohomology, all the standard sources for classical sheaf theory restrict attention to paracompact Hausdorff spaces of finite cohomological dimension. On the other hand, the manifolds occurring in the context of foliations and their holonomy groupoids are usually non-Hausdorff. For this reason, we begin this chapter with a preliminary section, where we show how the usual operations involving compact supports (Γ_c, $f_!$, ...) can be extended in an essentially unique way to non-Hausdorff manifolds.

Readers familiar with these operations in the Hausdorff case, who are prepared to believe that such an extension exists, may prefer to skip this section on a first reading of this chapter.

So, to fix the scope of this section, we assume that any space X has an open cover by subsets $U \subset X$ each of which are paracompact, Hausdorff, locally compact, and of cohomological dimension bounded by a number d (depending on X but not on U).

Let X be a space satisfying these general assumptions. An abelian sheaf A on X is said to be *c-soft* if for any Hausdorff open $U \subset X$ its restriction $A|_U$ is a c-soft sheaf on U in the usual sense [32]. By the same property for Hausdorff spaces, it follows that c-softness is a local property; i.e. a sheaf A is c-soft if and only if there is an open cover $X = \bigcup U_i$ such that each $A|_U$ is a c-soft sheaf on A.

Let A be a c-soft sheaf on X and let $\mathcal{G}(A)$ be its Godement resolution (see Example 4.8 (2)). For any Hausdorff open set $W \subset X$, let $\Gamma_c(W, A)$ be the usual set of compactly supported sections. If $W \subset U$ and U is non-Hausdorff, the usual 'extension by zero' may map an element of $\Gamma_c(W)$ to a discontinuous section over U. (For example, consider the non-Hausdorff manifold $U = \mathbb{R} \cup_{\mathbb{R}-\{0\}} \mathbb{R}$, the line with one double point

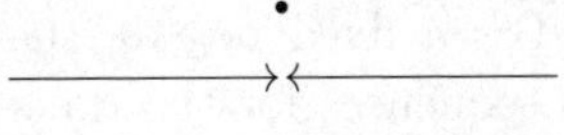

and let W be the upper copy of $\mathbb{R}$. If $\sigma \in \Gamma_c(W)$ is a section with

$\sigma(0) \neq 0$, then its extension to U is discontinuous.) Thus, extension by zero is a map $\Gamma_c(W, A) \to \Gamma_c(U, \mathcal{G}(A))$. For any (not necessarily Hausdorff) open set $U \subset X$, we define $\Gamma_c(U, A))$ to be the image of the map:

$$\bigoplus_W \Gamma_c(W, A) \longrightarrow \Gamma(U, \mathcal{G}(A)),$$

where W ranges over all Hausdorff open subsets $W \subset U$. An alternative definition follows by choosing a Hausdorff open cover in Proposition 5.2 below.

Observe that $\Gamma_c(U, A)$ so defined is evidently functorial in A, and that for any inclusion $U \subset U'$ we have an 'extension by zero' monomorphism $\Gamma_c(U, A) \to \Gamma_c(U', A)$.

The following lemma shows that in the definition of $\Gamma_c(U, A)$, it is enough to let W range over a Hausdorff open cover of U; in particular, it shows that the definition agrees with the usual one if U itself is Hausdorff.

Lemma 5.1 *Let A be a c-soft sheaf on X. For any open cover $U = \bigcup W_i$, where each W_i is Hausdorff, the sequence $\bigoplus_i \Gamma_c(W_i, A) \to \Gamma_c(U, A) \to 0$ is exact.*

Proof It suffices to show that for any Hausdorff open $W \subset U$, the map $\bigoplus_i \Gamma_c(W \cap W_i, A) \to \Gamma_c(W, A)$ is surjective. This is standard ([19, 32]). □

Proposition 5.2 (Mayer-Vietoris sequence) *Let $X = \bigcup_i U_i$ be an open cover indexed by an ordered set I, and let A be a c-soft sheaf on X. Then there is a long exact sequence*

$$\ldots \longrightarrow \bigoplus_{i_0 < i_1} \Gamma_c(U_{i_0 i_1}, A) \longrightarrow \bigoplus_{i_0} \Gamma_c(U_{i_0}, A) \longrightarrow \Gamma_c(X, A) \longrightarrow 0 \quad (1)$$

Here $U_{i_0 \ldots i_n} = U_{i_0} \cap \ldots \cap U_{i_n}$, as usual. (There is of course a similar exact sequence if I is not ordered.)

Proof The proposition is of course well known in the case where X is a paracompact Hausdorff space ([5]). We first reduce the proof to the case where each of the U_i is Hausdorff, as follows. Let $X = \bigcup_{j \in J} W_j$ be a cover by Hausdorff open sets, and consider the double complex $C_{p,q} = \bigoplus \Gamma_c(W_{j_0 \ldots j_p} \cap U_{i_0 \ldots i_q}, A)$, where the sum is over all $j_0 < \ldots < j_p$ and $i_0 < \ldots < i_q$. For a fixed $p \geq 0$, the column $C_{p,\bullet}$ is a sum of exact Mayer-Vietoris sequences for the Hausdorff open sets $W_{j_0 \ldots j_p}$, augmented by $C_{p,-1} = \bigoplus_{j_0 < \ldots < j_p} \Gamma_c(W_{j_0 \ldots j_p}, A)$. Keeping the notation $U_{i_0 \ldots i_q} = X = W_{j_0 \ldots j_p}$ if $q = -1 = p$, we observe that for a fixed $q \geq -1$, the row $C_{\bullet,q}$ is a sum of Mayer-Vietoris sequences for the spaces $U_{i_0..i_q}$ with respect to the open covers $\{W_j \cap U_{i_0..i_q}\}$. So, if the proposition would hold for covers by Hausdorff sets, each row $C_{\bullet,q}$ ($q \geq -1$) is also exact. By a standard double complex argument it follows that the augmentation column $C_{-1,\bullet}$ is also exact, and this column is precisely the sequence in the statement of the proposition. This shows that it suffices to prove the proposition in the special case where each U_i is Hausdorff.

So assume each $U_i \subset X$ is Hausdorff. Observe first that exactness of the sequence (1) at $\Gamma_c(X, A)$ now follows by Lemma 5.1. To show exactness elsewhere, consider for each finite subset $I_0 \subset I$ the space $U^{I_0} = \bigcup_{i \in I_0} U_i$ and the following subsequence:

$$\dots \longrightarrow \bigoplus_{i_0, i_1 \in I_0,\ i_0 < i_1} \Gamma_c(U_{i_0 i_1}, A) \longrightarrow \bigoplus_{i_0 \in I_0} \Gamma_c(U_{i_0}, A) \longrightarrow \Gamma_c(U^{I_0}, A) \longrightarrow 0 \tag{2}$$

of (1). Clearly (1) is the directed union of the sequences of the form (2), where $I_0 \subset I$ ranges over all finite subsets of I. So exactness of (1) follows from exactness of each such sequence of the form (2). Thus, it remains to prove the proposition in the special case of a *finite* cover $\{U_i\}$ of X by Hausdorff open sets.

So assume $X = U_1 \cup ... \cup U_n$ where each U_i is Hausdorff. For $n = 1$, there is nothing to prove. For $n = 2$, the sequence has the form

$$0 \longrightarrow \Gamma_c(U_1 \cap U_2, A) \longrightarrow \Gamma_c(U_1, A) \oplus \Gamma_c(U_2, A) \longrightarrow \Gamma_c(U_1 \cap U_2, A) \longrightarrow 0\,.$$

This sequence is exact at $\Gamma_c(X, A)$ by Lemma 5.1, and evidently exact at other places. Exactness for $n = 3$ can be proved using exactness for $n = 2$. Indeed, consider the large diagram below, whose upper two rows are the sequences for $n = 2, 3$ and whose third row is constructed by taking vertical cokernels, so that all columns are exact (we delete the sheaf A from the notation) (compare to [3, p. 187]). To show that the middle row is exact, it thus suffices to prove that the lower row is exact. This row can be decomposed into a Mayer-Vietoris sequence for the case $n = 2$, already shown to be exact,

$$0 \longrightarrow \Gamma_c(U_{123}) \longrightarrow \Gamma_c(U_{13}) \oplus \Gamma_c(U_{23}) \longrightarrow \Gamma_c(U_3 \cap (U_1 \cup U_2)) \longrightarrow 0$$

and the sequence $0 \to \Gamma_c(U_3 \cap (U_1 \cup U_2)) \to \Gamma_c(U_3) \to C \to 0$. The exactness of the latter sequence is easily proved by a diagram chase, using exactness of the right-hand column.

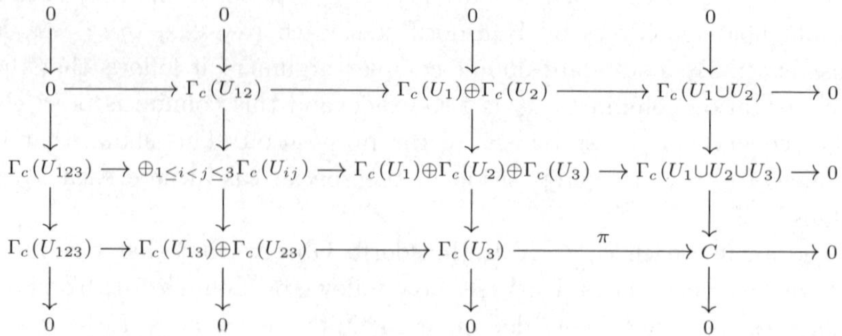

An identical argument will show that the exactness for a cover by $n+1$ opens follows from exactness for one by n opens, so the proof is completed by induction. □

Remarks 5.3 Proposition 5.2 is our main tool for transferring standard facts from sheaf theory on Hausdorff spaces to the non-Hausdorff case, as illustrated by the following consequences:

(i) Let $Y \subset X$ be a closed subspace, and let A be a c-soft sheaf on X. There is an exact sequence

$$0 \to \Gamma_c(X - Y, A) \xrightarrow{i} \Gamma_c(X, A) \xrightarrow{r} \Gamma_c(Y, A) \to 0$$

(i is extension by zero, r is the restriction). This (including the fact that the map r is well defined) follows by elementary homological algebra from the fact that this holds for Hausdorff spaces, by using Proposition 5.2 for a cover of X by Hausdorff open sets U_i, and for the induced covers of Y by $\{U_i \cap Y\}$ and $X - Y$ by $\{U_i - Y\}$.

(ii) For a family A_i of c-soft sheaves on X the direct sum $\oplus A_i$ is again c-soft, and $\Gamma_c(X, \oplus A_i) \cong \oplus \Gamma_c(X, A_i)$. In particular, when working over $\mathbb{R}$, we have for any c-soft sheaf S of $\mathbb{R}$-vector spaces and any vector space V that the tensor product $S \otimes_{\mathbb{R}} V$ (here V is the constant sheaf) is again c-soft, and the familiar formula $\Gamma_c(X, S \otimes_{\mathbb{R}} V) \cong \Gamma_c(X, S) \otimes_{\mathbb{R}} V$.

(iii) Let $A \to B$ be a quasi-isomorphism between chain complexes of c-soft sheaves on X. Then $\Gamma_c(X, A) \to \Gamma_c(X, B)$ is again a quasi-isomorphism. (By a 'mapping cone argument' [60, p. 19] we may assume that $B = 0$. In other words, we have to show that $\Gamma_c(X, A)$ is acyclic whenever A is. This follows from the Mayer-Vietoris sequence of Proposition 5.2 together with the Hausdorff case. (We remark that it is necessary to assume that the chain complexes are bounded below if X does not have finite cohomological dimension.)

Next, we consider the operation $f_!$ for sheaves of R-modules ([32]) in the non-Hausdorff case:

Proposition 5.4 *Let $f : Y \to X$ be a continuous map. There is a functor $f_!$ from c-soft sheaves on Y to c-soft sheaves on X with the following properties, for any c-soft sheaf B on Y:*

(i) *For any open $U \subset X$ there is a natural isomorphism*

$$\Gamma_c(U, f_! B) \cong \Gamma_c(f^{-1}(U), B) .$$

(ii) *For any point $x \in X$ we have $f_!(B)_x = \Gamma_c(f^{-1}(x), B)$.*

(iii) *For any fibered product*

$$\begin{array}{ccc} Z \times_X Y & \xrightarrow{p} & Y \\ {\scriptstyle q}\downarrow & & \downarrow{\scriptstyle f} \\ Z & \xrightarrow{e} & X \end{array}$$

along an étale map e there is a natural isomorphism $q_! p^* B \cong e^* f_! B$. (see Proposition 5.5 below for the case where e is not étale).

Proof We first construct $f_!$ with the stated properties in the special case where X is Hausdorff. For general X, the construction of $f_!$ is then obtained by gluing the construction over a cover of X by Hausdorff open sets.

So assume that X is Hausdorff. Consider a c-soft sheaf B on Y. For any open set $V \subset Y$, denote by B_V the sheaf on Y obtained by extending $B|_V$ by zero. Thus B_V is evidently c-soft, and $\Gamma_c(Y, B_V) = \Gamma_c(V, B)$. Moreover, an inclusion $V \subset W$ induces an evident map $B_V \hookrightarrow B_W$.

Now let $Y = \bigcup W_i$ be a cover by Hausdorff open sets. This cover induces a long exact sequence:

$$\cdots \longrightarrow \bigoplus_{i_0 < i_1} B_{W_{i_0 i_1}} \longrightarrow \bigoplus_{i_0} B_{W_{i_0}} \longrightarrow B \longrightarrow 0$$

of c-soft sheaves on Y. By the Remark 5.3 (iii), the functor $\Gamma_c(Y, -)$ applied to this long exact sequence again yields an exact sequence, and this is precisely the Mayer-Vietoris sequence of Proposition 5.2. For each $i_0, \ldots, i_n$ let $f_{i_0,\ldots,i_n} : W_{i_0,\ldots,i_n} \to X$ be the restriction of f; this is a map between Hausdorff spaces, so we have $(f_{i_0,\ldots,i_n})_!(B|_{W_{i_0,\ldots,i_n}})$ defined as usual. Define $f_!(B)$ as the cokernel fitting into a long exact sequence:

$$\cdots \longrightarrow \bigoplus_{i_0 < i_1} (f_{i_0 i_1})_!(B|_{W_{i_0 i_1}}) \longrightarrow \bigoplus_{i_0} (f_{i_0})_!(B|_{W_{i_0}}) \longrightarrow f_!(B) \longrightarrow 0 . \tag{3}$$

Note that $f_!(B)$ is c-soft because, by the Hausdorff case, each of the

$$(f_{i_0,\ldots,i_n})_!(B|_{W_{i_0,\ldots,i_n}})$$

is. Property (i) now easily follows from the Hausdorff case by Mayer-Vietoris arguments. To prove (ii), note that for $x \in X$, we have

$$(f_{i_0})_!(B_{W_{i_0}})_x = \Gamma_c(f^{-1}(x) \cap W_{i_0}; B)$$

by the Hausdorff case. So taking stalks of the long exact sequence in (3) at x and using the Mayer-Vietoris sequence of Proposition 5.2 for the space $f^{-1}(x)$, we find $f_!(B)_x = \Gamma_c(f^{-1}(x), B)$ as required. Finally, property (iii) is clear from the local nature of the construction. □

For an arbitrary (not-necessarily c-soft) sheaf A on Y, we define $f_!(A)$ as a complex of c-soft sheaves on X, by first taking the resolution $0 \to A \to S^0 \to S^1 \to \ldots$ of A of c-soft sheaves on Y, and then define $f_!(A)$ to be the complex $f_!(S^0) \to f_!(S^1) \to \ldots$. Notice that, in case A is itself already c-soft, by Proposition 5.4 (ii) and Remark 5.3 (iii), this complex is quasi-isomorphic to the single sheaf $f_!(A)$ viewed as a complex concentrated in degree 0.

More generally, we define $f_!$ as a functor $D(Y) \to D(X)$ at the level of derived categories, by first taking a c-soft resolution of a given complex and then applying the functor of Proposition 5.4 to each sheaf separately. This is well-defined by Remark 5.3 (iii) and Proposition 5.4 (ii).

Proposition 5.5 (Change of base) *For any pull-back diagram*

$$\begin{array}{ccc} Z \times_X Y & \xrightarrow{p} & Y \\ {\scriptstyle q}\downarrow & & \downarrow{\scriptstyle f} \\ Z & \xrightarrow{e} & X \end{array}$$

and any sheaf B on Y, there is a canonical quasi-isomorphism $q_!p^*B \simeq e^*f_!B$.

Proof Using Mayer-Vietoris for covers of X and Z by Hausdorff open sets, it suffices to consider the case where X and Z are both Hausdorff. Clearly it also suffices to prove the lemma in the special case where B is c-soft.

Let $Y = \bigcup W_i$ as in the remark above, so that $f_!(B)$ fits into a long exact sequence (3) of c-soft sheaves on X. Applying the exact functor e^* to this sequence and using the proposition in the Hausdorff case, one obtains a long exact sequence of the form:

$$\cdots \longrightarrow \bigoplus_{i_0<i_1} q_! p^*(B_{W_{i_0 i_1}}) \longrightarrow \bigoplus_{i_0} q_! p^*(B_{W_{i_0}}) \to e^* f_!(B) \longrightarrow 0\,. \qquad (4)$$

Now let $p^*(B) \to S^\bullet$ be a c-soft resolution over the pull-back $Z \times_X Y$. Then for any open $U \subset Y$, $S^\bullet_{p^{-1}(U)}$ is a c-soft resolution of $p^*(B_U)$, so $q_!(S^\bullet_{p^{-1}(U)})$ is a c-soft resolution of $q_! p^*(B)$. The lemma now follows by comparing the sequence (4) to the defining sequence

$$\cdots \longrightarrow \bigoplus_{i_0<i_1} q_!(S^\bullet_{p^{-1}W_{i_0 i_1}}) \longrightarrow \bigoplus_{i_0} q_!(S^\bullet_{p^{-1}W_{i_0}}) \longrightarrow q_!(S^\bullet) \longrightarrow 0$$

where $q_!(p^*(B)) = q_!(S^\bullet)$ by definition. □

Example 5.6 ($f_!$ on étale maps) Let $f\colon Y \to X$ be an étale map, i.e. a local homeomorphism. The pull-back functor $f^*: \mathrm{Sh}_R(X) \to \mathrm{Sh}_R(Y)$ has an exact left-adjoint $f_!\colon \mathrm{Sh}_R(Y) \to \mathrm{Sh}_R(X)$, described on the stalks by $f_!(B)_x = \bigoplus_{y \in f^{-1}(x)} B_y$. This construction agrees with the one described above. In particular, for étale f, the counit of the adjunction defines a map $\Sigma_f : f_! f^*(B) \to B$, 'summation along the fiber', for any sheaf B on X.

Example 5.7 ($f_!$ on proper maps) Define a map $f : Y \to X$ between (non-necessarily Hausdorff) spaces to be *proper* if

(i) the diagonal $Y \to Y \times_X Y$ is closed, and
(ii) for any Hausdorff open $U \subset X$ and any compact $K \subset U$, the set $f^{-1}(K)$ is compact.

It is easy to see that if f is proper then $f_! = f_*$, as in the Hausdorff case. Furthermore, for any c-soft sheaf A on X, there is a natural map $\Gamma_c(X, A) \to \Gamma_c(Y, f^*A)$ defined by pull-back, as in the Hausdorff case.

5.2 Compactly supported cohomology of étale groupoids

Throughout this chapter we shall work with sheaves over $\mathbb{R}$. (But let us remark that the arguments also apply to sheaves of R-modules for any ring R of finite cohomological dimension, provided all spaces are assumed to have finite cohomological dimension with respect to R.) For an étale groupoid G and a G-sheaf A (of R-modules), we will define in this section cohomology groups

$$H^i_c(G, A)\ , \qquad i \in \mathbb{Z}\ ,$$

to be referred to as the cohomology with compact supports of G with coefficients in the sheaf A. It will be proved that these cohomology groups are invariant under Morita equivalence. As a consequence, there are also well defined groups $H^i_c(G, A)$ for any foliation groupoid G; in particular, the theory applies to the holonomy and monodromy groupoids of foliations. But, as for cohomology in Chapter 4, we will not attempt to develop the theory for more general Lie groupoids.

So, consider an étale groupoid G and a G-sheaf A. Recall that from G we have constructed a simplicial space, the nerve of G (Section 4.2),

$$G_0 \leftleftarrows G_1 \ \overset{\leftarrow}{\underset{\leftarrow}{\leftarrow}}\ G_2 \ \cdots$$

in which each of the face maps $d_i : G_n \to G_{n-1}$, $i = 1, \ldots, n$, is étale. Recall also that each G-sheaf A induces sheaves $A_{(n)}$ on G_n,

$$A_{(n)} = \lambda_n^*(A)\ ,$$

where λ_n maps a string $(g_1, \ldots, g_n)$ to $s(g_n)$. Each face $d_i \colon G_n \to G_{n-1}$ induces a map

$$\Sigma_{d_i} : \Gamma_c(G_n, A_{(n)}) \longrightarrow \Gamma_c(G_{n-1}, A_{(n-1)}) \tag{5}$$

defined by the condition that it makes the following diagram commutative.

$$\begin{array}{ccc} \Gamma_c(G_n, A_{(n)}) & \xrightarrow{\ \cong\ } & \Gamma_c(G_n, d_i^* A_{(n-1)}) \\ {\scriptstyle \Sigma_{d_i}}\downarrow & & \downarrow{\scriptstyle \cong} \\ \Gamma_c(G_{n-1}, A_{(n-1)}) & \xleftarrow{\ \Sigma_{d_i}\ } & \Gamma_c(G_{n-1}, (d_i)_! d_i^* A_{(n-1)}) \end{array}$$

Here the top horizontal map comes from the canonical isomorphism $A_{(n)} \cong d_i^*(A_{(n-1)})$ (cf. (5)), the right hand vertical isomorphism is the one of Proposition 5.4 (i), and the bottom horizontal map is summation along the fiber (Example 5.6). Thus, modulo canonical isomorphisms, the left hand vertical map is summation along the fibers of d_i, and therefore we denote this map again by Σ_{d_i}, as in (5).

In this way, one obtains a simplicial abelian group

$$\Gamma_c(G_0, A_{(0)}) \leftleftarrows \Gamma_c(G_1, A_{(1)}) \ \overset{\leftarrow}{\underset{\leftarrow}{\leftarrow}}\ \Gamma_c(G_2, A_{(2)}) \ \cdots \tag{6}$$

and hence, by taking alternating sums of the Σ_{d_i} in the usual way, a chain complex $\Gamma_c(G_\bullet, A_{(\bullet)})$. We will view this as a bounded above cochain complex concentrated in negative degrees.

For a complex S of G-sheaves, this construction yields a double complex

$$\Gamma_c(G_{-p}, S^q_{(-p)})$$

concentrated in degrees $p \leq 0$. If S is bounded above, then so are the rows and columns of this double complex, as well as its total complex given in degree n by

$$\bigoplus_{q-p=n} \Gamma_c(G_p, S^q_{(p)}) \,. \tag{7}$$

Let us call a G-sheaf S c-soft if it is c-soft as a sheaf on G_0. Observe that, since G is étale, this implies that each induced sheaf $S_{(n)}$ on G_n is again c-soft. If A is an arbitrary G-sheaf (over $\mathbb{R}$), there always exists a bounded above resolution of A by c-soft G-sheaves; indeed if $d = \dim(G)$, there exists such a resolution

$$0 \longrightarrow A \longrightarrow S^0 \longrightarrow S^1 \longrightarrow \ldots \longrightarrow S^d \longrightarrow 0 \,.$$

For example, one can take the functorial resolution by differential forms

$$0 \longrightarrow A \longrightarrow A \otimes \Omega^0 \longrightarrow A \otimes \Omega^1 \longrightarrow \ldots \longrightarrow A \otimes \Omega^d \longrightarrow 0 \,;$$

or one can take the Godement resolution or an injective resolution, and truncate it at degree d (cf. [4, p. 116]).

Definition 5.8 For an étale groupoid G and a G-sheaf A (over $\mathbb{R}$), the groups $H^i_c(G, A)$ are the cohomology groups of the total complex (7), for any bounded above resolution S of A by c-soft G-sheaves.

Remarks 5.9 (1) The groups $H^i_c(G, A)$ are well defined, and do not depend on the resolution S. Indeed, it is easy to see that any two resolutions map into a third; moreover, if $S \to T$ is a quasi-isomorphism between bounded above complexes of c-soft sheaves, then by Remark 5.3 (iii), the map of double complexes

$$\Gamma_c(G_{-p}, S^q_{(-p)}) \longrightarrow \Gamma_c(G_{-p}, T^q_{(-p)})$$

is a quasi-isomorphism for each fixed p, and hence [60, pp. 59–60] induces a quasi-isomorphism of total complexes.

(2) Note that for any étale groupoid G of dimension d, the groups $H^i_c(G, A)$ are concentrated in degrees $i \leq d$.

(3) The dual groups

$$H_i(G, A) = H_c^{-i}(G, A)^\vee$$

could be called the Borel-Moore homology groups of G, analogously to the case of topological spaces [32, p. 374].

(4) The cohomology groups $H_c^*(G, A)$ depend covariantly on A in the evident way, and a short exact sequence

$$0 \longrightarrow A \longrightarrow B \longrightarrow C \longrightarrow 0$$

of G-sheaves induces a long exact sequence in compactly supported cohomology,

$$\ldots \longrightarrow H_c^n(G, A) \longrightarrow H_c^n(G, B) \longrightarrow H_c^n(G, C) \longrightarrow H_c^{n+1}(G, A) \longrightarrow \ldots$$

as usual.

Examples 5.10 As first examples, we mention some extreme cases.

(1) Consider a manifold M and a sheaf A on M. We can view M as an étale groupoid, the unit groupoid $u(M)$ (which we also denote simply by M), and A as a $u(M)$-sheaf. Then $H_c^*(u(M), A)$ is the usual cohomology with compact supports $H_c^*(M, A)$ of the manifold M with coefficients in the sheaf A.

(2) Consider a discrete group Γ, and a right $\mathbb{R}[\Gamma]$-module A. We can view Γ as an étale groupoid of dimension 0, and A as a Γ-sheaf. Then $H_c^n(\Gamma, A)$ is the reindexed group homology $H_{-n}(\Gamma, A)$.

(3) Suppose Γ is a discrete group acting (from the right, say) on a manifold M, and suppose A is an equivariant sheaf on M. Then the action groupoid $M \rtimes \Gamma$ is an étale groupoid and A is an $M \rtimes \Gamma$-sheaf. For a bounded c-soft resolution S of A, the double complex computing $H_c^*(M \rtimes \Gamma, A)$ takes the form

$$\bigoplus_{(\gamma_1, \ldots, \gamma_p)} \Gamma_c(M, S^q)$$

in bidegree $(-p, q)$, and we find a spectral sequence

$$H_p(\Gamma, H_c^q(M, A)) \Rightarrow H_c^{q-p}(M \rtimes \Gamma, A)\ .$$

In other words, the compactly supported cohomology of the groupoid $M \rtimes \Gamma$ is some mixture of compactly supported cohomology of the manifold M and the group homology of Γ.

Next, we consider hypercohomology with compact supports. Let G be an étale groupoid, and let A be a bounded above complex of G-sheaves.

Then there is quasi-isomorphism (a resolution) $A \to S$ into a bounded above complex of c-soft sheaves, and one can define the hypercohomology groups with compact supports $H_c^i(G, A)$ in the same way, as the cohomology of the total complex $\Gamma_c(G_{-p}, S^q_{(-p)})$. By the same argument as in Remark 5.9 (1), this does not depend on the choice of a resolution S of A.

Proposition 5.11 (Hypercohomology spectral sequence) *Let A be a bounded above cochain complex of G-sheaves, as above, and consider for each $q \in \mathbb{Z}$ the cohomology sheaf $H^q(A)$. There is a natural spectral sequence*

$$E_2^{p,q} = H_c^p(G, H^q(A)) \Rightarrow H_c^{q+p}(G, A) .$$

Proof Consider the resolution $0 \to A \to A \otimes \Omega^0 \to \ldots \to A \otimes \Omega^d \to 0$ by differential forms, and write C for the triple cochain complex

$$C^{p,q,r} = \Gamma_c(G_{-p}, A^q \otimes \Omega^r)$$

(we omit the lower index on $(A^q \otimes \Omega^r)_{(-p)} = A^q_{(-p)} \otimes \Omega^r$ from the notation). Let D be the double complex

$$D^{n,q} = \bigoplus_{p+r=q} C^{p,q,r} .$$

The total complex of C, and hence also that of D, computes the hypercohomology $H_c^*(G, A)$. The complex C is bounded in r and bounded above in p and q. Thus there is a spectral sequence of double complexes

$$II^n H^q(D) \Rightarrow H_c^{n+q}(G, A) .$$

But, for fixed p and r, one has

$$H^q(C^{p,\bullet,r}) = \Gamma_c(G_{-p}, H^q(A) \otimes \Omega^r) . \qquad (8)$$

(Indeed, for the G-sheaves $Z^q(A)$, $B^q(A)$ and $H^q(A)$ of cycles, boundaries and cohomology of A, the tensor products $Z^q(A) \otimes \Omega^r$, $B^q(A) \otimes \Omega^r$ and $H^q(A) \otimes \Omega^r$ are all c-soft. From this and exactness of - $\otimes\, \Omega^r$ the isomorphism (8) follows.) Thus

$$H^q(D^{n,\bullet}) = \bigoplus_{p+r=n} \Gamma_c(G_{-p}, H^q(A) \otimes \Omega^r) ,$$

and hence $H^n H^q(D) = H_c^n(G, H^q(A))$. □

5.3 The operation $\phi_!$

In this section we will construct for any homomorphism $\phi : K \to G$ between étale groupoids an operation $\phi_!$, from bounded above complexes of c-soft K-sheaves to such complexes of G-sheaves. In the next section, we will then derive a Leray (-Hochschild-Serre) spectral sequence for the map ϕ, and use it to prove the Morita invariance of compactly supported cohomology.

So, let G and K be étale groupoids, and let $\phi : K \to G$ be a homomorphism. Let G_0/ϕ be the comma category: its objects are pairs $(y, g : x \to \phi(y))$ with $y \in K_0$ and $g \in G_1$, and its arrows from $(y, g : x \to \phi(y))$ to $(y', g' : x' \to \phi(y'))$ are arrows $h : y \to y'$ in H such that $\phi(h)g = g'$ (so there are such arrows only if $x = x'$). Note that G_0/ϕ is again an étale groupoid (cf. Section 2.4). It has the obvious projection homomorphisms

$$K \xleftarrow{\tilde{\phi}} G_0/\phi \xrightarrow{\pi} G_0 \,. \tag{9}$$

(We view π as a homomorphism into the manifold G_0 viewed as the unit groupoid.)

The nerve of G_0/ϕ is a simplicial manifold

$$(G_0/\phi)_0 \leftleftarrows (G_0/\phi)_1 \,\substack{\leftarrow\\[-1ex]\leftarrow\\[-1ex]\leftarrow}\, (G_0/\phi)_2 \;\cdots \tag{10}$$

and the functor π in (9) induces projections

$$\pi_n : (G_0/\phi)_n \longrightarrow G_0 \tag{11}$$

compatible with the simplicial structure. The fiber of the functor $\pi : G_0/\phi \to G_0$ in (9) over a point $x \in G_0$ is the familiar comma groupoid x/ϕ. If $g' : x' \to x$ is an arrow in G, then precomposition with g' induces an obvious homomorphism $x/\phi \to x'/\phi$. Thus, the groupoid G acts from the right on the comma groupoid G_0/ϕ along the map π (cf. (9)). Consequently, each of the spaces $(G_0/\phi)_n$ carries a right G-action along π_n (cf. (11)), and these actions are compatible with the simplicial structure. In other words, Nerve(G_0/ϕ) is a simplicial G-space. Explicitly, a point of $(G_0/\phi)_n$ is a string

$$(y_0 \xleftarrow{k_1} y_1 \xleftarrow{k_1} \cdots \xleftarrow{k_n} y_n, \phi(y_n) \xleftarrow{g} x) \tag{12}$$

and G acts by precomposing on the last coordinate,

$$(k_1, \ldots, k_n, g)g' = (k_1, \ldots, k_n, gg') \,.$$

Now suppose S is a c-soft K-sheaf. Then S pulls back along the projection $\tilde{\phi} : G_0/\phi \to K$ to a sheaf $\tilde{\phi}^*(S)$ on G_0/ϕ, which induces sheaves

$$\tilde{\phi}^*(S)_{(n)}$$

on $(G_0/\phi)_{(n)}$ as before: the stalk of $\tilde{\phi}^*(S)_{(n)}$ at a point of the form (12) is S_{y_n}. Observe that $\tilde{\phi}^*(S)$ and each of the $\tilde{\phi}^*(S)_{(n)}$ are c-soft because $\tilde{\phi}\colon G_0/\phi \to K$ is étale.

Using the map π_n of (11), we now obtain a c-soft sheaf

$$(\pi_n)_!(\tilde{\phi}^*(S)_{(n)}) \tag{13}$$

on G_0, the stalk of which over a point $x \in G_0$ is given by the formula (cf. Proposition 5.4 (ii))

$$(\pi_n)_!(\tilde{\phi}^*(S)_{(n)})_x = \Gamma_c((x/\phi)_n, \tilde{\phi}^*(S)) \tag{14}$$

(where on the right we have denoted the restriction of $\tilde{\phi}^*(S)$ to x/ϕ again by $\tilde{\phi}^*(S)$). Since $(G_0/\phi)_n$ is a G-space, this sheaf $(\pi_n)_!(\tilde{\phi}^*(S)_{(n)})$ on G_0 is in fact a G-sheaf. At the level of stalks (14), the action by an arrow $g' : x' \to x$ is the map

$$\Gamma_c((x/\phi)_n, \tilde{\phi}^*(S)) \longrightarrow \Gamma_c((x'/\phi)_n, \tilde{\phi}^*(S))$$

induced by the diffeomorphism $(x/\phi)_n \to (x'/\phi)$ given by composition with g'. Also, since the $(G_0/\phi)_n$ form a simplicial G-space, these G-sheaves $(\pi_n)_!(\tilde{\phi}^*(S)_{(n)})$ in fact form a simplicial G-sheaf: its stalk at $x \in G_0$ is exactly the simplicial abelian group (6) of Section 5.2, associated to the étale groupoid x/ϕ and its sheaf $\tilde{\phi}^*(S)$. By taking alternating sums, we thus obtain a chain complex of G-sheaves, and by reindexing, a cochain complex concentrated in negative degrees. This complex is denoted by

$$\phi_!(S) .$$

More generally, for a bounded above cochain complex of c-soft K-sheaves S, we define $\phi_!(S)$ as the total complex associated to the double complex

$$(\pi_{-p})_!(\tilde{\phi}^*(S^q)_{(-p)}) .$$

The stalk of this complex at $x \in G_0$ is the complex

$$\Gamma_c((x/\phi)_{-p}, \tilde{\phi}^*(S^q)_{(-p)})$$

which defines the hypercohomology $H_c^*(x/\phi, \tilde{\phi}^*(S))$. In particular, if

$S \to T$ is a quasi-isomorphism between bounded above cochain complexes of c-soft K-sheaves, then by Remark 5.9 (1) it induces isomorphism

$$H^*_c(x/\phi, \tilde{\phi}^*(S)) \longrightarrow H^*_c(x/\phi, \tilde{\phi}^*(T))$$

for any point $x \in G_0$. In other words, the induced map $\phi_!(S) \to \phi_!(T)$ is a quasi-isomorphism between bounded above cochain complexes of G-sheaves.

If A is a bounded above complex of arbitrary (not necessarily c-soft) K-sheaves, we define $\phi_!(A)$ to be $\phi_!(S)$ where $A \to S$ is a quasi-isomorphism (a resolution) by a bounded above complex S of c-soft K-sheaves. Then by the observation above, $\phi_!(A)$ is well defined up to quasi-isomorphism. Moreover, a map $A \to B$ induces a map of resolutions and hence a map $\phi_!(A) \to \phi_!(B)$, again well define up to quasi-isomorphism. In other words, the construction provides a functor $\phi_!\colon D^-(K) \to D^-(G)$.

We summarize the discussion in the following theorem.

Theorem 5.12 *Let $\phi : K \to G$ be a homomorphism between étale groupoids. Then ϕ induces a functor*

$$\phi_!\colon D^-(K) \longrightarrow D^-(G)\,.$$

For any bounded above complex A of K-sheaves and any point $x \in G_0$, the stalk of $\phi_!(A)$ at x computes the compactly supported cohomology of x/ϕ with coefficients in the (x/ϕ)-sheaf $\tilde{\phi}^*(A)$ obtained by pulling back A along $x/\phi \to K$,

$$H^*(\phi_!(A)_x) = H^*_c(x/\phi, \tilde{\phi}^*(A))\,.$$

For a complex A as in the theorem, we will write

$$R^q\phi_!(A) = H^q(\phi_!(A))\,, \qquad q \in \mathbb{Z}\,,$$

for the cohomology sheaves of the complex $\phi_!(A)$. Then the theorem gives the formula

$$R^q\phi_!(A)_x = H^*_c(x/\phi, \tilde{\phi}^*(A))$$

for the stalks of $R^q\phi_!(A)$.

Examples 5.13 (1) If $\phi\colon N \to M$ is a smooth map between manifolds, we can view ϕ as a homomorphism between unit groupoids. In this case, the functor $\phi_! : D^-(N) \to D^-(M)$ constructed above for groupoids agrees with the usual one for spaces.

(2) Let $\phi : K \to G$ be a homomorphism between discrete groups, viewed as one-object étale (discrete) groupoids. Then G-sheaves are $\mathbb{R}[G]$-modules, and $D^-(G)$ is the derived category of bounded above cochain complexes of $\mathbb{R}[G]$-modules; similarly for K. For the unique object $x \in G_0$, the comma groupoid x/ϕ is a sum of connected discrete groupoids indexed by the right G-set of cosets $\mathrm{Im}(\phi)/G$, and x/ϕ is Morita equivalent to the sum of groups

$$\sum_{\xi \in \mathrm{Im}(\phi)/G} \mathrm{Ker}(\phi) ,$$

where G acts by acting on the cosets. Thus, the functor $\phi_! : D^-(K) \to D^-(G)$ sends a complex A to the complex

$$\bigoplus_{\xi \in \mathrm{Im}(\phi)/G} i^*(A)$$

of G-modules, where $i^*(A)$ denotes the restriction of A along the inclusion $i : \mathrm{Ker}(\phi) \to K$. In particular, is ϕ is a surjection and A is concentrated in degree 0, then

$$R^q\phi_!(A) = H_{-q}(\mathrm{Ker}(\phi), i^*(A)) .$$

(3) Let G be an étale groupoid, and let E be a right G-space, with action along a map $p_0 : E \to G_0$. Let $E \rtimes G$ be the semi-direct product groupoid, and write $p : E \rtimes G \to G$ for the projection. This is a homomorphism between étale groupoids. If S is a c-soft $E \rtimes G$-sheaf, then the sheaf $(p_0)_!(S)$ carries a natural G-action: the stalk of $(p_0)_!(S)$ at a point $x \in G_0$ is $\Gamma_c(p^{-1}(x), S)$, and any arrow $g : x \to y$ in G induces an action map $\rho_g : p_0^{-1}(y) \to p_0^{-1}(x)$ and an isomorphism of sheaves $S|_{p_0^{-1}(y)} \to \rho_g^*(S|_{p_0^{-1}(x)})$, hence an isomorphism $\Gamma_c(p_0^{-1}(y), S) \to \Gamma_c(p_0^{-1}(x), S)$. This shows that $(p_0)_!$ lifts to a functor from c-soft $E \rtimes G$-sheaves to c-soft G-sheaves. This functor agrees with the functor $p_!(S)$ constructed above. Indeed, $p_!(S)$ is a complex of G-sheaves whose stalk at x computes $H_c^*(x/\phi, S)$. But the étale groupoid x/ϕ is (strongly) equivalent to the space $p_0^{-1}(x)$ (viewed as a unit groupoid), via the functors

$$x/\phi \rightleftarrows p_0^{-1}(x)$$

sending an object $g : x \to p(e)$ to $eg \in p_0^{-1}(x)$ and a point $e \in p_0^{-1}(x)$ to $1_x : x \to p_0(e)$, respectively. Thus there is a simplicial homotopy contracting $\mathrm{Nerve}(x/\phi)$ to the constant simplicial space $p_0^{-1}(x)$, and

$H_c^*(x/\phi, S) \cong H_c^*(p_0^{-1}(x), S)$. (This is in fact a special case of Morita invariance, to be proved in the next section.) In this way, we find that for the square

$$\begin{array}{ccc} E & \xrightarrow{v} & E \rtimes G \\ {\scriptstyle p_0}\downarrow & & \downarrow{\scriptstyle p} \\ G_0 & \xrightarrow{u} & G \end{array}$$

we have $u^* p_! = (p_0)_! v^*$ as functors $D^-(E \rtimes G) \to D^-(G_0)$. This is a special case of the change-of-base formula in the next section.

5.4 Leray spectral sequence, Morita invariance and change-of-base

In this section we will discuss some of the basic properties of the operation $\phi_!$ which has just been introduced. To begin with, we will derive the Leray-Hochschild-Serre type spectral sequence involving $\phi_!$. Next, as a first application of this spectral sequence ,we will prove the Morita invariance of compactly supported cohomology. We will also prove a base change formula for (weak) pull-back squares of étale groupoids. As special cases of these general properties, we discuss compactly supported cohomology of orbifolds, and prove a comparison result for foliations similar to Theorem 4.18.

Proposition 5.14 (Leray spectral sequence) *Let $\phi : K \to G$ be a homomorphism between étale groupoids, and let A be a bounded above complex of K-sheaves. There is a natural spectral sequence*

$$E_2^{p,q} = H_c^p(G, R^q\phi_!(A)) \Rightarrow H_c^{p+q}(K, A) .$$

Proof The hypercohomology spectral sequence of Proposition 5.11 takes the form $H_c^p(G, R^q\phi_!(A)) \Rightarrow H_c^{p+q}(G, \phi_!(A))$, so it is sufficient to establish a natural isomorphism

$$H_c^p(G, \phi_!(A)) = H_c^p(K, A) . \tag{15}$$

By the usual double complex arguments, it suffices to do this for a single c-soft K-sheaf A (rather than a complex A). So, let A be such a sheaf. Spelling out the definitions, we find that we need to establish a quasi-isomorphism

$$\Gamma_c(G_\bullet, \lambda_\bullet^* \phi_!(A)) = \Gamma_c(K_\bullet, \lambda_\bullet'^*(A)) \tag{16}$$

where $\lambda_n : G_n \to G_0$ and $\lambda'_n : K_n \to K_0$ are the 'last vertex' maps, cf. Section 4.2. The complex $\phi_!(A)$ is given in degree $-m$ by applying $(\pi_m)_!$, for the projection $\pi_m : (G_0/\phi)_m \to G_m$, cf. (11). Consider the fibered product $S_{n,m}$ as in the following diagram.

$$\begin{array}{ccccccc} K_0 & \xleftarrow{\lambda'_m} & K_m & \xleftarrow{q_{n,m}} & S_{n,m} & \xrightarrow{r_{n,m}} & (G_0/\phi)_m \\ & & & & {\scriptstyle p_{n,m}}\downarrow & & \downarrow{\scriptstyle \pi_m} \\ & & & & G_n & \xrightarrow{\lambda_n} & G_0 \end{array}$$

Thus $S_{n,m}$ is the space of strings

$$(y_0 \xleftarrow{k_1} y_1 \xleftarrow{k_2} \dots \xleftarrow{k_m} y_m, \phi(y_m) \xleftarrow{g} x_n \xrightarrow{g_n} x_{n-1} \xrightarrow{g_{n-1}} \dots \xrightarrow{g_1} x_0)\,,$$

and we have written $p_{n,m}$, $q_{n,m}$ and $r_{n,m}$ for the evident projections. Notice that, since G is an étale groupoid, the maps λ_n, $r_{n,m}$ and $q_{n,m}$ are all étale.

The sheaf $\tilde{\phi}^*(A)_{(m)}$ on $(G_0/\phi)_m$ occurring in the definition of $\phi_!(A)$ has stalk A_{y_m} at $(y_0 \leftarrow y_1 \leftarrow \dots \leftarrow y_m, \phi(y_m) \leftarrow x)$. Thus

$$r^*_{n,m}\tilde{\phi}^*(A)_{(m)} = q^*_{n,m}\lambda'^*_m(A) = q^*_{n,m}(A_{(m)})\,,$$

and this is a c-soft sheaf on $S_{n,m}$ because $q_{n,m}$ is étale. So, using the base change formula $\lambda^*_n(\pi_m)_! = (p_{n,m})_! r^*_{n,m}$ for spaces (cf. Proposition 5.4 (iii)), we find a natural isomorphism

$$\lambda^*_n(\pi_m)_!(\tilde{\phi}^*(A)_{(m)}) = (p_{n,m})_! r^*_{n,m}(\tilde{\phi}^*(A)_{(m)}) = (p_{n,m})_! q^*_{n,m}(A_{(m)})\,.$$

Thus

$$\begin{aligned} \Gamma_c(G_n, \lambda^*_n(\pi_m)_!(\tilde{\phi}^*(A)_{(m)})) &= \Gamma_c(G_n, (p_{n,m})_! q^*_{n,m}(A_{(m)})) \\ &= \Gamma_c(S_{n,m}, q^*_{n,m}(A_{(m)})) \\ &= \Gamma_c(K_m, (q_{n,m})_! q^*_{n,m}(A_{(m)}))\,. \end{aligned}$$

For a fixed m, the stalk of the complex $(q_{\bullet,m})_! q^*_{\bullet,m}(A_{(m)})$ at a point $(y_0 \leftarrow y_1 \leftarrow \dots \leftarrow y_m)$ of K_m computes the compactly supported cohomology of the comma groupoid $G/\phi(y_m)$ with coefficients in the vector space A_{y_m}. Since G is étale, the groupoid $G/\phi(y_m)$ is discrete, and $H^*_c(G/\phi(y_m), A_{y_m})$ is the usual homology of discrete groupoids or categories. This homology vanishes in positive degrees because $G/\phi(y_m)$ has a terminal object. This shows that the 'summation along fibers' map

$$(q_{\bullet,m})_! q^*_{\bullet,m}(A_{(m)}) \longrightarrow A_{(m)}$$

is a quasi-isomorphism of complexes of c-soft sheaves on K_m. Thus

$$\Gamma_c(K_m, (q_{n,m})_! q_{n,m}^*(A_{(m)})) \longrightarrow \Gamma_c(K_m, A_{(m)})$$

is a quasi-isomorphism for any fixed m, and hence a quasi-isomorphism of double complexes. Since the domain of this quasi-isomorphism has been shown to be isomorphic to the complex $\Gamma_c(G_n, \lambda_n^*(\pi_m)_!(\tilde{\phi}^*(A)_{(m)}))$ which computes $H_c^*(G, \phi_!(A))$, this gives the desired quasi-isomorphism (16). □

Corollary 5.15 (Morita invariance) *Let $\phi\colon K \to G$ be a weak equivalence of étale groupoids. Then for any bounded above complex A of G-sheaves, ϕ induces an isomorphism*

$$H_c^*(K, \phi^*(A)) \longrightarrow H_c^*(G, A) .$$

Proof The map ϕ is necessarily étale by [48, Exercise 5.16 (4)]. For a point $x \in G_0$, the comma groupoid x/ϕ is discrete and equivalent to x/G. Hence $H_c^*(x/\phi, \tilde{\phi}^*(\phi^*(A)))$ is the homology of the discrete groupoid x/G with coefficients in A_x, hence vanishes in non-zero degrees. This shows that the Leray spectral sequence collapses, to give the stated isomorphism. □

Example 5.16 (Orbifolds) The Leray spectral sequence equally applies to topological groupoids. In particular, if G is a proper étale groupoid, its orbit space $M = |G|$ is a paracompact Hausdorff space, and we have a Leray spectral sequence for the quotient map $\phi\colon G \to M$,

$$H_c^p(M, R^q\phi_!(A)) \Rightarrow H_c^{p+q}(G, A) ,$$

with stalks of $R^q\phi_!(A)$ given by the homology of the isotropy groups

$$(R^q\phi_!(A))_{\phi(x)} = H_{-q}(G_x, A_x)$$

(analogous to Corollary 4.12). When working over $\mathbb{R}$ (or more generally, over a ring R with the property that $\#G_x$ is a unit in R), the spectral sequence collapses, to give an isomorphism

$$H_c^*(G, A) = H_c^*(M, A_G)$$

where A_G denotes the sheaf of coinvariants, with stalk $(A_G)_{\phi(x)} = H_0(G_x, A_x)$ at any point $\phi(x) \in M$.

As another application of the Leray spectral sequence, we state a comparison theorem for foliations analogous to Theorem 4.17.

Corollary 5.17 *Let M be a manifold of dimension n, equipped with a* *foliation $\mathcal{F}$ of dimension p. Suppose there is a number d such that the holonomy cover $\tilde{L}$ of each leaf L of $\mathcal{F}$ has the property that $H_c^i(\tilde{L}, \mathbb{R}) = 0$ for $p - d \leq i < p$. Then for each* $\mathrm{Hol}(M, \mathcal{F})$*-sheaf A there is a natural isomorphism*

$$H_c^s(M, u^*(A)) \longrightarrow H_c^{s-p}(\mathrm{Hol}_T(M, \mathcal{F}), A)$$

for any $s = n - d, \ldots, n$, where $u \colon M \to \mathrm{Hol}(M, \mathcal{F})$ denotes the inclusion of units and $T \subset M$ is a complete transversal section of $(M, \mathcal{F})$.

Remark. We have simply written A for the restriction of A to the étale holonomy groupoid $\mathrm{Hol}_T(M, \mathcal{F})$. Note that, by Poincaré duality (see also Section 5.5) the hypothesis on $H_c^i(\tilde{L}, \mathbb{R})$ is equivalent to the requirement that $H^j(\tilde{L}, \mathbb{R}) = 0$ for any $0 < j \leq d$ in case $\tilde{L}$ is orientable. Furthermore, $H_c^p(\tilde{L}, \mathbb{R}) = \mathbb{R}$. Also, by Remark 5.3 (ii), the hypothesis implies that $H_c^i(\tilde{L}, V) = 0$, $p - d \leq i < p$, for any vector space V.

Proof (Proof of Corollary 5.17) Consider the pull-back square

$$\begin{array}{ccc} P & \xrightarrow{k} & M \\ {\scriptstyle v}\downarrow & & \downarrow{\scriptstyle u} \\ \mathrm{Hol}_T(M, \mathcal{F}) & \xrightarrow{j} & \mathrm{Hol}(M, \mathcal{F}) \end{array}$$

of the proof of Theorem 4.17 in Section 4.4, in which the horizontal maps are weak equivalences. In particular, the homomorphism k induces isomorphisms

$$H_c^*(P, k^*u^*A) = H_c^*(M, u^*A)\ ,$$

so the Leray spectral sequence for the map v can be written as

$$E_2^{s,t} = H_c^s(\mathrm{Hol}_T(M, \mathcal{F}), R^t v_! v^* A) \Rightarrow H_c^{s+t}(M, u^*A)$$

(where on the left, we have written A for j^*A again). Recall that for a point $x \in T$, the comma groupoid x/v is Morita equivalent to the holonomy cover $\tilde{L}_x$ of the leaf L_x through x (cf. Section 4.4). Thus, as in Example 5.13 (2), the stalks of $R^t v_! v^* A$ are given by

$$(R^t v_! v^* A)_x = H_c^t(\tilde{L}_x, A_x)\ ,$$

where on the right the coefficients are in the vector space A_x. In particular, since $H_c^p(\tilde{L}_x, A_x) = A_x$, we have

(i) $E_2^{s-p,p} = H_c^{s-p}(\mathrm{Hol}_T(M, \mathcal{F}), A)$.

Also, since T has dimension q and $\tilde{L}_x$ has dimension p, we have

(ii) $E_2^{s,t} = 0$ for $s > p$, or $t > p$, or $p - d \leq t < p$,

the last case being the hypothesis of the corollary. Thus, the corollary would follow if we show that (ii) implies that

(iii) $E_2^{s-p,p} = E_\infty^{s-p,p}$ and
(iv) $E_2^{s-i,i} = 0$ for $i \neq p$.

This is simply a matter of bookkeeping the degrees of differentials

$$E_2^{s-r,t+r-1} \xrightarrow{d_r} E_q^{s,t} \xrightarrow{d_r} E_r^{s+r,t-r+1}, \qquad r \geq 2,$$

in the spectral sequence. Indeed, the image of d_r inside $E_r^{s-p,p}$ is zero because $E_r^{s',t'} = 0$ for $t' > p$. Also, the kernel of d_r inside $E_r^{s-p,p}$ is all of $E_r^{s-p,p}$ as soon as $E_r^{s-p+r,p-r+1} = 0$, and this holds for $s \geq n - d - 1$ because then either $p - d \leq p - r + 1 < p$ or $s - p + r > q$. Thus (iii) holds if $s \geq n - d - 1$. Next, for (iv), notice that (ii) already gives that $E_2^{s-i,i} = 0$ for $p - d < i < p$ and for $i > p$, so that we only need to check that $E_2^{s-i,i} = 0$ for $0 \leq i < p - d$. But in this case, if $s \geq n - d$ then $s - i > n - p = q$. □

Next, we turn to the change-of-base formula.

Proposition 5.18 *Let*

$$\begin{array}{ccc} P & \xrightarrow{b} & H \\ {\scriptstyle\psi}\downarrow & & \downarrow{\scriptstyle\phi} \\ K & \xrightarrow{a} & G \end{array}$$

be a weak pull-back square of étale groupoids. Then

$$a^*\phi_! = \psi_! b^* : D^-(H) \longrightarrow D^-(K).$$

Proof We will prove that for any c-soft H-sheaf B, there is a quasi-isomorphism

$$a^*\phi_!(B) \cong \psi_! b^*(B),$$

natural in B. It then follows by the usual double complex arguments that there also is a similar quasi-isomorphism for any such bounded below complex B of H-sheaves, since any such complex has a resolution by c-soft H-sheaves.

So, fix a c-soft H-sheaf B, and a bounded resolution $b^*(B) \to S$ by c-soft P-sheaves. The complex $\phi_!(B)$ of G-sheaves has as its stalk as $x \in G_0$ the complex $\Gamma_c((x/\phi)_n, B)$ computing the cohomology $H_c^*(x/\phi, B)$.

(Here we have abused notation, and we have simply written B for $\tilde{\phi}^*(B)_n$ and for $\tilde{\phi}^*(B)$, respectively.) Thus $a^*\phi_!(B)$ is the complex with stalk

$$\Gamma_c((a(z)/\phi)_n, B) \tag{17}$$

computing $H_c^*(a(z)/\phi, B)$ at any point $z \in K_0$.

On the other hand, $\psi_! b^*(B)$ is the complex of K-sheaves whose stalk at a point $z \in K_0$ is the complex (total complex of double complex)

$$\Gamma_c((z/\psi)_n, S^m) \tag{18}$$

computing $H_c^*(z/\psi, S) = H_c^*(z/\psi, b^*(B))$.

For a fixed $z \in K_0$, the quasi-isomorphism $b^*(B) \to S$ restricts along the inclusion

$$i_z : a(z)/\phi \longrightarrow P$$

to a quasi-isomorphism $i_z^* b^*(B) \to i_z^*(S)$ over $a(z)/\phi$, and this gives a quasi-isomorphism of complexes

$$\Gamma_c((a(z)/\phi)_n, i_z^* b^*(B)) \longrightarrow \Gamma_c((a(z)/\phi)_n, i_z^*(S^m)) \tag{19}$$

Now consider the homomorphism of étale groupoids

$$\pi_z : z/\psi \longrightarrow a(z)/\phi$$

sending an object $(z \xrightarrow{k} z', a(z') \xrightarrow{g} \phi(y))$ of z/ψ to the object $ga(k) : z \to \phi(y)$ of z/ψ, and with the evident effect on arrows. Since this homomorphism π_z is a Morita equivalence, Corollary 5.15 (or more precisely, the proof of Proposition 5.14) provides a canonical quasi-isomorphism

$$\Gamma_c((a(z)/\phi)_n, i_z^*(S^m)) \longrightarrow \Gamma_c((z/\psi)_n, \pi_z^* i_z^*(S^m)) \tag{20}$$

Furthermore, since the composite homomorphism $i_z\pi_z : z/\psi \to a(z)/\psi \to P$ is naturally isomorphic (but not equal) to the projection $\tilde{\psi} : z/\psi \to P$, the complex $\pi_z^* i_z^*(S^\bullet)$ in (20) is isomorphic to the complex $\tilde{\psi}^*(S)_{(n)}$, which we simply denoted by S in (18). Since, moreover, (granted our abuse of notation) the sheaf B on $(a(z)/\phi)_n$ in (17) is the same as the sheaf $i_z^* b^*(B)$ in (19), the quasi-isomorphisms (19) and (20) together give an explicit quasi-isomorphism

$$\Gamma_c((a(z)/\phi)_n, B) = \Gamma_c((z/\psi)_n, S^m)$$

which is the stalk of the desired quasi-isomorphism

$$a^*\phi_!(B) = \psi_! b^*(B).$$

5.5 Homology of the embedding category

The purpose of this section is to prove an analogue of Theorem 4.21 for compactly supported cohomology. As an application, we will derive a form of Poincaré duality for étale groupoids. This will in particular apply to leaf spaces of foliations, as modelled by étale groupoids.

In Section 4.5 we described the cohomology groups $H^n(\mathcal{C}, A)$ of a small category $\mathcal{C}$ with coefficients in a contravariant functor (a presheaf) from $\mathcal{C}$ into the category of abelian groups (or R-modules). Dually, for a covariant such functor A, the homology groups $H_n(\mathcal{C}, A)$ are defined as the homology groups of the complex $C_\bullet(\mathcal{C}, A)$, defined by

$$C_n(\mathcal{C}, A) = \bigoplus_{(f_1,\ldots,f_n)\in\mathcal{C}_n} A(\mathrm{s}(f_n)) \,,$$

where the sum ranges over all composable strings $c_0 \xleftarrow{f_1} c_1 \xleftarrow{f_2} \ldots \xleftarrow{f_n} c_n$. The differential is the alternating sum of the face maps $d_i : C_n \to C_{n-1}$, where d_i maps the summand $A(c_n)$ for $c_0 \xleftarrow{f_1} c_1 \xleftarrow{f_1} \ldots \xleftarrow{f_n} c_n$ into the summand $A(c_n)$ for $d_i(f_1, \ldots, f_n)$ by the identity map for $i < n$, and into the summand $A(c_{n-1})$ for $d_n(f_1, \ldots, f_n) = (f_1, \ldots, f_{n-1})$ by $A(f_n)$ for $i = n$.

Similarly, if A is a covariant functor from $\mathcal{C}$ into the category of bounded below chain complexes, we define the hypercohomology $H_*(\mathcal{C}, A)$ as the homology of the total complex associated to the double complex

$$C_p(\mathcal{C}, A^q) \,.$$

If $A \to B$ is a map (a natural transformation) between two such functors, then by the usual argument, the induced map $H_*(\mathcal{C}, A) \to H_*(\mathcal{C}, B)$ is an isomorphism whenever $A(c) \to B(c)$ is a quasi-isomorphism for each object c of $\mathcal{C}$.

We recall that there is an obvious duality between homology and cohomology. For a covariant functor A from $\mathcal{C}$ into (say) real vector spaces, let us write $A^\vee$ for the dual contravariant functor defined by $A^\vee(c) = A(c)^\vee = \mathrm{Hom}(A(c), \mathbb{R})$. Then the following proposition follows immediately from the definition.

Proposition 5.19 *For any small category $\mathcal{C}$ and any functor A from $\mathcal{C}$ into real vector spaces, there is a natural isomorphism*

$$H^n(\mathcal{C}, A^\vee) = H_n(\mathcal{C}, A)^\vee \,.$$

There is of course a similar isomorphism for hyper(co)homology, for a bounded below chain complex A and its dual cochain complex $A^\vee$, defined by $(A^\vee)^q = (A_q)^\vee$.

Now let G be an étale groupoid of dimension d, let $\mathcal{U}$ be a basis for G_0, and let $\mathrm{Emb}_{\mathcal{U}}(G)$ be the associated embedding category. Any G-sheaf A induces a covariant functor $\Gamma_c(A)$ on $\mathrm{Emb}_{\mathcal{U}}(G)$ with values in vector spaces, defined on objects by

$$\Gamma_c(A)(U) = \Gamma_c(U, A) .$$

The action of this functor on an arrow $\sigma\colon U \to V$ in $\mathrm{Emb}_{\mathcal{U}}(G)$ is defined by the map $\sigma_* : \Gamma_c(U, A) \to \Gamma_c(V, A)$, which is the composition of the isomorphism $\Gamma_c(U, A) = \Gamma_c(\mathrm{t}\sigma(U), A)$ given by the action of G on A and the extension by zero: $\Gamma_c(\mathrm{t}\sigma(U), A) \to \Gamma_c(V, A)$.

Theorem 5.20 *Let G be an étale groupoid and B be a c-soft sheaf. Then there is a natural isomorphism*

$$H_c^{-p}(G, B) = H_p(\mathrm{Emb}_{\mathcal{U}}(G), \Gamma_c(B)) \tag{21}$$

for any $p \geq 0$.

Remark. By the usual arguments, the theorem extends to the case where B is a bounded above cochain complex of G-sheaves with the property that $H_c^i(U, B^q) = 0$ for each $i > 0$, each q and each $U \in \mathcal{U}$. In this case we obtain an isomorphism of the form (21 where $\Gamma_c(B)$ is the chain complex $\Gamma_c(B)_q = \Gamma_c(B^{-q})$.

Proof (Proof of Theorem 5.20) The proof is actually quite similar to that of Theorem 4.21, and the reader may well prefer to skip it. For the present proof, it will be convenient to consider the chain complex

$$C_p(G, B) = \Gamma_c(G_p, B_{(p)})$$

and its homology $H_p(G, B)$, so that

$$H_p(G, B) = H_c^{-p}(G, B) ,$$

since B is assumed to be c-soft. Recall now the bisimplicial space $S_{p,q}$ from the proof of Theorem 4.21, with p, q-simplices

$$U_0 \xleftarrow{\sigma_1} U_1 \xleftarrow{\sigma_2} \cdots \xleftarrow{\sigma_q} U_q \xleftarrow{g} x_0 \xleftarrow{g_1} x_1 \xleftarrow{g_2} \cdots \xleftarrow{g_p} x_p ,$$

and the c-soft sheaf $B^{p,q}$ on $S_{p,q}$ with stalk B_{x_p} at such a p, q-simplex. Consider the bisimplicial abelian group

$$C_{p,q}(B) = \Gamma_c(S_{p,q}, B^{p,q})$$

and the associated double complex, which we also denote by $C_{p,q}(B)$. One could also write

$$C_{p,q}(B) = \bigoplus_{U_0 \leftarrow U_1 \leftarrow \ldots \leftarrow U_q} \Gamma_c((G/U_q)_p, B_{(p)}) ,$$

where on the right G/U_q is the comma groupoid, and $B_{(p)}$ stands for the sheaf on $(G/U_q)_p$ with stalk B_{x_p} at a point $(x \xleftarrow{g} x_0 \xleftarrow{g_1} \ldots \xleftarrow{g_p} x_p)$ with $x \in U_q$.

For a fixed q, the complex $\Gamma_c((G/U_q)_p, B_{(p)})$ computes the compactly supported cohomology of the comma groupoid G/U_q. Since this groupoid is Morita equivalent to the space U_q, Corollary 5.15 yields that, for a fixed q, one has $H_p(C_{\bullet,q}(B)) = H_c^{-p}(U_q, B)$. Since B is assumed c-soft, we find

$$H_p(C_{\bullet,q}(B)) = 0 , \qquad p > 0 ,$$
$$H_0(C_{\bullet,q}(B)) = \Gamma_c(U_q, B) .$$

This shows that the homology of the total complex $\mathrm{Tot}(C_{\bullet,\bullet}(B))$ is isomorphic to $H_*(\mathrm{Emb}_{\mathcal{U}}(G), \Gamma_c(B))$.

On the other hand, for a fixed p, we consider the projection $\mathrm{pr} = \mathrm{pr}_{p,q} \colon S_{p,q} \to G_p$, again as in the proof of Theorem 4.21. This projection is an étale map, and at a point $(x_0 \xleftarrow{g_1} x_1 \xleftarrow{g_2} \ldots \xleftarrow{g_p} x_p)$ of G_p, the stalk of $\mathrm{pr}_!(B)$ is

$$\bigoplus_{U_0 \leftarrow U_1 \leftarrow \ldots \leftarrow U_q} \bigoplus_{g} B_{x_p} ,$$

where g ranges over all arrows in G from x_0 to some point x in U_q. This stalk is a chain complex in q, and its homology is that of the colimit of comma categories $U/\mathrm{Emb}_{\mathcal{U}}(G)$ where $U \in \mathcal{U}$ ranges over all neighbourhoods of x_0 (as in the proof of Theorem 4.21). Since these comma categories each have an initial object and hence vanishing homology in positive degrees, we find that $H_q(\mathrm{pr}_!(B)_{x_0}) = 0$ for $q > 0$ while

$$H_0(\mathrm{pr}_!(B)_{x_0}) = \varinjlim_{x \in U} H_0(U/\mathrm{Emb}_{\mathcal{U}}(G), B_{x_p}) \cong B_{x_p} .$$

Thus, for a fixed p, the complex $\mathrm{pr}_!(B_{p,\bullet})$ is quasi-isomorphic to the sheaf $B_{(p)}$ on G_p (seen as a complex concentrated in degree 0). Since all sheaves involved are c-soft, we conclude that $\Gamma_c(S_{p,\bullet}, B^{p,\bullet}) = \Gamma_c(G_p, \mathrm{pr}_!(B^{p,\bullet}))$ is quasi-isomorphic to $\Gamma_c(G_p, B_{(p)})$. Thus, for a fixed p, the homology of the total complex of $C_{p,\bullet}(B)$ is isomorphic to $H_*(G, B)$, and the theorem is proved. □

Corollary 5.21 *Let G be an étale groupoid, and let A be an arbitrary G-sheaf. Then there is a natural isomorphism*

$$H_c^{-p}(G, A) = H_p(\mathrm{Emb}_{\mathcal{U}}(G), \Gamma_c(A \otimes \Omega^{-\bullet})) \; .$$

Remark. On the right hand, this is the hyperhomology of the usual complex of differential forms, reindexed so as to form a chain complex.

Corollary 5.22 *Let G be an étale groupoid of dimension d, and let A be a locally constant G-sheaf. Suppose that the basis $\mathcal{U}$ of G_0 consists of contractible open sets. Then there is a natural isomorphism*

$$H_c^p(G, A) = H_{d-p}(\mathrm{Emb}_{\mathcal{U}}(G), H_c^d(\,\text{-}\,, A)) \; .$$

Proof The chain complex $\Gamma_c(U, A \otimes \Omega^{-\bullet})$ computes $H_c^{-\bullet}(U, A)$, which is zero except in degree $-d$. So the hypercohomology spectral sequence of small categories collapses, and the previous corollary yields

$$\begin{aligned} H_c^{-p}(G, A) &= H_p(\mathrm{Emb}_{\mathcal{U}}(G), H_c^d(\,\text{-}\,, A)[d]) \\ &= H_{d+p}(\mathrm{Emb}_{\mathcal{U}}(G), H_c^d(\,\text{-}\,, A)) \; , \end{aligned}$$

where $H_c^d(\,\text{-}\,, A)[d]$ denotes the functor $H_c^d(\,\text{-}\,, A)$ concentrated in degree $-d$. Replacing p by $-p$ yields the result. □

Next, we apply the duality for small categories (Proposition 5.19) to étale groupoids. Let S be a c-soft G-sheaf. We define a dual G-sheaf $D(S)$ as follows. For each open subset U of G_0, consider the vector space $\Gamma_c(U, S)$ and its dual $\Gamma_c(U, S)^\vee$. For two open subsets $U \subset V$ of G_0, the extension-by-zero map dualizes to a restriction map $\Gamma_c(V, S)^\vee \to \Gamma_c(U, S)^\vee$. With these restriction maps, the assignment $U \mapsto \Gamma_c(U, S)^\vee$ actually defines an injective sheaf on U, since S is c-soft. Indeed, the gluing property for covers is the dual of the Mayer-Vietoris sequence of Proposition 5.2 (see also [32]). This sheaf on G_0 has a natural action by the étale groupoid G, defined in the same way as the action of the embedding category on the functor $\Gamma_c(S)$ earlier in this section. The G-sheaf obtained in this way is denoted $D(S)$. Its definition can be summarised by the identity

$$\Gamma(U, D(S)) = \Gamma_c(U, S)^\vee \qquad (22)$$

for all open subsets U of G_0. More generally, for an arbitrary G-sheaf A, we take its resolution by differential forms, $0 \longrightarrow A \longrightarrow A \otimes \Omega^0 \longrightarrow \ldots \longrightarrow A \otimes \Omega^d \longrightarrow 0$ (where $d = \dim(G)$). Then for open subsets U of G_0, the assignment $U \mapsto \Gamma_c(U, A \otimes \Omega^{-\bullet})$ defines a covariant functor into chain complexes, and, since each of the $A \otimes \Omega^q$ is c-soft, the dual functor

$U \mapsto \Gamma_c(U, A \otimes \Omega^{-\bullet})^\vee$ defines a cochain complex of G-sheaves, which we denote by $D(A)$. Thus

$$\Gamma(U, D(A)^q) = \Gamma_c(U, A \otimes \Omega^{-q})^\vee . \tag{23}$$

This notation is consistent with the earlier one (22), for c-soft sheaves S, because if $A = S$ is c-soft, then the double complex in (23) is quasi-isomorphic to the complex $D(S)$ (concentrated in degree 0) of (22).

Corollary 5.23 *Let G be an étale groupoid, and let A be a G-sheaf. There is a natural isomorphism*

$$H_c^{-p}(G, A)^\vee = H^p(G, D(A)) .$$

Proof By Corollary 5.21 we have

$$H_c^{-p}(G, A)^\vee = H_p(\mathrm{Emb}_{\mathcal{U}}(G), \Gamma_c(A \otimes \Omega^{-\bullet}))^\vee ,$$

while, by the analogue of Theorem 4.21 for cohomology,

$$H^p(G, D(A)) = H^p(\mathrm{Emb}_{\mathcal{U}}(G), \Gamma D(A)) .$$

But $\Gamma D(A)(U) = \Gamma_c(U, A \otimes \Omega^{-\bullet})^\vee$, so the duality for small categories (Proposition 5.19) gives the result. □

As an example, consider the case where A is the constant sheaf $\mathbb{R}$. Then $D(\mathbb{R})(U)$ is the complex $\Gamma_c(U, \Omega^{-q})^\vee$. Since $H_c^i(U, \mathbb{R}) = \mathbb{R}$ for $i = d$ and zero otherwise, $D(\mathbb{R})$ is quasi-isomorphic to $H_c^d(U, \mathbb{R})^\vee$ concentrated in degree d. Recall that $U \mapsto H_c^d(U, \mathbb{R})^\vee$ is by definition the *orientation sheaf* on G_0 (cf. [32]). It is in fact again a G-sheaf. If we denote this orientation G-sheaf by $\mathcal{O}$, then we conclude that there is a quasi-isomorphism

$$D(\mathbb{R}) \longrightarrow \mathcal{O}[d]$$

where $\mathcal{O}[d]$ is the cochain complex of sheaves given by $\mathcal{O}$ concentrated in degree $-d$.

Thus, as a special case of the previous corollary, we obtain the following version of *Poincaré duality* for étale groupoids.

Corollary 5.24 *Let G be an étale groupoid of dimension d. There is a natural isomorphism*

$$H_c^p(G, \mathbb{R})^\vee = H^{d-p}(G, \mathcal{O}) .$$

Bibliography

1. J. Bernstein, V. Lunts, Equivariant sheaves and functors. *Lecture Notes in Math.* **1578**, Springer-Verlag, Berlin (1994).
2. R. Bott, Characteristic classes and foliations. *Lecture Notes in Math.* **279**, Springer-Verlag, Berlin (1972), 1–94.
3. R. Bott, L. W. Tu, *Differential Forms in Algebraic Topology.* Springer-Verlag, New York (1982).
4. A. Borel, *Intersection cohomology.* Birkhäuser, Boston (1984).
5. G. Bredon, *Sheaf Theory.* McGraw-Hill, New York-Toronto (1967).
6. M. R. Bridson, A. Haefliger, *Metric spaces of non-positive curvature.* Springer-Verlag, Berlin (1999).
7. J.-L. Brylinski, V. Nistor, Cyclic cohomology of étale groupoids. *K-Theory* **8** (1994), no. 4, 341–365.
8. H. Bursztyn, A. Weinstein, Picard groups in Poisson geometry. *Preprint* (2003).
9. H. Bursztyn, A. Weinstein, Poisson geometry and Morita equivalence. *This volume* (2004).
10. Cannas da Silva, A. Weinstein, *Geometric Models for Noncommutative Algebras.* Berkeley Mathematics Lecture Notes **10** (American Mathematical Society, Providence, Rhode Island) (1999).
11. H. Cartan, S. Eilenberg, *Homological algebra.* Princeton University Press, Princeton (1999).
12. A. Connes, *Noncommutative geometry.* Academic Press, San Diego (1994).
13. M. Crainic, Cyclic cohomology of étale groupoids: the general case. *K-Theory* **17** (1999), no. 4, 319–362.
14. M. Crainic, I. Moerdijk, A Homology Theory for Etale Groupoids. *J. Reine Angew. Math.* **521** (2000), 25–46.

15. M. Crainic, I. Moerdijk, Foliation groupoids and their cyclic homology. *Adv. Math.* **157** (2001), no. 2, 177–197.
16. M. Crainic, I. Moerdijk, Cech-De Rham theory for leaf spaces of foliations. *Math. Annalen* **328** (2004), 59–85.
17. P. Deligne, D. Mumford, The irreducibility of the space of curves of given genus. *Inst. Hautes Etudes Sci. Publ. Math.* **36** (1969), 75–109.
18. W. T. van Est, Rapport sur les S-atlas. *Astérisque* **116** (1984), 235–292.
19. R. Godement, Topologie Algébrique et Théorie des Faisceaux. Hermann, Paris (1958).
20. M. Golubitsky, V. Guillemin, *Stable Mappings and Their Singularities.* Springer-Verlag, New York (1973).
21. A. Gorokhovsky, J. Lott, Local index theory over étale groupoids. *J. Reine Angew. Math.* **560** (2003), 151–198.
22. A. Haefliger, Sur les feuilletages analytiques. *C. R. Acad. Sci. Paris* **242** (1956), 2908–2910.
23. A. Haefliger, Structures feuilletées et cohomologie à valeur dans un faisceau de groupoïdes, *Comment. Math. Helv.* **32** (1958), 248–329.
24. A. Haefliger, Differential cohomology. *Differential topology (Varenna, 1976)*, 19–70, Liguori, Naples (1979).
25. A. Haefliger, Some remarks on foliations with minimal leaves. *J. Differential Geom.* **15** (1980), no. 2, 269–284.
26. A. Haefliger, Groupoïdes d'holonomie et classifiants. Transversal structure of foliations (Toulouse, 1982), *Astérisque* **116** (1984), 70–97.
27. J.L. Heitsch, A cohomology for foliated manifolds. *Bull. Amer. Math. Soc.* **79** (1973), no. 6, 1283–1285.
28. A. Henriques, D. Metzler, Presentations of Noneffective Orbifolds. To appear in *Trans. Amer. Math. Soc.*
29. P. J. Higgins and K. C. H. Mackenzie, Algebraic constructions in the category of Lie algebroids. *J. Algebra* **129** (1990), 194–230.
30. M. Hilsum, G. Skandalis, Morphismes K-orientés d'espaces de feuilles et fonctorialité en théorie de Kasparov (d'après une conjecture d'A. Connes). *Ann. Sci. Ecole Norm. Sup. (4)* **20** (1987), no. 3, 325–390.
31. P. J. Hilton, U. Stammbach, *A course in homological algebra.* Springer-Verlag, New York (1971).
32. B. Iversen, *Sheaf Theory.* Springer-Verlag, New York (1986).
33. S. Jekel, On two theorems of A. Haefliger concerning foliations. *Topology* **15** (1976), 267–271.
34. N.P. Landsman, Bicategories of operator algebras and Poisson man-

ifolds. Mathematical physics in mathematics and physics (Siena, 2000), 271–286, *Fields Inst. Commun.* **30**, Amer. Math. Soc., Providence (2001).

35. N.P. Landsman, Quantization as a functor. *Quantization, Poisson brackets and beyond (Manchester, 2001)*, 9–24, Contemp. Math. **315**, Amer. Math. Soc., Providence (2002).
36. E. Lupercio, B. Uribe, Loop groupoids, gerbes, and twisted sectors on orbifolds. *Orbifolds in mathematics and physics (Madison, WI, 2001)*, 163–184, Contemp. Math. **310**, Amer. Math. Soc., Providence (2002).
37. K.C.H. Mackenzie, *Lie Groupoids and Lie Algebroids in Differential Geometry.* London Mathematical Society Lecture Notes Series **124**, Cambridge University Press (1987).
38. S. Mac Lane, *Homology.* Springer-Verlag, Berlin (1963).
39. S. Mac Lane, *Categories for the working mathematician.* Springer-Verlag, New York (1998).
40. D. Metzler, Topological and Smooth Stacks. *Preprint* (2003).
41. I. Moerdijk, The classifying topos of a continuous groupoid. I. *Trans. Amer. Math. Soc.* **310** (1988), no. 2, 629–668.
42. I. Moerdijk, Toposes and groupoids. *Springer Lecture Notes in Math.* **1348** (1988), 280–29.
43. I.Moerdijk, Classifying toposes and foliations. *Ann. Inst. Fourier (Grenoble)* **41** (1991), no. 1, 189–209.
44. I. Moerdijk, On the weak homotopy type of étale groupoids. *Integrable systems and foliations/Feuilletages et systèmes intégrables (Montpellier, 1995)*, 147–156, Progr. Math. **145**, Birkhäuser Boston, Boston (1997).
45. I. Moerdijk, Etale groupoids, derived categories, and operations. *Groupoids in analysis, geometry, and physics (Boulder, CO, 1999)*, 101–114, Contemp. Math. **282**, Amer. Math. Soc., Providence (2001).
46. I. Moerdijk, Orbifolds as groupoids: an introduction. *Orbifolds in mathematics and physics (Madison, WI, 2001)*, 205–222, Contemp. Math. **310**, Amer. Math. Soc., Povidence (2002).
47. I. Moerdijk, Lie groupoids, gerbes, and non-abelian cohomology. *K-Theory* **28** (2003), no. 3, 207–258.
48. I. Moerdijk, J. Mrčun, *Introduction to Foliations and Lie Groupoids.* Cambridge University Press, Cambridge (2003).
49. C.C. Moore, C. Schochet, *Global analysis on foliated spaces.* Springer-

Verlag, New York (1988).
50. J. Mrčun, *Stability and invariants of Hilsum-Skandalis maps.* Ph. D. Thesis, Utrecht (1996).
51. J. Mrčun, Functoriality of the bimodule associated to a Hilsum-Skandalis map. *K-Theory* **18** (1999), no. 3, 235–253.
52. P.S. Muhly, J.N. Renault, D.P. Williams, Equivalence and isomorphism for groupoid C^*-algebras. *J. Operator Theory* **17** (1987), no. 1, 3–22.
53. J. Phillips, The holonomic imperative and the homotopy groupoid of a foliated manifold. *Rocky Mountain J. Math.* **17** (1987), 151–165.
54. D.A. Pronk, Etendues and stacks as bicategories of fractions. *Compositio Math.* **102** (1996), no. 3, 243–303
55. I. Satake, On a generalization of the notion of manifold. *Proc. Nat. Acad. Sci. U.S.A.* **42** (1956), 359–363.
56. V. Sergiescu, Cohomologie basique et dualité des feuilletages riemanniens. *Ann. Inst. Fourier (Grenoble)* **35** (1985), no. 3, 137–158.
57. E. H. Spanier, *Algebraic Topology.* McGraw-Hill, New York (1966).
58. J. Tapia, Sur la pente du feuilletage de Kronecker et la cohomologie étale de l'espace des feuilles. *C. R. Acad. Sci. Paris Sér. I Math.* **305** (1987), no. 10, 427–429.
59. J.-L. Tu, P. Xu, C. Laurent, Twisted K-theory of differentiable stacks. *Preprint* (2003).
60. C. A. Weibel, *An introduction to homological algebra.* Cambridge University Press, Cambridge (1994).
61. H. E. Winkelnkemper, The graph of a foliation, *Ann. Global Anal. Geom.* **1**(1983), no. 3, 51–75.

PART FOUR

Geometric Methods in Representation Theory

Wilfried Schmid[1]
schmid@math.harvard.edu

Harvard University
Department of Mathematics
One Oxford Street
Cambridge, MA 02138, USA

Lecture Notes Taken by Matvei Libine
matvei@math.umass.edu

Department of Mathematics & Statistics
710 North Pleasant Street
University of Massachusetts
Amherst, MA 01003-9305 USA

[1] Supported in part by NSF grant DMS-0070714

1

Reductive Lie Groups: Definitions and Basic Properties

The results stated in this section are fairly standard. Proofs and further details can be found in [23], for instance.

1.1 Basic Definitions and Examples

In these notes "Lie algebra" means finite dimensional Lie algebra over $\mathbb{R}$ or $\mathbb{C}$. These arise as Lie algebras of, respectively, Lie groups and complex Lie groups. We begin by recalling some basic definitions:

Definition 1.1 A Lie algebra $\mathfrak{g}$ is *simple* if it has no proper ideals and $\dim \mathfrak{g} > 1$. A Lie algebra $\mathfrak{g}$ is *semisimple* if it can be written as a direct sum of simple ideals $\mathfrak{g}_i$,

$$\mathfrak{g} = \oplus_{1 \leq i \leq N}\ \mathfrak{g}_i \,.$$

One calls a Lie algebra $\mathfrak{g}$ *reductive* if it can be written as a direct sum of ideals

$$\mathfrak{g} = \mathfrak{s} \oplus \mathfrak{z} \,,$$

with $\mathfrak{s}$ semisimple and $\mathfrak{z}$ = center of $\mathfrak{g}$. A Lie group is *simple*, respectively *semisimple*, if it has finitely many connected components and if its Lie algebra is simple, respectively semisimple. A Lie subgroup $G \subset GL(n, \mathbb{R})$ or $G \subset GL(n, \mathbb{C})$ is said to be *reductive* if it has finitely many connected components, its Lie algebra is reductive and Z_{G^0} = center of the identity component G^0 of G consists of semisimple linear transformations – equivalently, if Z_{G^0} is conjugate, possibly after extension of scalars from $\mathbb{R}$ to $\mathbb{C}$, to a subgroup of the diagonal subgroup in the ambient $GL(n, \mathbb{R})$ or $GL(n, \mathbb{C})$.

Remark 1.2 Our definition of a reductive Lie algebra is not the one most commonly used, but is equivalent to it. The semisimple ideal $\mathfrak{s}$ is uniquely determined by $\mathfrak{g}$ since $\mathfrak{s} = [\mathfrak{g}, \mathfrak{g}]$.

In the definition of a simple Lie algebra we require $\dim \mathfrak{g} > 1$ because we want to exclude the one-dimensional abelian Lie algebra, which is reductive.

We shall talk about reductive Lie group only in the context of linear groups, i.e., for Lie subgroups of $GL(n,\mathbb{R})$ or $GL(n,\mathbb{C})$. Note that the Lie groups

$$\left\{ \begin{pmatrix} e^t & 0 \\ 0 & e^{-t} \end{pmatrix} \,\middle|\, t \in \mathbb{R} \right\}, \qquad \left\{ \begin{pmatrix} 1 & x \\ 0 & 1 \end{pmatrix} \,\middle|\, x \in \mathbb{R} \right\}$$

are both isomorphic to the additive group of real numbers, and hence to each other, but only the first is reductive in the sense of our definition.

Example 1.3 The Lie groups $SL(n,\mathbb{R})$, $SL(n,\mathbb{C})$ are simple; the Lie groups $GL(n,\mathbb{R})$, $GL(n,\mathbb{C})$ are reductive. Any compact real Lie group is a linear group, as can be deduced from the Peter-Weyl Theorem, and is moreover reductive (see for example Proposition 1.59, Theorem 4.20 and its Corollary 4.22 in [23]).

Remark 1.4 We will show in Example 1.13 that the universal covering $\widetilde{SL(n,\mathbb{R})}$ of $SL(n,\mathbb{R})$, $n \geq 2$, is not a linear group.

One studies reductive Lie groups because these are the groups that naturally arise in geometry, physics and number theory. The notion of semisimple Lie group is a slight generalization of the notion of simple Lie group, and the notion of reductive Lie group in turn generalizes the notion of semisimple Lie group.

1.2 Maximal Compact Subgroups and the Cartan Decomposition

From now on we shall use the symbol $G_{\mathbb{R}}$ to denote a *linear reductive* Lie group; we reserve the symbol G for the complexification of $G_{\mathbb{R}}$ – see section 1.3. As a standing assumption, we suppose that $Z_{G^0_{\mathbb{R}}}$ = center of the identity component $G^0_{\mathbb{R}}$ can be expressed as a direct product

$$\begin{aligned} Z_{G^0_{\mathbb{R}}} &= C \cdot A, \text{ with } C \text{ compact, } A \cong (\mathbb{R}^k, +) \text{ for some } k \geq 0, \text{ and} \\ \mathfrak{a}_{\mathbb{R}} &= \text{Lie algebra of } A \ \text{ has a basis } \{\xi_1, \ldots, \xi_k\} \text{ such that each } \xi_j \\ &\qquad \text{is diagonalizable, with rational eigenvalues.} \end{aligned} \tag{1}$$

The ξ_j are regarded as matrices via the embedding $G_{\mathbb{R}} \subset GL(n, \mathbb{R})$ or $G_{\mathbb{R}} \subset GL(n, \mathbb{C})$ that exhibits $G_{\mathbb{R}}$ as linear group. In the present section, real eigenvalues would suffice, but the rationality of the eigenvalues will become important when we complexify $G_{\mathbb{R}}$.

Example 1.5 The groups $GL(n, \mathbb{R})$, $GL(n, \mathbb{C})$ satisfy the condition (1): for $G_{\mathbb{R}} = GL(n, \mathbb{R})$, $Z_{G^0_{\mathbb{R}}} = \{c \cdot 1_{n\times n} \mid c \in \mathbb{R},\, c^n > 0\}$ and $C = \{\pm 1 \cdot 1_{n\times n}\}$ or $C = \{1_{n\times n}\}$, depending on whether n is even or odd. For $G_{\mathbb{R}} = GL(n, \mathbb{C})$, $Z_{G^0_{\mathbb{R}}}$ is the group of non-zero complex multiples of the identity matrix and $C = \{c \cdot 1_{n\times n} \mid |c| = 1\}$. In both cases, $\mathfrak{a}_{\mathbb{R}}$ consists of all real multiples of the identity matrix, and $A = \exp(\mathfrak{a}_{\mathbb{R}}) \cong (\mathfrak{a}_{\mathbb{R}}, +) \cong (\mathbb{R}, +)$.

We let $K_{\mathbb{R}} \subset G_{\mathbb{R}}$ denote a *maximal compact subgroup*. Maximal compact subgroups exist for dimension reasons. The following general facts can be found in [16], for example.

Proposition 1.6 *Under the stated hypotheses on* $G_{\mathbb{R}}$*,*

a) *any compact subgroup of* $G_{\mathbb{R}}$ *is contained in some maximal compact subgroup* $K_{\mathbb{R}}$*, and* $\dim_{\mathbb{R}} K_{\mathbb{R}} \geq 1$ *unless* $G_{\mathbb{R}}$ *is abelian;*
b) *any two maximal compact subgroups of* $G_{\mathbb{R}}$ *are conjugate by an element of* $G_{\mathbb{R}}$*;*
c) *the inclusion* $K_{\mathbb{R}} \hookrightarrow G_{\mathbb{R}}$ *induces an isomorphism of component groups* $K_{\mathbb{R}}/K^0_{\mathbb{R}} \simeq G_{\mathbb{R}}/G^0_{\mathbb{R}}$*;*
d) *if* $G_{\mathbb{R}}$ *is semisimple, the normalizer of* $K_{\mathbb{R}}$ *in* $G_{\mathbb{R}}$ *coincides with* $K_{\mathbb{R}}$.

Since the maximal compact subgroups are all conjugate, the choice of any one of them is non-essential. At various points, we shall choose a maximal compact subgroup; the particular choice will not matter.

Let $\mathfrak{g}_{\mathbb{R}}$ and $\mathfrak{k}_{\mathbb{R}}$ denote the Lie algebras of $G_{\mathbb{R}}$ and $K_{\mathbb{R}}$, respectively. Then $K_{\mathbb{R}}$ acts on $\mathfrak{g}_{\mathbb{R}}$ via the restriction of the adjoint representation Ad. We recall the *Cartan decomposition* of $\mathfrak{g}_{\mathbb{R}}$:

Proposition 1.7 *There exists a unique $K_{\mathbb{R}}$-invariant linear complement $\mathfrak{p}_{\mathbb{R}}$ of $\mathfrak{k}_{\mathbb{R}}$ in $\mathfrak{g}_{\mathbb{R}}$ (so $\mathfrak{g}_{\mathbb{R}} = \mathfrak{k}_{\mathbb{R}} \oplus \mathfrak{p}_{\mathbb{R}}$ as direct sum of vector spaces), with the following properties:*
a) *The linear map $\theta : \mathfrak{g}_{\mathbb{R}} \to \mathfrak{g}_{\mathbb{R}}$, defined by*

$$\theta\xi = \begin{cases} \xi, & \text{if } \xi \in \mathfrak{k}_{\mathbb{R}}; \\ -\xi, & \text{if } \xi \in \mathfrak{p}_{\mathbb{R}}, \end{cases}$$

is an involutive automorphism *of $\mathfrak{g}_{\mathbb{R}}$; equivalently, $[\mathfrak{p}_{\mathbb{R}}, \mathfrak{p}_{\mathbb{R}}] \subset \mathfrak{k}_{\mathbb{R}}$ and $[\mathfrak{k}_{\mathbb{R}}, \mathfrak{p}_{\mathbb{R}}] \subset \mathfrak{p}_{\mathbb{R}}$.*
b) *Every $\xi \in \mathfrak{p}_{\mathbb{R}} \subset \mathfrak{gl}(n, \mathbb{C})$ is diagonalizable, with real eigenvalues; here $\mathfrak{p}_{\mathbb{R}} \subset \mathfrak{g}_{\mathbb{R}} \subset \mathfrak{gl}(n, \mathbb{C})$ refers to the inclusion induced by the inclusion of Lie groups $G_{\mathbb{R}} \subset GL(n, \mathbb{R}) \subset GL(n, \mathbb{C})$ or $G_{\mathbb{R}} \subset GL(n, \mathbb{C})$.*

Remark 1.8 Analogously to b), every $\xi \in \mathfrak{k}_{\mathbb{R}} \subset \mathfrak{gl}(n, \mathbb{C})$ is diagonalizable, with purely imaginary eigenvalues. Indeed, $\{t \mapsto \exp t\xi\}$ is a one-parameter subgroup of $K_{\mathbb{R}}$, and must therefore have bounded matrix entries; that is possible only when ξ is diagonalizable over $\mathbb{C}$, with purely imaginary eigenvalues.

One calls $\theta : \mathfrak{g}_{\mathbb{R}} \to \mathfrak{g}_{\mathbb{R}}$ the *Cartan involution* of $\mathfrak{g}_{\mathbb{R}}$. It lifts to an involutive automorphism of $G_{\mathbb{R}}$, which we also denote by θ.

Example 1.9 The group $G_{\mathbb{R}} = SL(n, \mathbb{R})$ contains $K_{\mathbb{R}} = SO(n)$ as maximal compact subgroup. In this situation

$$\begin{aligned} \mathfrak{g}_{\mathbb{R}} &= \{\xi \in \operatorname{End}(\mathbb{R}^n) \mid \operatorname{tr}\xi = 0\}, \\ \mathfrak{k}_{\mathbb{R}} &= \{\xi \in \operatorname{End}(\mathbb{R}^n) \mid {}^t\xi = -\xi,\ \operatorname{tr}\xi = 0\}, \\ \mathfrak{p}_{\mathbb{R}} &= \{\xi \in \operatorname{End}(\mathbb{R}^n) \mid {}^t\xi = \xi,\ \operatorname{tr}\xi = 0\}. \end{aligned}$$

On the Lie algebra level, $\theta\xi = -{}^t\xi$, and on the group level, $\theta g = {}^t g^{-1}$. The group $K_{\mathbb{R}}$ can be characterized as the fixed point set of θ, i.e., $K_{\mathbb{R}} = \{g \in G_{\mathbb{R}};\ \theta g = g\}$.

The Cartan decomposition of $\mathfrak{g}_{\mathbb{R}}$ has a counterpart on the group level, the so-called *global Cartan decomposition*:

Proposition 1.10 *The map $K_{\mathbb{R}} \times \mathfrak{p}_{\mathbb{R}} \to G_{\mathbb{R}}$, defined by $(k, \xi) \mapsto k \cdot \exp \xi$, determines a diffeomorphism of manifolds.*

In the setting of the above example the proposition is essentially equivalent to the well-known assertion that any invertible real square matrix can be expressed uniquely as the product of an orthogonal matrix and a positive definite symmetric matrix.

Remark 1.11 As a consequence of the proposition, $K_\mathbb{R} \hookrightarrow G_\mathbb{R}$ is a strong deformation retract. In particular this inclusion induces isomorphisms of homology and homotopy groups.

1.3 Complexifications of Linear Groups

We continue with the hypotheses stated in the beginning of the last subsection. Like any linear Lie group, $G_\mathbb{R}$ has a *complexification* – a complex Lie group G, with Lie algebra

$$\mathfrak{g} =_{\text{def}} \mathbb{C} \otimes_\mathbb{R} \mathfrak{g}_\mathbb{R}\,, \tag{2}$$

containing $G_\mathbb{R}$ as a Lie subgroup, such that

a) the inclusion $G_\mathbb{R} \hookrightarrow G$ induces $\mathfrak{g}_\mathbb{R} \hookrightarrow \mathfrak{g}$, $\xi \mapsto 1 \otimes \xi$, and

b) $G_\mathbb{R}$ meets every connected component of G.

To construct a complexification, one regards $G_\mathbb{R}$ as a subgroup[1] of $GL(n,\mathbb{R})$, so that $\mathfrak{g}_\mathbb{R} \subset \mathfrak{gl}(n,\mathbb{R})$. That makes $\mathfrak{g}$ a Lie subalgebra of $\mathfrak{gl}(n,\mathbb{C}) = \mathbb{C} \otimes_\mathbb{R} \mathfrak{gl}(n,\mathbb{R})$. Then G^0, the connected Lie subgroup of $GL(n,\mathbb{C})$ with Lie algebra $\mathfrak{g}$, satisfies the condition a). By construction $G^0_\mathbb{R} \subset G^0$, and $G_\mathbb{R}$ normalizes G^0, hence $G = G_\mathbb{R} \cdot G^0$ is a complex Lie group with Lie algebra $\mathfrak{g}$, which contains $G_\mathbb{R}$ and satisfies both a) and b). When G is a complexification of $G_\mathbb{R}$, one calls $G_\mathbb{R}$ a *real form* of G. We do not exclude the case of a Lie group $G_\mathbb{R}$ which happens to be a complex Lie group; in the case of $G_\mathbb{R} = GL(n,\mathbb{C})$, for example, $G \cong GL(n,\mathbb{C}) \times GL(n,\mathbb{C}) \subset GL(2n,\mathbb{C})$.

In general, the complexification of a linear Lie group depends on its realization as a linear group. In our situation, the complexification G inherits the property (1) from $G_\mathbb{R}$. It implies that G is determined by $G_\mathbb{R}$ up to isomorphism, but the *embedding* does depend on the realization as real group, unless $G^0_\mathbb{R}$ has a compact center.

One can complexify the Cartan decomposition: let $\mathfrak{g} = \mathbb{C} \otimes_\mathbb{R} \mathfrak{g}_\mathbb{R}$ as before, $\mathfrak{k} = \mathbb{C} \otimes_\mathbb{R} \mathfrak{k}_\mathbb{R}$, and $\mathfrak{p} = \mathbb{C} \otimes_\mathbb{R} \mathfrak{p}_\mathbb{R}$; then $\mathfrak{g} = \mathfrak{k} \oplus \mathfrak{p}$ as vector spaces. The complexification G of $G_\mathbb{R}$ naturally contains K = complexification of $K_\mathbb{R}$, as complex Lie subgroup. Observe that K cannot be compact unless $K_\mathbb{R} = \{e\}$, which does not happen unless $G_\mathbb{R}$ is abelian; indeed, any nonzero $\xi \in \mathfrak{k}_\mathbb{R}$ is diagonalizable over $\mathbb{C}$, with purely imaginary eigenvalues, not all zero, so the complex one-parameter subgroup $\{z \mapsto \exp(z\xi)\}$ of

[1] If $G_\mathbb{R}$ is presented as a linear group $G_\mathbb{R} \subset GL(m,\mathbb{C})$, one uses the usual inclusion $GL(m,\mathbb{C}) \hookrightarrow GL(2m,\mathbb{R})$ to exhibit $G_\mathbb{R}$ as subgroup of $GL(n,\mathbb{R})$, with $n = 2m$.

K is unbounded. By construction, the Lie algebras $\mathfrak{g}_{\mathbb{R}}$, $\mathfrak{k}_{\mathbb{R}}$, $\mathfrak{g}$, $\mathfrak{k}$ and the corresponding groups satisfy the following containments:

$$\begin{array}{ccc} \mathfrak{g}_{\mathbb{R}} & \subset & \mathfrak{g} \\ \cup & & \cup \\ \mathfrak{k}_{\mathbb{R}} & \subset & \mathfrak{k} \end{array} \qquad\qquad \begin{array}{ccc} G_{\mathbb{R}} & \subset & G \\ \cup & & \cup \\ K_{\mathbb{R}} & \subset & K \end{array} \tag{3}$$

Since $[\mathfrak{p}_{\mathbb{R}}, \mathfrak{p}_{\mathbb{R}}] \subset \mathfrak{k}_{\mathbb{R}}$ and $[\mathfrak{k}_{\mathbb{R}}, \mathfrak{p}_{\mathbb{R}}] \subset \mathfrak{p}_{\mathbb{R}}$,

$$\mathfrak{u}_{\mathbb{R}} =_{\text{def}} \mathfrak{k}_{\mathbb{R}} \oplus i\mathfrak{p}_{\mathbb{R}} \quad \text{is a real Lie subalgebra of } \mathfrak{g}. \tag{4}$$

We let $U^0_{\mathbb{R}}$ denote the connected Lie subgroup of G with Lie algebra $\mathfrak{u}_{\mathbb{R}}$. If $G_{\mathbb{R}}$ is semisimple, one knows that $U^0_{\mathbb{R}}$ is compact [16]; as a consequence of our hypotheses, $U^0_{\mathbb{R}}$ is compact even in the reductive case. Thus $U^0_{\mathbb{R}}$ lies in a maximal compact subgroup of G, which we denote by $U_{\mathbb{R}}$. Not only is $U_{\mathbb{R}} \subset G$ a maximal compact subgroup, but also

$$\begin{aligned} &\text{a) } U_{\mathbb{R}} \text{ is a real form of } G; \\ &\text{b) } K_{\mathbb{R}} = U_{\mathbb{R}} \cap G_{\mathbb{R}}. \end{aligned} \tag{5}$$

Both assertions are well known in the semisimple case, to which the general case can be reduced.

The process of complexification establishes a bijection, up to isomorphism, between compact Lie groups and linear, reductive, complex Lie groups[2] satisfying (1); in the opposite direction, the correspondence is given by the passage to a maximal compact subgroup, which is then a *compact real form* of the complex group. The groups $U_{\mathbb{R}}$ and G are related in this fashion: the former is a compact real form of the latter. In view of Remark 1.11, $U_{\mathbb{R}} \hookrightarrow G$ is a strong deformation retract, which induces isomorphisms of homology and homotopy.

Example 1.12 The pair $G_{\mathbb{R}} = SL(n, \mathbb{R})$, $K_{\mathbb{R}} = SO(n)$, has complexifications $G = SL(n, \mathbb{C})$, $K = SO(n, \mathbb{C})$; the corresponding compact real form of $G = SL(n, \mathbb{C})$ is $U_{\mathbb{R}} = SU(n)$.

Since $\mathfrak{g} = \mathbb{C} \otimes_{\mathbb{R}} \mathfrak{u}_{\mathbb{R}}$, these two Lie algebras have the same representations over $\mathbb{C}$ – representations can be restricted from $\mathfrak{g}$ to $\mathfrak{u}_{\mathbb{R}}$, and in the opposite direction, can be extended complex linearly. On the global

[2] Any connected complex Lie group with a reductive Lie algebra can be realized as a linear group; we mention the hypothesis of linearity to signal that we want the linear realization to be reductive in the sense of our definition.

level, these operations induce a canonical bijection

$$\left\{\begin{matrix}\text{holomorphic}\\ \text{finite dimensional}\\ \text{representations of } G\end{matrix}\right\} \simeq \left\{\begin{matrix}\text{finite dimensional}\\ \text{continuous complex}\\ \text{representations of } U_{\mathbb{R}}\end{matrix}\right\}, \tag{6}$$

a bijection because $U_{\mathbb{R}} \hookrightarrow G$ induces an isomorphism of the component group and the fundamental group. Of course we are also using the well known fact that continuous finite dimensional representations of Lie groups are necessarily real analytic, and are determined on the identity component by the corresponding infinitesimal representations of the Lie algebra.

We had mentioned earlier that the universal covering group of $G = SL(n, \mathbb{R})$, $n \geq 2$, is not a linear group. We can now sketch the argument:

Example 1.13 Let $\widetilde{SL(n, \mathbb{R})}$ be the universal covering group of $SL(n, \mathbb{R})$, $n \geq 2$. Since

$$\pi_1(SL(n, \mathbb{R})) = \pi_1(SO(n)) = \begin{cases} \mathbb{Z}, & \text{if } n = 2; \\ \mathbb{Z}/2\mathbb{Z}, & \text{if } n \geq 3, \end{cases}$$

the universal covering $\widetilde{SL(n, \mathbb{R})} \to SL(n, \mathbb{R})$ is a principal $\mathbb{Z}$-bundle when $n = 2$ and a principal $\mathbb{Z}/2\mathbb{Z}$-bundle when $n \geq 3$. If $\widetilde{SL(n, \mathbb{R})}$ were linear, its complexification would have to be a covering group of $SL(n, \mathbb{C})$ = complexification of $SL(n, \mathbb{R})$, of infinite order when $n = 2$ and of order (at least) two when $n \geq 3$. But $SU(n)$, $n \geq 2$, is simply connected, as can be shown by induction on n. But then $SL(n, \mathbb{C})$ = complexification of $SU(n)$ is also simply connected, and therefore cannot have a non trivial covering. We conclude that $\widetilde{SL(n, \mathbb{R})}$ is not a linear group.

2

Compact Lie Groups

In this section we consider the well understood case of a connected, compact Lie group. As was remarked in Example 1.3, any such group is automatically linear and reductive. In the setting of chapter 1, the groups $K_{\mathbb{R}}$ and $U_{\mathbb{R}}$ are compact, but not necessarily connected. In any case, knowing the representations of the identity component explicitly gives considerable information about the representations of a non-connected compact group – modulo knowledge of the representations of the component group, of course. In section 4, we shall suppose that the group $G_{\mathbb{R}}$ has a connected complexification G; in that case, $U_{\mathbb{R}}$ will indeed be connected. Let us state the hypotheses of the current section explicitly:

$$U_{\mathbb{R}} \text{ is a connected compact Lie group.} \tag{7}$$

As in the previous section $\mathfrak{g} = \mathbb{C} \otimes_{\mathbb{R}} u_{\mathbb{R}}$ denotes the complexified lie algebra of $U_{\mathbb{R}}$.

2.1 Maximal Tori, the Unit Lattice, and the Weight Lattice

A common strategy in mathematics is to study properties of a new class of objects by looking for sub-objects whose properties are already known, but which are "large enough" to convey some useful information about the objects to be studied. In representation theory this means studying representations of compact groups by restricting them to maximal tori, while representations of noncompact linear reductive groups are studied by restricting them to maximal compact subgroups.

With $U_{\mathbb{R}}$ connected and compact, as we are assuming, let $T_{\mathbb{R}} \subset U_{\mathbb{R}}$

denote a *maximal torus*. It is not difficult to see that maximal tori exist and are nontrivial – i.e., $\dim T_{\mathbb{R}} > 0$ – unless $U_{\mathbb{R}} = \{e\}$. Moreover,

Proposition 2.1

a) *Any two maximal tori in $U_{\mathbb{R}}$ are conjugate by an element of $U_{\mathbb{R}}$;*
b) *any $g \in U_{\mathbb{R}}$ is conjugate to some $t \in T_{\mathbb{R}}$;*
c) *$T_{\mathbb{R}}$ coincides with its own centralizer in $U_{\mathbb{R}}$, i.e., $T_{\mathbb{R}} = Z_{U_{\mathbb{R}}}(T_{\mathbb{R}})$;*
d) *$T_{\mathbb{R}}$ is the identity component of its own normalizer in $U_{\mathbb{R}}$, i.e., $T_{\mathbb{R}} = N_{U_{\mathbb{R}}}(T_{\mathbb{R}})^0$.*

We fix a particular maximal torus $T_{\mathbb{R}}$, with Lie algebra $\mathfrak{t}_{\mathbb{R}}$ and complexified Lie algebra $\mathfrak{t}$. In view of the Proposition, the particular choice will not matter. Since $T_{\mathbb{R}}$ is abelian and connected, the exponential $\exp : \mathfrak{t}_{\mathbb{R}} \to T_{\mathbb{R}}$ is a surjective homomorphism, relative to the additive structure of $\mathfrak{t}_{\mathbb{R}}$.

Remark 2.2 The exponential mapping of a general connected Lie group need not be surjective. For instance, $g = \begin{pmatrix} -1 & 1 \\ 0 & -1 \end{pmatrix} \in SL(2,\mathbb{R})$ cannot lie in the image of the exponential map. Indeed, if this element of g could be expressed as $g = \exp\xi$, for some $\xi \in \mathfrak{sl}(2,\mathbb{R})$, then ξ is not diagonalizable, even over $\mathbb{C}$, since g is not diagonalizable over $\mathbb{C}$. That forces ξ to have two equal eigenvalues, necessarily eigenvalues zero because $\operatorname{tr}\xi = 0$. Contradiction: $g = \exp(\xi)$ does not have eigenvalues 1.

The exponential map $\exp : \mathfrak{t}_{\mathbb{R}} \to T_{\mathbb{R}}$ is not only a surjective homomorphism, but also locally bijective, hence a covering homomorphism,

$$\exp : \mathfrak{t}_{\mathbb{R}}/L \xrightarrow{\sim} T_{\mathbb{R}} \qquad (\, L = \{\xi \in t_{\mathbb{R}} \mid \exp\xi = e\}\,)\,. \tag{8}$$

That makes $L \subset t_{\mathbb{R}}$ a discrete, cocompact subgroup. In other words, L is a lattice, the so called *unit lattice*. Let $\widehat{T}_{\mathbb{R}}$ denote the group of characters, i.e. the group of homomorphisms from $T_{\mathbb{R}}$ to the unit circle $S^1 = \{z \in \mathbb{C} \mid |z| = 1\}$. Then

$$\Lambda =_{\text{def}} \{\,\lambda \in it^*_{\mathbb{R}} \mid \langle \lambda, L\rangle \subset 2\pi i\mathbb{Z}\,\} \;\xrightarrow{\sim}\; \widehat{T}_{\mathbb{R}}\,, \tag{9}$$

$$\Lambda \ni \lambda \longmapsto e^{\lambda} \in \widehat{T}_{\mathbb{R}}\,,$$

with $e^{\lambda} : T_{\mathbb{R}} \to \{z \in \mathbb{C} \mid |z| = 1\}$ defined by $e^{\lambda}(\exp\xi) = e^{\langle\lambda,\xi\rangle}$, for any $\xi \in \mathfrak{t}_{\mathbb{R}}$. One calls $\Lambda \subset i\mathfrak{t}^*_{\mathbb{R}}$ the *weight lattice*; except for the factor $2\pi i$ in its definition, it is the lattice dual of the unit lattice $L \subset \mathfrak{t}_{\mathbb{R}}$.

2.2 Weights, Roots, and the Weyl Group

Let π be a representation of $U_{\mathbb{R}}$ on a finite-dimensional complex vector space V – in other words, a continuous homomorphism $\pi : U_{\mathbb{R}} \to GL(V)$. Since $T_{\mathbb{R}}$ is compact and the field $\mathbb{C}$ algebraically closed, the action of any $t \in T_{\mathbb{R}}$ can be diagonalized, and since $T_{\mathbb{R}}$ is abelian, the action of all the $t \in T_{\mathbb{R}}$ can be diagonalized simultaneously. Thus we obtain the *weight space decomposition*

$$V = \oplus_{\lambda \in \Lambda} V^{\lambda}, \tag{10}$$

where

$$\begin{aligned} V^{\lambda} &= \{ v \in V \mid \pi(t)v = e^{\lambda}(t)v \ \ \forall t \in T_{\mathbb{R}} \} \\ &= \{ v \in V \mid \pi(\xi)v = \langle \lambda, \xi \rangle v \ \ \forall \xi \in \mathfrak{t}_{\mathbb{R}} \}. \end{aligned}$$

If $V^{\lambda} \neq \{0\}$, one calls λ, V^{λ}, and $\dim V^{\lambda}$ respectively a *weight* of π, the *weight space* corresponding to λ, and the *multiplicity* of the weight λ.

In the case of the adjoint representation of $U_{\mathbb{R}}$ on the complexified Lie algebra $\mathfrak{g} = \mathbb{C} \otimes_{\mathbb{R}} \mathfrak{u}_{\mathbb{R}}$, one singles out the weight zero: $\mathfrak{g} = \mathfrak{g}^0 \oplus (\oplus_{\lambda \neq 0} \mathfrak{g}^{\lambda})$. Evidently $\mathfrak{g}^0 \supset \mathfrak{t}$, since $T_{\mathbb{R}}$ acts trivially on $\mathfrak{t}$. In fact, $\mathfrak{g}^0 = \mathfrak{t}$, for one could otherwise show that $T_{\mathbb{R}}$ is not a maximal torus. Nonzero weights of the adjoint representation are called *roots*, hence

$$\mathfrak{g} = \mathfrak{t} \oplus (\oplus_{\alpha \in \Phi} \mathfrak{g}^{\alpha}), \quad \text{with} \quad \Phi = \Phi(\mathfrak{g}, \mathfrak{t}) = \text{set of all roots}. \tag{11}$$

This is the *root space decomposition* of $\mathfrak{g}$ relative to the action of $T_{\mathbb{R}}$, or equivalently, relative to the action of $\mathfrak{t}$. One refers to Φ as the *root system*. Very importantly,

$$\alpha \in \Phi \implies \dim \mathfrak{g}^{\alpha} = 1, \tag{12}$$

i.e., all roots have multiplicity one. Since $\Phi \subset \Lambda - \{0\} \subset i\mathfrak{t}_{\mathbb{R}}^* \subset \mathfrak{t}^*$, roots take purely imaginary values on the real Lie algebra $\mathfrak{t}_{\mathbb{R}}$, which implies

$$\overline{\mathfrak{g}^{\alpha}} = \mathfrak{g}^{\overline{\alpha}} = \mathfrak{g}^{-\alpha}, \quad \text{and hence} \quad \Phi = -\Phi; \tag{13}$$

here $\overline{\mathfrak{g}^{\alpha}}$ denotes the complex conjugate of $\mathfrak{g}^{\alpha}$ with respect to the real form $\mathfrak{u}_{\mathbb{R}} \subset \mathfrak{g}$.

Let $\pi : \mathfrak{u}_{\mathbb{R}} \to \mathfrak{gl}(V)$ denote the infinitesimal representation induced by the global representation π considered at the beginning of this subsection, and $\pi : \mathfrak{g} \to \mathfrak{gl}(V)$ its complex extension. The latter may be interpreted as a linear map $\mathfrak{g} \otimes_{\mathbb{C}} V \to V$, which is $U_{\mathbb{R}}$-invariant – therefore also $T_{\mathbb{R}}$-invariant – when $U_{\mathbb{R}}$ is made to act on $\mathfrak{g}$ via Ad. Hence, for every $\alpha \in \Phi$ and every weight λ of π,

$$\pi(\mathfrak{g}^{\alpha})\, V^{\lambda} \;\subset\; V^{\lambda+\alpha}\,;$$

in particular

$$\pi(\mathfrak{g}^{\alpha})\, V^{\lambda} \;=\; 0 \quad \text{if } \lambda+\alpha \text{ is not a weight.}$$

Applied to the adjoint representation, this means

$$[\mathfrak{g}^{\alpha}, \mathfrak{g}^{\beta}] \;\subset\; \begin{cases} \mathfrak{g}^{\alpha+\beta} & \text{if } \alpha+\beta \in \Phi \\ \mathfrak{t} & \text{if } \alpha+\beta = 0 \\ 0 & \text{if } \alpha+\beta \notin \Phi \cup \{0\}\,, \end{cases} \tag{14}$$

for all roots $\alpha, \beta \in \Phi$.

An element $\xi \in \mathfrak{t}$ is said to be *singular* if $\langle \alpha, \xi \rangle = 0$ for some root α, and otherwise *regular*. The set

$$i\mathfrak{t}'_{\mathbb{R}} \;=_{\mathrm{def}}\; \{\, \mathrm{i}\mathfrak{t}_{\mathbb{R}} \mid \xi \text{ is regular}\,\} \tag{15}$$

breaks up into a finite, disjoint union of open, convex cones, the so-called *Weyl chambers.* If $C \subset i\mathfrak{t}_{\mathbb{R}}$ is a particular Weyl chamber and ξ an element of C, the subset

$$\Phi^{+} \;=\; \{\alpha \in \Phi \mid \langle \alpha, \xi \rangle > 0\} \;\subset\; \Phi \tag{16}$$

depends only on C, not on the choice of $\xi \in C$; by definition, Φ^{+} is a *system of positive roots.* Essentially by construction,

$$\begin{aligned} &\text{a)} \quad \Phi \;=\; \Phi^{+} \cup (-\Phi^{+}) \quad \text{(disjoint union);} \\ &\text{b)} \quad \alpha, \beta \in \Phi^{+}\,, \quad \alpha+\beta \in \Phi \quad \Longrightarrow \quad \alpha+\beta \in \Phi^{+}. \end{aligned} \tag{17}$$

The Weyl chamber C can be recovered from the system of positive roots Φ^{+},

$$C \;=\; \{\, \xi \in i\mathfrak{t}_{\mathbb{R}} \mid \langle \alpha, \xi \rangle > 0\,\}\,. \tag{18}$$

In fact, $C \longleftrightarrow \Phi^{+}$ establishes a bijection between Weyl chambers and positive root systems.

Via the adjoint action, the normalizer $N_{U_{\mathbb{R}}}(T_{\mathbb{R}})$ acts on $\mathfrak{t}_{\mathbb{R}}$, on $i\mathfrak{t}^{*}_{\mathbb{R}}$, on Λ, on Φ, and on the set of Weyl chambers. Since $T_{\mathbb{R}} = Z_{U_{\mathbb{R}}}(T_{\mathbb{R}}) = N_{U_{\mathbb{R}}}(T_{\mathbb{R}})^{0}$ by Proposition 2.1, the *Weyl group*

$$W \;=\; W(U_{\mathbb{R}}, T_{\mathbb{R}}) \;=_{def}\; N_{U_{\mathbb{R}}}(T_{\mathbb{R}})/T_{\mathbb{R}} \tag{19}$$

is a finite group, which acts on $\mathfrak{t}_{\mathbb{R}}$, on $i\mathfrak{t}^{*}_{\mathbb{R}}$, on the weight lattice Λ, and on Φ. This action permutes the Weyl chambers, hence also the positive root systems.

Proposition 2.3 *The Weyl group $W(U_{\mathbb{R}}, T_{\mathbb{R}})$ acts faithfully on $\mathfrak{t}_{\mathbb{R}}$, and acts simply transitively on the set of Weyl chambers $\{C\}$, as well as on the set of positive root systems $\{\Phi^+\}$.*

In particular, if some $n \in N_{U_{\mathbb{R}}}(T_{\mathbb{R}})$ fixes a root system Φ^+, then $n \in T_{\mathbb{R}}$. In the following, we shall fix a positive root system Φ^+. The particular choice will not matter since they are all conjugate to each other. The Weyl chamber C that corresponds to Φ^+ is called the *dominant Weyl chamber.*

2.3 The Theorem of the Highest Weight

We consider finite dimensional representations of $U_{\mathbb{R}}$ on complex vector spaces, as in the previous subsection. Recall that a finite dimensional representation (π, V) is *irreducible* if the representation space V contains no $U_{\mathbb{R}}$-invariant subspaces other than $\{0\}$ and V itself; it is *completely reducible* if it can be written as the direct sum of irreducible subrepresentations. If (π, V) is unitary – i.e., if V comes equipped with $U_{\mathbb{R}}$-invariant inner product – the orthogonal complement of an invariant subspace is again invariant. One can therefore successively split off one minimal invariant subspace at a time. Since minimal invariant subspaces are irreducible, this shows that finite dimensional, unitary representations are completely reducible. Any representation (π, V) of the compact group $U_{\mathbb{R}}$ can be made unitary: one puts an arbitrary inner product on the space V, and then uses Haar measure to average the g-translates of this inner product for all $g \in U_{\mathbb{R}}$; the averaged inner product is $U_{\mathbb{R}}$-invariant. This implies the well known fact that

finite dimensional representations of
compact groups are completely reducible.

In particular, to understand the finite dimensional representations of $U_{\mathbb{R}}$, it suffices to understand the finite dimensional, irreducible representations.

The preceding discussion applies to any compact Hausdorff group. We now return to the case of a connected, compact Lie group $U_{\mathbb{R}}$. Our definitions and statements will involve the choice of a maximal torus $T_{\mathbb{R}}$ and a positive root system $\Phi^+ \subset i\mathfrak{t}^*_{\mathbb{R}}$. As was remarked earlier, these are not essential choices. Since $U_{\mathbb{R}}$ is compact, there exists a negative definite, $\mathrm{Ad}(U_{\mathbb{R}})$-invariant, symmetric bilinear form

$$S : \mathfrak{u}_{\mathbb{R}} \times \mathfrak{u}_{\mathbb{R}} \longrightarrow \mathbb{R}, \tag{20}$$

$\mathrm{Ad}(U_{\mathbb{R}})$-invariant in the sense that $S(\mathrm{Ad}\, g(\xi), \mathrm{Ad}\, g(\eta)) = S(\xi,\eta)$, for all $\xi, \eta \in \mathfrak{u}_{\mathbb{R}}$ and $g \in U_{\mathbb{R}}$, or equivalently, on the infinitesimal level,

$$S(\,[\zeta,\xi]\,,\,\eta\,) + S(\,\xi\,,\,[\zeta,\eta]\,) \;=\; 0\,, \qquad \text{for all } \zeta,\xi,\eta \in \mathfrak{u}_{\mathbb{R}}\,. \tag{21}$$

One way to construct S is to take a finite-dimensional representation (π, V) which is faithful, or at least faithful on the level of the Lie algebra, and define

$$S(\xi,\eta) \;=\; \mathrm{tr}\big(\pi(\xi)\pi(\eta)\big)\,. \tag{22}$$

Indeed, S is symmetric, bilinear, $\mathrm{Ad}(U_{\mathbb{R}})$-invariant by construction, and any nonzero $\xi \in \mathfrak{u}_{\mathbb{R}}$ acts on V dagonalizably, with purely imaginary eigenvalues, not all zero, so $\mathrm{tr}\big(\pi(\xi)\pi(\xi)\big) < 0$. If $U_{\mathbb{R}}$ is semisimple, one can let the adjoint representation play the role of π; in that case one calls S the *Killing form.*

The bilinear form S is far from uniquely determined by the required properties. However, its restriction to the various simple ideals in $\mathfrak{u}_{\mathbb{R}}$ *is unique*, up to scaling, as follows from Schur's lemma. This partial uniqueness suffices for our purposes.

Extending scalars from $\mathbb{R}$ to $\mathbb{C}$, we obtain an $\mathrm{Ad}(U_{\mathbb{R}})$-invariant, symmetric, complex bilinear form $S : \mathfrak{g} \times \mathfrak{g} \to \mathbb{C}$, which is positive definite on $i\mathfrak{u}_{\mathbb{R}}$. By restriction it induces a positive definite inner product $(\,.\,,\,.\,)$ on $i\mathfrak{t}_{\mathbb{R}}$, and by duality also on $i\mathfrak{t}^*_{\mathbb{R}}$. The Weyl group $W = W(U_{\mathbb{R}}, T_{\mathbb{R}})$ preserves these inner products, since they are obtained by restriction of an *Ad*$(U_{\mathbb{R}})$-invariant bilinear form.

Definition 2.4 An element $\lambda \in i\mathfrak{t}^*_{\mathbb{R}}$ is said to be *dominant* if $(\lambda, \alpha) \geq 0$ for all $\alpha \in \Phi^+$, and *regular* if $(\lambda,\alpha) \neq 0$ for all $\alpha \in \Phi$.

These notions apply in particular to any λ in the weight lattice Λ. The inner product identifies $i\mathfrak{t}^*_{\mathbb{R}}$ with its own dual $i\mathfrak{t}_{\mathbb{R}}$, and via this identification, the set of all dominant, regular $\lambda \in i\mathfrak{t}^*_{\mathbb{R}}$ corresponds precisely to the dominant Weyl chamber $C \subset \mathfrak{t}_{\mathbb{R}}$, i.e., the Weyl chamber determined by Φ^+. Since W acts simply transitively on the set of Weyl chambers, every regular $\lambda \in i\mathfrak{t}^*_{\mathbb{R}}$ is W-conjugate to exactly one dominant, regular $\lambda' \in i\mathfrak{t}^*_{\mathbb{R}}$. In fact, this statement remains correct if one drops the condition of regularity; this can be seen by perturbing a singular $\lambda \in i\mathfrak{t}^*_{\mathbb{R}}$

slightly, so as to make it regular. The action of W preserves the weight lattice, hence

$$\text{every } \lambda \in \Lambda \text{ is } W\text{-conjugate to a unique dominant } \lambda' \in \Lambda\,; \tag{23}$$

in other words, $\{\lambda \in \Lambda \mid \lambda \text{ is dominant}\} \cong W\backslash\Lambda$.

Theorem 2.5 (Theorem of the Highest Weight) *For an irreducible, finite dimensional, complex representation π, the following conditions on a weight λ of π are equivalent:*

1. *$\lambda + \alpha$ is not a weight, for any positive root $\alpha \in \Phi^+$;*
2. *there exists a non-zero $v_0 \in V^\lambda$ such that $\pi(\mathfrak{g}^\alpha)\, v_0 = 0$ for all $\alpha \in \Phi^+$;*
3. *any weight of π can be expressed as $\lambda - A$, where A is a sum of positive roots (possibly empty; repetitions are allowed).*

There exists exactly one weight λ of π with these (equivalent) properties, the so-called highest weight *of π. The highest weight is dominant, has multiplicity one (i.e. $\dim V^\lambda = 1$), and determines the representation π up to isomorphism. Every dominant $\lambda \in \Lambda$ arises as the highest weight of an irreducible representation π.*

The assertion that property *2* implies property *3* can be deduced from the Poincaré-Birkhoff-Witt theorem, and the other implications among the three properties can be established by elementary arguments. Among the remaining statements, only the existence of an irreducible finite dimensional representation with a given regular dominant weight requires some effort. One can prove this existence statement analytically, via the Weyl character formula, algebraically, by realizing the representation in question as a quotient of a Verma module, or geometrically, as will be sketched in subsection 2.5.

In effect, the theorem parameterizes the isomorphism classes of irreducible finite dimensional representations over $\mathbb{C}$ in terms of their highest weights,

$$\left\{\begin{array}{c}\text{irreducible}\\ \text{finite-dimensional complex}\\ \text{representations of } U_{\mathbb{R}}\\ \text{up to isomorphism}\end{array}\right\} \longleftrightarrow \{\lambda \in \Lambda \mid \lambda \text{ dominant}\} \longleftrightarrow W\backslash\Lambda\,. \tag{24}$$

Beyond the enumeration of irreducible finite dimensional representations, the theorem also provides structural information about these representations. In fact, most of the general structural properties that are used in applications are consequences of the Theorem of the Highest Weight.

2.4 Borel Subalgebras and the Flag Variety

We now also consider the complexification G of the connected compact Lie group $U_{\mathbb{R}}$. In principle, its Lie algebra $\mathfrak{g} = \mathbb{C} \otimes_{\mathbb{R}} \mathfrak{u}_{\mathbb{R}}$ can be any reductive Lie algebra over $\mathbb{C}$, and G any connected, complex, reductive Lie group.

Definition 2.6 A *Borel subalgebra* of $\mathfrak{g}$ is a maximal solvable subalgebra. A *Borel subgroup* of G is a connected complex Lie subgroup of G whose Lie algebra is a Borel subalgebra of $\mathfrak{g}$.

Proposition 2.7 *Any two Borel subgroups of G, respectively Borel subalgebras of $\mathfrak{g}$, are conjugate under the action of G. Any Borel subgroup $B \subset G$ coincides with its own normalizer, i.e., $N_G(B) = B$. In particular, Borel subgroups are closed.*

Remark 2.8 The property $N_G(B) = B$ implies that *any* complex subgroup of G whose Lie algebra is a Borel subalgebra is automatically connected, and hence is a Borel subgroup. In other words, the requirement of connectedness in Definition 2.6 can be dropped without changing the notion of a Borel subgroup.

The linear subspaces of $\mathfrak{g}$ of a fixed dimension constitute a smooth projective variety, a so-called Grassmann variety. Being a subalgebra, or more specifically a solvable subalgebra, amounts to an algebraic condition on an arbitrary point in this Grassmannian. Since all Borel subalgebras have the same dimension, call it d, a solvable subalgebra of $\mathfrak{g}$ is maximal solvable if and only if it has dimension d. We conclude that

$$X \ = \ \text{set of all Borel subalgebras of } \mathfrak{g} \tag{25}$$

is a closed subvariety of the Grassmannian. This gives X the structure of a complex projective variety. By definition, X is the *flag variety* of $\mathfrak{g}$. We already know that G acts transitively on X via the adjoint action, which is algebraic. In particular X is smooth.

The flag variety X can be characterized by a universal property: it dominates all the complex projective varieties with a transitive, algebraic action of G – any other variety Y with these properties is a G-equivariant quotient of X, i.e., the image of X under a surjective algebraic map that relates the G-actions on X and Y. Such G-equivariant quotients of X are called *generalized flag varieties*

As before, let $T_{\mathbb{R}} \subset U_{\mathbb{R}}$ be a maximal torus, $\mathfrak{t}$ its complexified Lie algebra, and Φ^+ a system of positive roots. One can show quite directly that

$$\mathfrak{b}_0 \;=\; \mathfrak{t} \,\oplus\, \left(\oplus_{\alpha\in\Phi^+}\,\mathfrak{g}^{-\alpha}\right) \tag{26}$$

is maximal solvable in $\mathfrak{g}$, hence a Borel subalgebra. Any other Borel subalgebra is conjugate to it under the action of G, and even under the action of $U_{\mathbb{R}}$, as we shall see. The corresponding Borel subgroup B_0 is also the normalizer of B_0, hence the normalizer of $\mathfrak{b}_0$ in G – in other words, B_0 is the isotropy subgroup at $\mathfrak{b}_0$ for the action of G on X. That implies $X \simeq G/B_0$, since G acts transitively on X.

Lemma 2.9 $U_{\mathbb{R}} \cap B_0 = T_{\mathbb{R}}$.

Proof. Complex conjugation with respect to the real form $\mathfrak{u}_{\mathbb{R}} \subset \mathfrak{g}$ maps $\mathfrak{g}^{-\alpha}$ to $\mathfrak{g}^{\alpha}$, hence $\mathfrak{b}_0 \cap \overline{\mathfrak{b}_0} = \mathfrak{t}$, hence

$$\mathfrak{u}_{\mathbb{R}} \cap \mathfrak{b}_0 \;=\; (\mathfrak{u}_{\mathbb{R}} \cap \mathfrak{b}_0) \cap \overline{(\mathfrak{u}_{\mathbb{R}} \cap \mathfrak{b}_0)} \;=\; \mathfrak{u}_{\mathbb{R}} \cap \mathfrak{b}_0 \cap \overline{\mathfrak{b}_0} \;=\; \mathfrak{u}_{\mathbb{R}} \cap \mathfrak{t} \;=\; \mathfrak{t}_{\mathbb{R}}\,. \tag{27}$$

On group level that means $(U_{\mathbb{R}} \cap B_0)^0 = T_{\mathbb{R}}$. Any $n \in U_{\mathbb{R}} \cap B_0$ therefore normalizes $T_{\mathbb{R}}$; n maps Φ^+ to itself, since otherwise $\operatorname{Ad} n(\mathfrak{b}_0)$ could not equal $\mathfrak{b}_0$, as it must. We had mentioned earlier that any $n \in N_{U_{\mathbb{R}}}(T_{\mathbb{R}})$ which fixes Φ^+ lies in $T_{\mathbb{R}}$. The lemma follows.

Because of the lemma, we can identify the $U_{\mathbb{R}}$-orbit through the identity coset in $G/B_0 \simeq X$ with $U_{\mathbb{R}}/T_{\mathbb{R}}$. This orbit is closed because $U_{\mathbb{R}}$ is compact, and is open, as can be seen by counting dimensions. Hence $U_{\mathbb{R}}$ acts transitively on X, and

$$X \;\simeq\; G/B_0 \;\simeq\; U_{\mathbb{R}}/T_{\mathbb{R}}\,. \tag{28}$$

The transitivity of the $U_{\mathbb{R}}$-action now implies that each Borel subgroup of G intersects $U_{\mathbb{R}}$ in a maximal torus. Arguing as in the proof of the lemma, one finds that $W = N_{U_{\mathbb{R}}}(T_{\mathbb{R}})/T_{\mathbb{R}}$ acts simply transitively on the set of Borel subgroups which contain $T_{\mathbb{R}}$. Thus each maximal torus in $U_{\mathbb{R}}$ lies in exactly N Borel subgroups, $N =$ cardinality of W.

Example 2.10 A *complete flag* in $\mathbb{C}^n$ is a nested sequence of subspaces

$$0 \subset F_1 \subset F_2 \subset \cdots \subset F_{n-1} \subset \mathbb{C}^n\,, \qquad \dim F_j = j\,.$$

The tautological action of $G = SL(n,\mathbb{C})$ on $\mathbb{C}^n$ induces a transitive action on the set of all such complete flags. As a consequence of Lie's theorem, any Borel subgroup of $SL(n,\mathbb{C})$ is the stabilizer of a complete flag. In the case of $G = SL(n,\mathbb{C})$, then, the flag variety X is the variety of all complete flags in $\mathbb{C}^n$ – hence the name *flag variety.*

2.5 The Borel-Weil-Bott Theorem

Recall the notion of a G-equivariant holomorphic line bundle over a complex manifold with a holomorphic G-action – in our specific case, a G-equivariant holomorphic line bundle over the flag variety X: it is a holomorphic line bundle $\mathcal{L} \to X$, equipped with a holomorphic action of G on $\mathcal{L}$ by bundle maps, which lies over the G-action on the base X. The isotropy group at any point $x_0 \in X$ then acts on the fibre $\mathcal{L}_{x_0}$ at x_0. In this way, one obtains a holomorphic representation $\varphi : B_0 \to GL(1,\mathbb{C}) = \mathbb{C}^*$ of B_0, the isotropy group at the identity coset in $G/B_0 \simeq X$. One dimensional representations are customarily called characters. Since G acts transitively on X, the passage from G-equivariant holomorphic line bundles – taken modulo isomorphism, as usual – to holomorphic characters $\varphi : B_0 \to \mathbb{C}^*$ can be reversed,

$$\left\{ \begin{matrix} \text{holomorphic } G\text{-equivariant} \\ \text{line bundles over } X \simeq G/B_0 \end{matrix} \right\} \simeq \left\{ \begin{matrix} \text{holomorphic} \\ \text{characters of } B_0 \end{matrix} \right\}. \tag{29}$$

By construction this is an isomorphism of groups, relative to operations of tensor product of G-equivariant holomorphic line bundles and multiplication of holomorphic characters, respectively.

Holomorphic characters $\varphi : B_0 \to \mathbb{C}^*$ drop to $B_0/[B_0, B_0]$, the quotient of B_0 modulo its commutator. Note that B_0 contains T, the complexification of the maximal torus $T_\mathbb{R}$. One can show that the inclusion $T \hookrightarrow B_0$ induces an isomorphism $T \sim B_0/[B_0, B_0]$. Thus B_0 and T have the same group of holomorphic characters. On the other hand,

$$\left\{ \begin{matrix} \text{holomorphic} \\ \text{characters of } T \end{matrix} \right\} \simeq \widehat{T_\mathbb{R}} \tag{30}$$

by restriction from T to its compact real form $T_\mathbb{R}$ (6). Combining these isomorphisms and identifying the dual group $\widehat{T_\mathbb{R}}$ with the weight lattice Λ as usual, we get a canonical isomorphism

$$\left\{ \begin{matrix} \text{group of holomorphic } G\text{-equivariant} \\ \text{line bundles over } X \end{matrix} \right\} \simeq \Lambda\,. \tag{31}$$

We write $\mathcal{L}_\lambda$ for the line bundle corresponding to $\lambda \in \Lambda$ under this isomorphism.

The action of G on X and $\mathcal{L}_\lambda$ determines a holomorphic, linear action on the space of global sections $H^0(X, \mathcal{O}(\mathcal{L}_\lambda))$ and, by functorality, also on the higher cohomology groups $H^p(X, \mathcal{O}(\mathcal{L}_\lambda))$, $p > 0$. These groups are finite dimensional since X is compact. The Borel-Weil-Bott theorem describes the resulting representations of the compact real form $U_\mathbb{R} \subset G$ and, in view of (6), also as holomorphic representations of G.

Theorem 2.11 (Borel-Weil [32]) *If λ is a dominant weight, the representation of $U_\mathbb{R}$ on $H^0(X, \mathcal{O}(\mathcal{L}_\lambda))$ is irreducible, of highest weight λ, and $H^p(X, \mathcal{O}(\mathcal{L}_\lambda)) = 0$ for $p > 0$. If $\lambda \in \Lambda$ fails to be dominant, $H^0(X, \mathcal{O}(\mathcal{L}_\lambda)) = 0$.*

In particular the theorem provides a concrete, geometric realization of every finite dimensional irreducible representation of $U_\mathbb{R}$. The description of $H^0(X, \mathcal{O}(\mathcal{L}_\lambda))$ can be deduced from the theorem of the highest weight, and the vanishing of the higher cohomology groups is a consequence of the Kodaira vanishing theorem.

Bott [9] extended the Borel-Weil theorem by identifying the higher cohomology groups as representations of $U_\mathbb{R}$. The description involves

$$\rho = \frac{1}{2} \sum\nolimits_{\alpha \in \Phi+} \alpha \in i\mathfrak{t}_\mathbb{R}^* . \tag{32}$$

Since $\Phi \subset \Lambda$, 2ρ is evidently a weight. In fact,

$$\mathcal{L}_{-2\rho} = \text{canonical bundle of } X \tag{33}$$

as can be shown quite easily. In general, ρ itself need not be a weight; if not a weight, it can be made a weight by going to a twofold covering group. Geometrically this means that the canonical line bundle of X has a square root, possibly as a G-equivariant holomorphic line bundle, and otherwise as an equivariant holomorphic line bundle for a twofold covering of G. Whether or not ρ is a weight,

$$w\rho - \rho \in \Lambda \quad \text{for all } w \in W . \tag{34}$$

Also, ρ has the following important property:

$$\text{for } \lambda \in \Lambda , \quad \lambda \text{ is dominant} \iff \lambda + \rho \text{ is dominant regular.} \tag{35}$$

It follows that for $\lambda \in \Lambda$, if $\lambda + \rho$ is regular, there exists a unique $w \in W$ which makes $w(\lambda + \rho)$ dominant regular, and for this w, $w(\lambda + \rho) - \rho$ is a dominant weight.

Theorem 2.12 (Borel-Weil-Bott) *If $\lambda \in \Lambda$ and if $\lambda + \rho$ is singular, then*

$$H^p(X, \mathcal{O}(\mathcal{L}_\lambda)) = 0 \quad \text{for all } p \in \mathbb{Z}.$$

If $\lambda + \rho$ is regular, let w be the unique element of W such that $w(\lambda+\rho)$ is dominant, and define $p(\lambda) = \#\{\alpha \in \Phi^+ \mid (\lambda+\rho, \alpha) < 0\}$. In this situation,

$$H^p(X, \mathcal{O}(\mathcal{L}_\lambda)) = \begin{cases} \text{is non-zero, irreducible, of highest} & \\ \qquad\qquad \text{weight } w(\lambda+\rho) - \rho & \text{if } p = p(\lambda); \\ 0 & \text{if } p \neq p(\lambda). \end{cases}$$

The description of the highest weight as $w(\lambda+\rho) - \rho$ – rather than $w\lambda$, for instance – makes the statement compatible with Serre duality, as it has to be.

Bott proved this result by reducing it to the Borel-Weil theorem. The mechanism is a spectral sequence which relates the cohomology groups of line bundles $\mathcal{L}_\lambda$, $\mathcal{L}_{s(\lambda+\rho)-\rho}$, corresponding to parameters related by a so-called simple Weyl reflection $s \in W$ [9]. An outline of Bott's argument can be found in [7].

The Borel-Weil theorem alone suffices to realize all irreducible representations of $U_{\mathbb{R}}$. Bott's contribution is important for other reasons. At the time, it made it possible to compute some previously unknown cohomology groups of interest to algebraic geometers. The Borel-Weil-Bott theorem made the flag varieties test cases for the fixed point formula for the index of elliptic operators. In representation theory, the theorem gave the first indication that higher cohomology groups might be useful in constructing representations geometrically. That turned out to be the case, for the representations of the discrete series of a non-compact reductive group, for example.

3
Representations of Reductive Lie Groups

We now return to the case of a not-necessarily-compact, connected, linear reductive group $G_{\mathbb{R}}$. The notation of sections 1.2-3 applies. In particular, $K_{\mathbb{R}} \subset G_{\mathbb{R}}$ denotes a maximal compact subgroup, and $\mathfrak{g}$ is the complexified Lie algebra of $G_{\mathbb{R}}$.

3.1 Notions of Continuity and Admissibility, $K_{\mathbb{R}}$-finite and C^{∞} Vectors

Interesting representations of noncompact groups are typically infinite dimensional. To apply analytic and geometric methods, it is necessary to have a topology on the representation space and to impose an appropriate continuity condition on the representation in question. In the finite dimensional case, there is only one "reasonable" topology and continuity hypothesis, but in the infinite dimensional case, choices must be made. One may want to study both complex and real representations. There is really no loss in treating only the complex case, since one can complexify a real representation and regard the original space as an $\mathbb{R}$-linear subspace of its complexification.

We shall consider representations on *complete locally convex Hausdorff topological vector spaces over* $\mathbb{C}$. That includes complex Hilbert spaces, of course. Unitary representations are of particular interest, and one might think that Hilbert spaces constitute a large enough universe of representation spaces. It turns out, however, that even the study of unitary representations naturally leads to the consideration of other types of topological spaces, such as Fréchet spaces and DNF spaces. Most analytic arguments depend on completeness and the Hausdorff property.

Local convexity is required to define the integral of vector-valued functions, which is a crucial tool in the study of representations of reductive groups – see (36), for example.

Let $\mathrm{Aut}(V)$ denote the group of continuous, continuously invertible, linear maps from a complete locally convex Hausdorff topological vector space V to itself; we do not yet specify a topology on this group. There are at least four reasonable notions of continuity one could impose on a homomorphism $G_{\mathbb{R}} \to \mathrm{Aut}(V)$:

a) *continuity*: the action map $G_{\mathbb{R}} \times V \to V$ is continuous, relative to the product topology on $G_{\mathbb{R}} \times V$;
b) *strong continuity*: for every $v \in V$, $g \mapsto \pi(g)v$ is continuous as map from $G_{\mathbb{R}}$ to V;
c) *weak continuity*: for every $v \in V$ and every φ in the continuous linear dual space V', the complex-valued function $g \mapsto \langle \varphi, \pi(g)v \rangle$ is continuous;
d) *continuity in the operator norm*, which makes sense only if V is a Banach space; in that case, $\mathrm{Aut}(V)$ can be equipped with the norm topology, and continuity in the operator norm means that $\pi : G_{\mathbb{R}} \to \mathrm{Aut}(V)$ is a continuous homomorphism of topological groups.

Remark 3.1 The following implications hold for essentially formal reason:

$$\text{continuity} \implies \text{strong continuity} \implies \text{weak continuity}\,,$$

and if V is a Banach space,

$$\text{continuity in the operator norm} \implies \text{continuity}\,.$$

Also, if V is a Banach space,

$$\text{continuity} \iff \text{strong continuity} \iff \text{weak continuity}\,.$$

In this chain of implications, *strong continuity* $\implies$ *continuity* follows relatively easily from the uniform boundedness principle, but the implication *weak continuity* $\implies$ *strong continuity* is more subtle – details can be found in [37].

Example 3.2 The translation action of $(\mathbb{R}, +)$ on $L^p(\mathbb{R})$ is continuous for $1 \leq p < \infty$, but *not* continuous in the operator norm; for $p = \infty$ the translation action fails to be continuous, strongly continuous, even weakly continuous.

Continuity in the operator norm is too much to ask for – most of the representations of interest involve translation. Thus, from now on, "representation" shall mean a continuous – continuous in the sense described above – linear action $\pi : G_{\mathbb{R}} \to \mathrm{Aut}(V)$ on a complete, locally convex Hausdorff space V. If π is continuous, the dual linear action of the topological dual space V', equipped with the strong dual topology[1], need not be continuous. However, when V is a reflexive Banach space, V and V' play symmetric roles in the definition of weak continuity; in this case, the dual action is also continuous, so there exists a "dual representation" π' of $G_{\mathbb{R}}$ on the dual Banach space V'.

An infinite dimensional representation (π, V) typically has numerous invariant subspaces $V_1 \subset V$, but the induced linear action of $G_{\mathbb{R}}$ on V/V_1 is a purely algebraic object unless V/V_1 is Hausdorff, i.e., unless $V_1 \subset V$ is a closed subspace. For this reasons, the existence of a non-closed invariant subspace should not be regarded as an obstacle to irreducibility: (π, V) is *irreducible* if V has no proper *closed* $G_{\mathbb{R}}$-invariant subspaces. A representation (π, V) has *finite length* if every increasing chain of closed $G_{\mathbb{R}}$-invariant subspaces breaks off after finitely many steps. One calls a representation (π, V) *admissible* if $\dim_{\mathbb{R}} \mathrm{Hom}_{K_{\mathbb{R}}}(U, V) < \infty$ for every finite-dimensional irreducible representation (τ, U) of $K_{\mathbb{R}}$. Informally speaking, admissibility means that the restriction of (π, V) to $K_{\mathbb{R}}$ contains any irreducible $K_{\mathbb{R}}$-representation only finitely often.

Theorem 3.3 (Harish-Chandra [15]) *Every irreducible unitary representation (π, V) of $G_{\mathbb{R}}$ is admissible.*

Harish-Chandra proved this theorem for a larger class of reductive Lie groups, not assuming linearity. Godement [12] gave a simplified, transparent argument for linear groups. Atiyah's lecture notes [1] include Godement's argument and many related results.

Heuristically, admissible representations of finite length constitute the smallest class that is invariant under "standard constructions" (in a very wide sense!) and contains the irreducible unitary representations. One should regard inadmissible irreducible representations as exotic. Indeed, the first example of an inadmissible irreducible representation on a Banach space – a representation of the group $G_{\mathbb{R}} = SL(2, \mathbb{R})$ – is relatively recent [33], and depends on a counterexample to the invariant subspace problems for Banach spaces. All irreducible representations which have

[1] In the case of a Banach space, this is the dual Banach topology; for the general case, see [34], for example.

come up naturally in geometry, differential equations, physics, and number theory are admissible.

Definition 3.4 Let (π, V) be a representation of $G_{\mathbb{R}}$. A vector $v \in V$ is

a) $K_{\mathbb{R}}$-*finite* if v lies in a finite-dimensional $K_{\mathbb{R}}$-invariant subspace;
b) a C^∞ *vector* if $g \mapsto \pi(g)v$ is a C^∞ map from $G_{\mathbb{R}}$ to V;
c) in the case of a Banach space V only, an *analytic vector*, if $g \mapsto \pi(g)v$ is a C^ω map (C^ω means real analytic);
d) a weakly analytic vector if, for every $\varphi \in V'$, the complex valued function $g \mapsto \langle \varphi, \pi(g)v \rangle$ is real analytic.

All reasonable notions of a real analytic V-valued map agree when V is a Banach space, but not for other locally convex topological vector spaces. That is the reason for defining the notion of an analytic vector only in the Banach case. Surprisingly perhaps, even weakly real analytic functions with values in a Banach space are real analytic in the usual sense, i.e., locally representable by absolutely convergent vector valued power series – see [24, appendix] for an efficient argument. In the Banach case, then, the notions of an analytic vector and of a weakly analytic coincide; for other representations, the former is not defined, but the latter still makes sense.

As a matter of self-explanatory notation, we write $V_{K_{\mathbb{R}}-\mathrm{finite}}$ for the space of $K_{\mathbb{R}}$-finite vectors in V.

Theorem 3.5 (Harish-Chandra [15]) *If (π, V) is an admissible representation,*

a) *$V_{K_{\mathbb{R}}-finite}$ is a dense subspace of V;*
b) *every $v \in V_{K_{\mathbb{R}}-finite}$ is both a C^∞ vector and a weakly analytic vector.*

Let us sketch the proof. For every $f \in C_c(G_{\mathbb{R}})$ = space compactly supported continuous functions on $G_{\mathbb{R}}$, the operator valued integral

$$\pi(f) \in \mathrm{End}(V), \qquad \pi(f)v = \int_{G_{\mathbb{R}}} f(g)\,\pi(g)\,v\,dg \qquad (v \in V), \quad (36)$$

is well defined and convergent. If $f \in C_c^\infty(G_{\mathbb{R}})$, for any $v \in V$, $\pi(f)v$ is a C^∞ vector, as was first observed by Gårding. When $f = f_n$ runs through an "approximate identity" in $C_c^\infty(G_{\mathbb{R}})$, the sequence $\pi(f_n)v$ converges to v. This much establishes Gårding's theorem – the space of C^∞ vectors is dense in V. In analogy to the operators $\pi(f)$, one can also define $\pi(h)$ for $h \in C(K_{\mathbb{R}})$. Because of the Stone-Weierstrass theorem, the space of left $K_{\mathbb{R}}$-finite functions is dense in $C(K_{\mathbb{R}})$. Letting $h = h_n$ run through an approximate identity in $C(K_{\mathbb{R}})$, consisting of left $K_{\mathbb{R}}$-finite functions h_n, one can approximate any $v \in V$ by the sequence of vectors $\pi(h_n)v$, which are all $K_{\mathbb{R}}$-finite. This proves the density of the $K_{\mathbb{R}}$-finite vectors. One can combine this latter argument with Gårding's, to approximate any $v \in V$ by a sequence of $K_{\mathbb{R}}$-finite C^∞ vectors; when v transforms according to a particular irreducible finite dimensional representation τ of $K_{\mathbb{R}}$, one can also make the approximating sequence lie in the space of τ-isotypic vectors. This space is finite dimensional because of the admissibility hypothesis. The density of the subspace of τ-isotypic C^∞ vectors in the space of all τ-isotypic vectors therefore implies that all τ-isotypic vectors are C^∞ vectors, for every τ, so all $K_{\mathbb{R}}$-finite vectors are C^∞ vectors. The functions $g \mapsto \langle \varphi, \pi(g)v \rangle$, for $v \in V_{K_{\mathbb{R}}-\text{finite}}$ and $\varphi \in V'$, satisfy elliptic differential equations with C^ω coefficients, which implies they are real analytic. Thus all $K_{\mathbb{R}}$-finite vectors are weakly analytic, as asserted by the theorem.

The theorem applies in particular to $K_{\mathbb{R}}$, considered as maximal compact subgroup of itself. Finite dimensional subspaces are automatically closed, so the density of $K_{\mathbb{R}}$-finite vectors forces any infinite dimensional representation (π, V) of $K_{\mathbb{R}}$ to have proper closed invariant subspaces. In other words,

Corollary 3.6 *Every irreducible representation of $K_{\mathbb{R}}$ is finite dimensional.*

3.2 Harish-Chandra Modules

We continue with the notation of the previous section, but (π, V) will now specifically denote an admissible representation of $G_{\mathbb{R}}$, and later an admissible representation of finite length. We write $V_{K_{\mathbb{R}}-\text{finite}}$ for the space of $K_{\mathbb{R}}$-finite vectors, and V^∞ for the space of C^∞ vectors. The latter is a $G_{\mathbb{R}}$-invariant subspace of V, which contains the former as $K_{\mathbb{R}}$-invariant subspace; both are dense in V. The Lie algebra $\mathfrak{g}_{\mathbb{R}}$ acts on V^∞ by differentiation, and we extend this action by $\mathbb{C}$-linearity to the complexified Lie algebra $\mathfrak{g}$ on V^∞ – equivalently, V^∞ has a natural structure of module over $U(\mathfrak{g})$, the universal enveloping algebra of $\mathfrak{g}$. One might think that the $U(\mathfrak{g})$-module V^∞ is the right notion of infinitesimal representation attached to the global representation (π, V). It has a very serious drawback, however: except in the finite dimensional case, V^∞ may be highly reducible as $U(\mathfrak{g})$-module even if (π, V) is irreducible.

The first hint of a solution to this problem appeared in Bargmann's description of the irreducible unitary representations of $SL(2,\mathbb{R})$ [2]. Later formalized and developed by Harish-Chandra, it starts with the observation that

$$V_{K_{\mathbb{R}}-\text{finite}} \text{ is a } U(\mathfrak{g})-\text{submodule of } V^\infty. \tag{37}$$

Indeed, the action map $U(\mathfrak{g}) \otimes V_{K_{\mathbb{R}}-\text{finite}} \to V^\infty$ is $K_{\mathbb{R}}$-invariant when $K_{\mathbb{R}}$ acts on $U(\mathfrak{g})$ via the adjoint action and on V and its subspaces via π. The image of the action map is therefore exhausted by finite dimensional, $K_{\mathbb{R}}$-invariant subspaces; in other words, the image of this action map lies in $V_{K_{\mathbb{R}}-\text{finite}}$.

Finite dimensional representations of the compact Lie group $K_{\mathbb{R}}$ extend naturally to its complexification (6). By definition, $V_{K_{\mathbb{R}}-\text{finite}}$ is the union of finite dimensional $K_{\mathbb{R}}$-invariant subspaces, so the $K_{\mathbb{R}}$-action on $V_{K_{\mathbb{R}}-\text{finite}}$ extends naturally to the complexification K of $K_{\mathbb{R}}$. Even though $V_{K_{\mathbb{R}}-\text{finite}}$ has no natural Hausdorff topology – it is not closed in V unless $\dim V < \infty$ – it makes sense to say that K acts holomorphically on $V_{K_{\mathbb{R}}-\text{finite}}$: like $K_{\mathbb{R}}$, K acts *locally finitely*, in the sense that every vector lies in a finite dimensional invariant subspace; the invariant finite dimensional subspaces do carry natural Hausdorff topologies, and K does act holomorphically on them. The Lie algebra $\mathfrak{k}$ has two natural actions on $V_{K_{\mathbb{R}}-\text{finite}}$, by differentiation of the K-action, and via the inclusion $\mathfrak{k} \subset \mathfrak{g}$ and the $U(\mathfrak{g})$-module structure. These two actions coincide, essentially by construction. To simplify the notation, we denote the actions on $V_{K_{\mathbb{R}}-\text{finite}}$ by juxtaposition. With this convention,

$$k\,(\xi v) = (\operatorname{Ad} k\,\xi)(kv), \quad \forall k \in K,\ \xi \in U(\mathfrak{g}),\ v \in V_{K_{\mathbb{R}}-\text{finite}}, \tag{38}$$

as can be deduced from the well-known formula

$$\exp(\operatorname{Ad} k\,\xi) = k \exp(\xi) k^{-1},$$

for $\xi \in \mathfrak{g}_{\mathbb{R}}$, $k \in K_{\mathbb{R}}$.

Definition 3.7 A $(\mathfrak{g}, K)$*-module* is a complex vector space M, equipped with the structure of $U(\mathfrak{g})$-module and with a linear action of K such that:

a) The action of K is locally finite, i.e., every $m \in M$ lies in a finite dimensional K-invariant subspace on which K acts holomorphically;

b) when the K-action is differentiated, the resulting action of the Lie algebra $\mathfrak{k}$ agrees with the action of $\mathfrak{k}$ on M via $\mathfrak{k} \hookrightarrow \mathfrak{g}$ and the $U(\mathfrak{g})$-module structure.
c) the identity (38) holds for all $k \in K$, $\xi \in U(\mathfrak{g})$, $v \in M$.

A *Harish-Chandra module* is a $(\mathfrak{g}, K)$-module M which is finitely generated over $U(\mathfrak{g})$ and admissible, in the sense that every irreducible K-representation occurs in M with finite multiplicity.

The discussion leading up to the definition shows that the space of $K_{\mathbb{R}}$-finite vectors $V_{K_{\mathbb{R}}-\text{finite}}$ of an admissible representation (π, V) is an admissible $(\mathfrak{g}, K)$-module. Very importantly,

$$\begin{aligned} &\text{the correspondence } \tilde{V} \mapsto \tilde{V}_{K_{\mathbb{R}}-\text{finite}} \text{ sets up a bijection} \\ &\{\text{closed } G_{\mathbb{R}}\text{-invariant subspaces } \tilde{V} \subset V\} \leftrightarrow \\ &\{(\mathfrak{g}, K)\text{-submodules } \tilde{V}_{K_{\mathbb{R}}-\text{finite}} \subset V_{K_{\mathbb{R}}-\text{finite}}\}, \end{aligned} \tag{39}$$

as follows from the weakly analytic nature of $K_{\mathbb{R}}$-finite vectors (3.5). When (π, V) is not only admissible but also of finite length, every ascending chain of $(\mathfrak{g}, K)$-submodules of $V_{K_{\mathbb{R}}-\text{finite}}$ breaks off eventually. Because of the Nötherian property of $U(\mathfrak{g})$, that implies the finite generation of $V_{K_{\mathbb{R}}-\text{finite}}$ over $U(\mathfrak{g})$. In short, the space of $K_{\mathbb{R}}$-finite vectors $V_{K_{\mathbb{R}}-\text{finite}}$ of an admissible representation of finite length is a Harish-Chandra module. The statement (39) also implies that (π, V) is irreducible if and only if $V_{K_{\mathbb{R}}-\text{finite}}$ is irreducible as $(\mathfrak{g}, K)$-module. This property of $V_{K_{\mathbb{R}}-\text{finite}}$ makes it the appropriate notion of the infinitesimal representation corresponding to (π, V).

From now on, we write $\mathrm{HC}(V)$ for the space of $K_{\mathbb{R}}$-finite vectors of an admissible representation (π, V) of finite length and call $\mathrm{HC}(V)$ *the Harish-Chandra module of* π. The next statement formalizes the properties of Harish-Chandra modules we have mentioned so far:

Theorem 3.8 (Harish-Chandra [15]) *The association* $V \mapsto \mathrm{HC}(V) = V_{K_{\mathbb{R}}-\mathit{finite}}$ *establishes a covariant, exact, faithful functor*

$$\left\{\begin{matrix}\textit{category of admissible } G_{\mathbb{R}}\textit{-representations} \\ \textit{of finite length and } G_{\mathbb{R}}\textit{-equivariant maps}\end{matrix}\right\}$$

$$\xrightarrow{\mathrm{HC}} \left\{\begin{matrix}\textit{category of Harish-Chandra modules} \\ \textit{and } (\mathfrak{g}, K)\textit{-equivariant linear maps}\end{matrix}\right\}.$$

Definition 3.9 Two finite length admissible representations (π_i, V_i), $i = 1, 2$, are *infinitesimally equivalent* if $HC(V_1) \simeq HC(V_2)$.

Loosely speaking, infinitesimal equivalence means that the two representations are the same except for the choice of topology. A concrete example may be helpful. The group

$$G_{\mathbb{R}} = SU(1,1) = \left\{ \begin{pmatrix} a & b \\ \bar{b} & \bar{a} \end{pmatrix} \;\middle|\; a, b \in \mathbb{C},\ |a|^2 - |b|^2 = 1 \right\} \tag{40}$$

has $G = SL(2,\mathbb{C})$ as complexification, and is conjugate in G to $SL(2,\mathbb{R})$. As maximal compact subgroup, we choose the diagonal subgroup, in which case its complexification also consists of diagonal matrices:

$$\begin{aligned} K_{\mathbb{R}} &= \left\{ k_\theta = \begin{pmatrix} e^{i\theta} & 0 \\ 0 & e^{-i\theta} \end{pmatrix} \;\middle|\; \theta \in \mathbb{R} \right\} \cong U(1), \\ K &= \left\{ \begin{pmatrix} a & 0 \\ 0 & a^{-1} \end{pmatrix} \;\middle|\; a \in \mathbb{C}^* \right\} \cong \mathbb{C}^*. \end{aligned} \tag{41}$$

By linear fractional transformations, $SU(1,1)$ acts transitively on $D =$ open unit disc in $\mathbb{C}$, with isotropy subgroup $K_{\mathbb{R}}$ at the origin. Left translation on $D \cong SU(1,1)/K_{\mathbb{R}}$ induces a linear action ℓ of $SU(1,1)$ on $C^\infty(D)$,

$$(\ell(g)f)(x) =_{\text{def}} f(g^{-1} \cdot z), \quad g \in SU(1,1),\ f \in C^\infty(D),\ z \in D, \tag{42}$$

and on the subspace

$$H^2(D) \;=_{\text{def}}\; \begin{array}{c}\text{space of holomorphic functions on } D \\ \text{with } L^2 \text{ boundary values,}\end{array} \tag{43}$$

topologized by the inclusion $H^2(D) \hookrightarrow L^2(S^1)$. One can show that both actions are representations, i.e., they are continuous with respect to the natural topologies of the two spaces.

Recall the definition of k_θ in (41). Since $\ell(k_\theta)z^n = e^{-2in\theta} z^n$, $f \in H^2(D)$ is $K_{\mathbb{R}}$-finite if and only if f has a finite Taylor series at the origin, i.e., if and only if f is a polynomial:

$$H^2(D)_{K_{\mathbb{R}}-\text{finite}} = \mathbb{C}[z]. \tag{44}$$

In particular, $(\ell, H^2(D))$ is admissible. This representation is not irreducible, since $H^2(D)$ contains the constant functions $\mathbb{C}$ as an obviously closed invariant subspace. It does have finite length; in fact, the quotient $H^2(D)/\mathbb{C}$ is irreducible, as follows from a simple infinitesimal calculation in the Harish-Chandra module $\mathrm{HC}(H^2(D)) = \mathbb{C}[z]$.

Besides $V = H^2(D)$, the action (42) on each of the following spaces, equipped with the natural topology in each case, defines a representation of $SU(1,1)$:

a) $H^p(D)$ = space of holomorphic functions on D with L^p boundary values, $1 \le p \le \infty$;
b) $H^\infty(D)$ = space of holomorphic functions on D with C^∞ boundary values;
c) $H^{-\infty}(D)$ = space of holomorphic functions on D with distribution boundary values;
d) $H^\omega(D)$ = space of holomorphic functions on D with real analytic boundary values;
e) $H^{-\omega}(D)$ = space of all holomorphic functions on D.

Taking boundary values, one obtains inclusions $H^p(D) \hookrightarrow L^p(S^1)$, which are equivariant with respect to the action of $SU(1,1)$ on $L^p(S^1)$ by linear fractional transformations. The latter fails to be continuous when $p = \infty$, but that is not the case for the image of $H^\infty(D)$ in $L^\infty(S^1)$. Essentially by definition, every holomorphic function on D has hyperfunction boundary values. This justifies the notation $H^{-\omega}(D)$; the superscript $-\omega$ stands for hyperfunctions. One can show that $H^\infty(D)$ is the space of C^∞ vectors for the Hilbert space representation $(\ell, H^2(D))$.

Arguing as in the case of $H^2(D)$, one finds that the representation ℓ of $SU(1,1)$ on each of the spaces a)-e) has $\mathbb{C}[z]$ as Harish-Chandra module, so all of them are infinitesimally equivalent. This is the typical situation, not just for $SU(1,1)$, but for all groups $G_\mathbb{R}$ of the type we are considering: every infinite dimensional admissible representation (π, V) of finite length lies in an infinite family of representations, all infinitesimally equivalent, but pairwise non-isomorphic. In the context of unitary representations the situation is different:

Theorem 3.10 (Harish-Chandra [15]) *If two irreducible unitary representations are infinitesimally equivalent, they are isomorphic as unitary representations.*

If we were dealing with finite dimensional representations, this would follow from an application of Schur's Lemma. Schur's lemma, it should be recalled, is a consequence of the existence of eigenvalues of endomorphisms of finite dimensional vector spaces over $\mathbb{C}$. In the setting of Harish-Chandra modules, endomorphisms are in particular $K_\mathbb{R}$-invariant, and must therefore preserve subspaces on which $K_\mathbb{R}$ acts according to any particular irreducible representation of $K_\mathbb{R}$. These subspaces are finite dimensional, and their direct sum, over all irreducible representations of $K_\mathbb{R}$ up to equivalence, is the Harish-Chandra module in question. Thus endomorphisms of Harish-Chandra modules can be put into Jordan canonical form, even though the modules are infinite dimensional. In short, there exists a version of Schur's lemma for Harish-Chandra modules; it plays the crucial role in the proof of the theorem.

According to results of Casselman [10], every Harish-Chandra module M has a *globalization*, meaning an admissible $G_{\mathbb{R}}$-representation (π, V) of finite length, such that $\mathrm{HC}(V) = M$. That makes the following two problems equivalent:

a) Classify irreducible admissible representations of $G_{\mathbb{R}}$, up to infinitesimal equivalence;
b) Classify irreducible Harish-Chandra modules for the pair $(\mathfrak{g}, K)$.

Under slight additional hypotheses, problem b) has been solved by Langlands [25], Vogan–Zuckerman [36], and Beilinson–Bernstein [3, 17], by respectively analytic, algebraic, and geometric means. The three solutions give formally different answers, which can be related most easily in terms of the Beilinson–Bernstein construction [18]. We shall discuss these matters in section 4.

A Harish-Chandra module corresponds to an irreducible unitary representation if and only if it carries an invariant, positive definite hermitian form – invariant in the sense that the action of $K_{\mathbb{R}}$ preserves it, and that every $\xi \in \mathfrak{g}$ acts as a skew-hermitian transformation. Whether a given irreducible Harish-Chandra module carries an invariant, nontrivial, possibly indefinite hermitian form is easy to decide: the Harish-Chandra module needs to be conjugate-linearly isomorphic to its own dual. When a nontrivial invariant hermitian form exists, it is unique up to scaling. The classification of the irreducible unitary representations of $G_{\mathbb{R}}$ comes down to determining which invariant hermitian forms have a definite sign. Many results in this direction exist, most of them due to Vogan and Barbasch, but a general answer is not in sight, not even a good general conjecture.

We mentioned already that every Harish-Chandra module M has a *globalization*. It is natural to ask if a globalization can be chosen in a functorial manner – in other words, whether the functor HC in Theorem 3.8 has a right inverse. Such functorial globalizations do exist. Four of them are of particular interest, the C^{∞} and $C^{-\infty}$ globalizations of Casselman-Wallach [11, 35], as well as the minimal globalization and the maximal globalization [29, 22]. All four are *topologically exact*, i.e., they map exact sequences of Harish-Chandra modules into exact sequences of representations in which every morphism has *closed range*. The main technical obstacle in constructing the canonical globalizations is to establish this closed range property. In the case of an admissible representation (π, V) of finite length, on a reflexive Banach space V, the C^∞ globalization of HC(V) is topologically isomorphic to the space of C^∞ vectors V^∞. Similarly the minimal globalization is topologically isomorphic to the space of analytic vectors V^ω; both have very naturally defined topologies. The other two constructions are dual to these: the $C^{-\infty}$ globalization is isomorphic to $((V')^\infty)'$, the strong dual of the space of C^∞ vectors of the dual representation (π', V'), and the maximal globalization is similarly isomorphic to $((V')^\omega)'$. In the case of the earlier example of $(\ell, H^2(D))$, the four globalizations of HC($H^2(D)$) can be identified with $H^\infty(D)$, $H^\omega(D)$, $H^{-\infty}(D)$, and $H^{-\omega}(D)$, respectively.

4
Geometric Constructions of Representations

In this section we shall freely use the notational conventions of the preceding sections. To simplify the discussion, we suppose

$$\text{the complexification } G \text{ of } G_{\mathbb{R}} \text{ is connected.} \tag{45}$$

That is the case for $G_{\mathbb{R}} = GL(n, \mathbb{R})$, for example: the group itself is not connected, but it does have a connected complexification. The hypothesis (45) in particular implies:

$$\text{for each } g \in G_{\mathbb{R}}\,,\ \operatorname{Ad} g\colon \mathfrak{g} \longrightarrow \mathfrak{g} \text{ is an inner automorphism.} \tag{46}$$

The latter condition is important; without it irreducible Harish-Chandra modules need not have infinitesimal characters – see definition (4.2) below. From a technical point of view, the weaker condition (46) suffices entirely for our purposes, at the cost of additional terminology and explanations. We assume (45) only to avoid these. The compact real form $U_{\mathbb{R}} \subset G$ is then also connected, as we had assumed in section 2.

Recall the notion of a Cartan subalgebra of the complex reductive Lie algebra $\mathfrak{g}$: a maximal abelian subalgebra $\mathfrak{h} \subset \mathfrak{g}$ such that $\operatorname{Ad}\xi : \mathfrak{g} \to \mathfrak{g}$ is diagonalizable, for every $\xi \in \mathfrak{h}$. Any two Cartan subalgebras are conjugate under the adjoint action of G. The complexified Lie algebra $\mathfrak{t}$ of a maximal torus $T_{\mathbb{R}} \subset U_{\mathbb{R}}$ is a particular example of a Cartan subalgebra. Since G acts on the set of compact real forms by conjugation, every Cartan subalgebra of $\mathfrak{g}$ arises as the complexified Lie algebra of a maximal torus in some compact real form of G. In particular the discussion in sections 2.1-2 applies to any Cartan subalgebra. With $\mathfrak{t}$ and $T_{\mathbb{R}} \subset U_{\mathbb{R}}$ as above, there are two potential notions of Weyl group, namely the "compact Weyl group" $W(U_{\mathbb{R}}, T_{\mathbb{R}}) = N_{U_{\mathbb{R}}}(T_{\mathbb{R}})/T_{\mathbb{R}}$ and the "complex Weyl group" $W(G,T) = N_G(T)/T$, with $T =$ complexification of $T_{\mathbb{R}}$. They coincide, in fact, and we shall denote both by the symbol W.

4.1 The Universal Cartan Algebra and Infinitesimal Characters

By definition, the flag variety X of the complex reductive Lie algebra $\mathfrak{g}$ parameterizes the Borel subalgebras of $\mathfrak{g}$:

$$X \ni x \quad \longleftrightarrow \quad \mathfrak{b}_x \subset \mathfrak{g}\,. \tag{47}$$

Define

$$\mathfrak{h}_x = \mathfrak{b}_x/[\mathfrak{b}_x, \mathfrak{b}_x]. \tag{48}$$

This quotient is independent of x in the following equivalent senses:

a) if $g \cdot x = y$ for some $g \in G$ and $x, y \in X$, the map $\mathfrak{h}_x \to \mathfrak{h}_y$ induced by $\operatorname{Ad} g : \mathfrak{b}_x \to \mathfrak{b}_y$ depends only on x and y, not on the particular choice of g;
b) $\mathfrak{h}_x$ is the fiber at x of a canonically flat holomorphic vector bundle over X;
c) let $\mathfrak{t} \subset \mathfrak{g}$ be a Cartan subalgebra and $\Phi^+ \subset \Phi(\mathfrak{g}, \mathfrak{t})$ a positive root system. Then $\mathfrak{b}_0 = \mathfrak{t} \oplus (\oplus_{\alpha\in\Phi^+} \mathfrak{g}^{-\alpha})$ is a Borel subalgebra, with $[\mathfrak{b}_0, \mathfrak{b}_0] = \oplus_{\alpha\in\Phi^+} g^{-\alpha}$. The resulting isomorphism $\mathfrak{h}_0 = \mathfrak{b}_0/[\mathfrak{b}_0, \mathfrak{b}_0] \cong \mathfrak{t}$ depends only on the choice of Φ^+.

We write $\mathfrak{h}$ instead of $\mathfrak{h}_x$ to signify independence of x. This is the *universal Cartan algebra.* It is *not a subalgebra* of $\mathfrak{g}$, but $\mathfrak{h}$ *is* canonically isomorphic to any *ordered Cartan subalgebra*, i.e., to any Cartan subalgebra $\mathfrak{t} \subset \mathfrak{g}$ with a specified choice of positive root system Φ^+. We use the canonical isomorphism between $\mathfrak{h}$ and any ordered Cartan subalgebra $(\mathfrak{t}, \Phi^+)$ to transfer from $\mathfrak{t}$ to $\mathfrak{h}$ the weight lattice, root system, positive root system, and Weyl group. In this way we get the *universal weight lattice* $\Lambda \subset \mathfrak{h}^*$, the *universal root system* $\Phi \subset \Lambda$, the *universal positive root system* $\Phi^+ \subset \Phi$, and the *universal Weyl group* W, which acts on $\mathfrak{h}$ and dually on $\mathfrak{h}^*$, leaving invariant both Λ and Φ. Moreover, there exists a W-invariant, positive inner product $(.,.)$ on the $\mathbb{R}$-linear subspace $\mathbb{R}\otimes_{\mathbb{Z}}\Lambda$ which depends on the choice of S in (20), but on nothing else. Going back to (31), we see that the parametrization of the G-equivariant holomorphic line bundles $\mathcal{L}_\lambda \to X$ in terms of λ becomes completely canonical when we regard λ as lying in universal weight lattice Λ.

We can use these ideas to characterize the so-called Harish-Chandra isomorphism. Let $Z(\mathfrak{g})$ denote the center of the universal enveloping algebra $U(\mathfrak{g})$, and $S(\mathfrak{h})^W$ the algebra of W-invariants in the symmetric algebra of $\mathfrak{h}$, or what comes to the same, the algebra of W-invariant polynomial functions on $\mathfrak{h}^*$. By differentiation of the G-action, $\mathfrak{g}$ acts on holomorphic sections of $\mathcal{L}_\lambda$ as a Lie algebra of vector fields. This induces an action of $U(\mathfrak{g})$, and therefore also of $Z(\mathfrak{g})$, on the sheaf of holomorphic sections $\mathcal{O}(\mathcal{L}_\lambda)$.

Theorem 4.1 (Harish-Chandra [14]) *There exists a canonical isomorphism*

$$\gamma : Z(\mathfrak{g}) \xrightarrow{\sim} S(\mathfrak{h})^W$$

such that, for any $\lambda \in \Lambda$, any $\zeta \in Z(\mathfrak{g})$ acts on the sheaf of holomorphic sections $\mathcal{O}(\mathcal{L}_\lambda)$ as multiplication by the scalar $\gamma(\zeta)(\lambda+\rho)$.

In this statement, $\gamma(\zeta)(\lambda+\rho)$ refers to the value of the W-invariant polynomial function $\gamma(\zeta)$ at the point $\lambda+\rho$. More generally this makes sense for elements of $\mathfrak{h}^*$: every $\lambda \in \mathfrak{h}^*$ determines a character[1]

$$\chi_\lambda : Z(\mathfrak{g}) \longrightarrow \mathbb{C}, \qquad \chi_\lambda(\zeta) = \gamma(z)(\zeta). \tag{49}$$

In view of Harish-Chandra's theorem, $\chi_\lambda = \chi_\mu$ if and only if $\lambda = w \cdot \mu$ for some $w \in W$, and every character of $Z(\mathfrak{g})$ is of this type.

Definition 4.2 One says that a Harish-Chandra module M has an *infinitesimal character* if $Z(\mathfrak{g})$ acts on M via a character.

When the infinitesimal character exists, it can of course be expressed as χ_λ, for some $\lambda \in \Lambda$. Applying Schur's lemma for Harish-Chandra modules, as was explained in section 3.2, one finds that every irreducible Harish-Chandra module does have an infinitesimal character.

4.2 Twisted $\mathcal{D}$-modules

The flag variety X is projective, and thus in particular has an algebraic structure. The complex linear reductive group G has an algebraic structure as well, and the action of G on X is algebraic. The G-equivariant

[1] in the present context, "character" means algebra homomorphism into the one dimensional algebra $\mathbb{C}$.

line bundles $\mathcal{L}_\lambda \to X$ are associated to algebraic characters of the structure group of the algebraic principal bundle $G \to G/B$, so they, too, have algebraic structures, and G acts algebraically also on these line bundles. In the following, we equip X with the Zariski topology, and all sheaves are understood to be sheaves relative to the Zariski topology. In the current setting, $\mathcal{O}$ denotes the sheaf of algebraic functions on X and $\mathcal{O}(\mathcal{L}_\lambda)$ the sheaf of algebraic sections of $\mathcal{L}_\lambda$.

The locally defined linear differential operators on X with algebraic coefficients constitute a sheaf of algebras, customarily denoted by $\mathcal{D}$. By definition, $\mathcal{D}$ acts on $\mathcal{O}$; in more formal language, $\mathcal{O}$ is a sheaf of modules over the sheaf of algebras $\mathcal{D}$. For $\lambda \in \Lambda$,

$$\mathcal{D}_\lambda \;=\; \mathcal{O}(\mathcal{L}_\lambda) \otimes_{\mathcal{O}} \mathcal{D} \otimes_{\mathcal{O}} \mathcal{O}(\mathcal{L}_{-\lambda}) \tag{50}$$

is also a sheaf of algebras, the sheaf of linear differential operators acting on sections of $\mathcal{L}_\lambda$. It is a so-called *twisted sheaf of differential operators.* Since we can think of $\mathcal{O}$ as differential operators of degree zero, there exists a natural inclusion $\mathcal{O} \hookrightarrow \mathcal{D}_\lambda$. By differentiation of the G-action, every $\xi \in \mathfrak{g}$ determines a globally defined first order differential operator acting on $\mathcal{O}(\mathcal{L}_\lambda)$. In this way we get a canonical homomorphism of Lie algebras

$$\mathfrak{g} \;\longrightarrow\; \Gamma\mathcal{D}_\lambda \;=\; H^0(X, \mathcal{D}_\lambda)\,; \tag{51}$$

here, as usual, we give the associative algebra $\Gamma\mathcal{D}_\lambda$ the additional structure of a Lie algebra by taking commutators of differential operators. When $\mathfrak{g}$ is semisimple, this morphism is injective. In any case, it induces

$$U(\mathfrak{g}) \;\longrightarrow\; \Gamma\mathcal{D}_\lambda \;=\; H^0(X, \mathcal{D}_\lambda)\,, \tag{52}$$

a homomorphism of associative algebras.

Until now we have supposed that λ lies in the weight lattice. However, there is a natural way to make sense of the sheaf of algebras $\mathcal{D}_\lambda$ for any $\lambda \in \mathfrak{h}^*$. In terms of local coordinates, the twisting operation (50) involves taking logarithmic derivatives, so the lattice parameter λ occurs polynomially; one can therefore let it take values in $\mathfrak{h}^*$. With this extended definition, the natural inclusion $\mathcal{O} \hookrightarrow \mathcal{D}_\lambda$ and the morphism (52) exist just as before.

Theorem 4.3 (Beilinson-Bernstein [3]) *For any $\lambda \in \mathfrak{h}^*$, the morphism (52) induces*

$$U_{\lambda+\rho} =_{def} U(\mathfrak{g})/\textit{ideal generated by } \{\zeta - \chi_{\lambda+\rho}(\zeta) \mid \zeta \in Z(\mathfrak{g})\} \xrightarrow{\sim} \Gamma\mathcal{D}_\lambda .$$

The higher cohomology groups of $\mathcal{D}_\lambda$ vanish: $H^p(X, \mathcal{D}_\lambda) = 0$ for $p > 0$.

Note that any $U_{\lambda+\rho}$-module can be regarded as a $U(\mathfrak{g})$-module with infinitesimal character $\chi_{\lambda+\rho}$, and vice versa. Following the usual custom, we shall use the terminology "$\mathcal{D}_\lambda$-module" as shorthand for "sheaf of $\mathcal{D}_\lambda$-modules". A $\mathcal{D}_\lambda$-module is said to be *coherent* if it is coherent over the sheaf of algebras $\mathcal{D}_\lambda$ – in other words, if locally around any point, it can be presented as the quotient of a free $\mathcal{D}_\lambda$-module of finite rank, modulo the image of some other free $\mathcal{D}_\lambda$-module of finite rank. Theorem 4.3 makes it possible to define the two functors

$$\left\{ \begin{array}{c} \text{category of finitely} \\ \text{generated } U_{\lambda+\rho}\text{-modules} \end{array} \right\} \underset{\Gamma}{\overset{\Delta}{\rightleftarrows}} \left\{ \begin{array}{c} \text{category of} \\ \text{coherent } \mathcal{D}_\lambda\text{-modules} \end{array} \right\}, \tag{53}$$

with $\Gamma = H^0(X, \,\cdot\,) =$ global sections functor, and

$$\Delta M \;=\; \mathcal{D}_\lambda \otimes_{U_{\lambda+\rho}} M\,; \tag{54}$$

Δ is called the "localization functor". Because of (4.1), the category on the left in (53) depends only on the W-orbit of $\lambda + \rho$; the category on the right, on the other hand, depends on λ itself.

Recall the definition of the W-invariant inner product $(\cdot,\cdot)$ on the real form $\mathbb{R} \otimes_{\mathbb{Z}} \Lambda \subset \mathfrak{h}^*$ in section 4.1. We shall use the same notation to denote the bilinear – not hermitian! – extension of the inner product to the complex vector space $\mathfrak{h}^*$. In analogy to our earlier terminology we call $\lambda \in \mathfrak{h}^*$ *regular* if $(\lambda, \alpha) \neq 0$ for all $\alpha \in \Phi$, and otherwise *singular.*

Definition 4.4 An element $\lambda \in \mathfrak{h}^*$ is said to be *integrally dominant* if $2\frac{(\lambda,\alpha)}{(\alpha,\alpha)} \notin \mathbb{Z}_{<0}$ for all $\alpha \in \Phi^+$.

Remark 4.5 For $\lambda \in \Lambda$ these quotients are integers, and for a generic $\lambda \in \mathfrak{h}^*$ all of them are non-integral. In every case there exists $w \in W$ such that $w(\lambda + \rho)$ is integrally dominant.

Theorem 4.6 (Beilinson-Bernstein [3]) *Let $\mathcal{S}$ be a coherent $\mathcal{D}_\lambda$-module.*

A) *If $\lambda + \rho$ is integrally dominant and regular, the global sections of $\mathcal{S}$ generate its stalks;*
B) *If $\lambda + \rho$ is integrally dominant, then $H^p(X, \mathcal{S}) = 0$ for all $p \neq 0$.*

The conclusion of A) means that the stalk $\mathcal{S}_x$ at any $x \in X$ is generated over the ring $(\mathcal{D}_\lambda)_x$ by the image of $\Gamma\mathcal{S}$ in $\mathcal{S}_x$. Note the formal analogy with Cartan's theorems A and B for coherent analytic sheaves on Stein manifolds – see [13], for example. The next statement follows quite directly from the theorem:

Corollary 4.7 *If $\lambda + \rho$ is integrally dominant and regular, the localization functor Δ defines an equivalence of categories, with inverse functor* Γ.

The hypothesis of regularity is essential. For example, when the canonical bundle of X has a G-equivariant square root $\mathcal{L}_{-\rho}$, the Borel-Weil-Bott theorem asserts that the $\mathcal{D}_{-\rho}$-module $\mathcal{O}(\mathcal{L}_{-\rho})$ has no non-zero global sections, nor even non-zero higher cohomology, but $-\rho + \rho = 0$ *is* integrally dominant. A refined version of the corollary does apply whenever $\lambda + \rho$ is integrally dominant but singular. The idea is to construct a quotient category of the category of coherent $\mathcal{D}_\lambda$-modules, by dividing out the subcategory of sheaves with trivial cohomology. The greater subtlety of this situation merely reflects, and even explains, a familiar fact: in the study of $U(\mathfrak{g})$-modules with singular infinitesimal character, one encounters difficulties not present in the regular case. The category of coherent $\mathcal{D}_\lambda$-modules "knows nothing" about regularity or singularity; for any $\mu \in \Lambda$,

$$\left\{\begin{matrix}\text{category of coherent}\\ \mathcal{D}_\lambda\text{-modules}\end{matrix}\right\} \ni \mathcal{S} \mapsto \mathcal{O}(\mathcal{L}_\mu) \otimes_{\mathcal{O}} \mathcal{S} \in \left\{\begin{matrix}\text{category of coherent}\\ \mathcal{D}_{\lambda+\mu}\text{-modules}\end{matrix}\right\} \tag{55}$$

defines an equivalence of categories, and μ can always be chosen so as to make $\lambda + \mu + \rho$ integrally dominant and regular. The existence of sheaves without cohomology carries sole responsibility for the more complicated nature of the singular case!

The corollary can be extended in another direction. When $\lambda + \rho$ fails to be integrally dominant, the localization functor Δ becomes an equivalence of categories on the level of derived categories. Heuristically, and quite imprecisely, that means replacing $\Gamma\mathcal{S}$ by the formal Euler characteristic $\sum_p (-1)^p H^p(X, \mathcal{S})$. In other words, the derived category of coherent $\mathcal{D}_\lambda$-modules is equivalent to the derived category of coherent $\mathcal{D}_{w(\lambda+\rho)-\rho}$-modules, for any $w \in W$. This latter equivalence can be described geometrically, in terms of Beilinson-Bernstein's *intertwining functors* [4].

The corollary and the first of the two extensions makes it possible to translate problems about finitely generated $U(\mathfrak{g})$-modules with an infinitesimal character into problems in algebraic geometry. What are advantages to working on the geometric side, rather than directly with $U(\mathfrak{g})$-modules? For those of us who think geometrically, the geometric arguments seem far more transparent than their algebraic counterparts. More importantly, some results on $U(\mathfrak{g})$-modules, which seem inaccessible by algebra, have been proved geometrically – the proofs of the Kazhdan-Lusztig conjectures [3, 5] and of the Barbasch-Vogan conjecture [30] are particular examples.

4.3 Construction of Harish-Chandra Modules

The equivalence of categories (4.7) persists when certain additional ingredients are fed in on both sides. In the case of a Harish-Chandra module M with infinitesimal character $\chi_{\lambda+\rho}$, the algebraic action[2] of the group K on M induces an algebraic action of K on the $\mathcal{D}_\lambda$-module ΔM. One might think that the admissibility of M puts an additional restriction on ΔM, but that is not the case: any $(\mathfrak{g}, K)$-module with an infinitesimal character is automatically admissible. By a *$(\mathcal{D}_\lambda, K)$-module* one means a sheaf of $\mathcal{D}_\lambda$-modules, equipped with an algebraic action of K that is compatible with the $\mathcal{D}_\lambda$-structure. Compatibility in the current setting is entirely analogous to the earlier notion of compatibility in the context of $(\mathfrak{g}, K)$-module. More precisely, the Lie algebra $\mathfrak{k}$ acts on M both via the differentiation of the K-action and via (51) and the inclusion $\mathfrak{k} \hookrightarrow \mathfrak{g}$. These two actions of $\mathfrak{k}$ must agree, and the analogue of (38) must also be satisfied. With these additional ingredients, the functors Δ and Γ in (53) induce

$$\left\{\begin{matrix}\text{category of Harish-Chandra modules} \\ \text{with infinitesimal character } \chi_{\lambda+\rho}\end{matrix}\right\} \underset{\Gamma}{\overset{\Delta}{\rightleftarrows}} \left\{\begin{matrix}\text{category of coherent} \\ (\mathcal{D}_\lambda, K)\text{-modules}\end{matrix}\right\}. \tag{56}$$

Corollary 4.7 has a counterpart for these restricted functors:

Corollary 4.8 *If $\lambda + \rho$ is integrally dominant and regular, the functor Δ in (56) defines an equivalence of categories, with inverse functor Γ.*

[2] All finite dimensional representations of a complex linear reductive group are algebraic.

This corollary, too, has refinements that apply when either or both of the hypotheses of integral dominance and regularity are dropped. Under an equivalence of categories, irreducible objects correspond to irreducible objects, hence

Corollary 4.9 *If $\lambda + \rho$ is integrally dominant and regular, the functor Δ establishes a bijection*

$$\left\{ \begin{matrix} \textit{irreducible Harish-Chandra modules} \\ \textit{with infinitesimal character } \chi_{\lambda+\rho} \end{matrix} \right\} \xrightarrow[\sim]{\Delta} \left\{ \begin{matrix} \textit{irreducible} \\ (\mathcal{D}_\lambda, K)\textit{-modules} \end{matrix} \right\}.$$

If $\lambda+\rho$ is integrally dominant but singular, Δ sets up a bijection between the set of all irreducible Harish-Chandra modules with infinitesimal character $\chi_{\lambda+\rho}$ on the one hand, and the set of irreducible $(\mathcal{D}_\lambda, K)$-modules with non-zero cohomology on the other.

This latter corollary describes the irreducible Harish-Chandra modules in terms of irreducible $(\mathcal{D}_\lambda, K)$-modules. But how does one construct such sheaves? Two properties of the K-action on X simplify the problem. First of all,

$$K \text{ acts on } X \text{ with finitely many orbits.} \tag{57}$$

Since K acts algebraically, these orbits are algebraic subvarieties. Moreover,

$$\text{all } K\text{-orbits in } X \text{ are affinely embedded,} \tag{58}$$

which means that they intersect any open affine subset $U \subset X$ in an affine set. As a negative example, we mention $\mathbb{CP}^n - \{\text{point}\}$, with $n \geq 2$, which is not affinely embedded in $\mathbb{CP}^n$. Such sets do arise as K-orbits in generalized flag varieties.

We now consider a particular irreducible $(\mathcal{D}_\lambda, K)$-module $\mathcal{S}$. For geometric reasons, the support of $\mathcal{S}$ must consist of the closure of a single K-orbit Q. We let ∂Q denote the boundary of Q in X, i.e., $\partial Q =$ (closure of Q) $- Q$. The inclusion $j : Q \hookrightarrow X$ factors as a product

$$j = j_o \circ j_c\,, \quad \text{with } j_c : Q \hookrightarrow X - \partial Q \text{ and } j_o : X - \partial Q \hookrightarrow X\,; \tag{59}$$

note that j_c is a smooth closed embedding and j_0 an open embedding. Since $\mathcal{S}|_{X-\partial Q}$ is K-equivariant, irreducible, and supported on the K-orbit Q, $\mathcal{L}_\lambda$ must exist at least as a K-equivariant line bundle on the formal neighborhood of Q in $X - \partial Q$, even if it does not exist as line bundle on all of X, and $\mathcal{S}|_{X-\partial Q}$ must be the $\mathcal{D}_\lambda$-module direct image of $\mathcal{O}_Q(\mathcal{L}_\lambda|_Q)$ under j_c – in formal notation,

$$\mathcal{S}|_{X-\partial Q} \;=\; j_{c+}\mathcal{O}_Q(\mathcal{L}_\lambda|_Q)\,. \tag{60}$$

Depending on the orbit Q, this forces certain integrality conditions on λ; if these integrality conditions do not hold, no irreducible $(\mathcal{D}_\lambda, K)$-module can have the closure of Q as support. The $\mathcal{D}_\lambda$-module direct image $j_{c+}\mathcal{O}_Q(\mathcal{L}_\lambda|_Q)$ is easy to describe because j_c is a smooth closed embedding: its sections can be expressed as normal derivatives, of any order, applied to sections of $\mathcal{L}_\lambda|_Q$ over Q. Since $\mathcal{S}$ is irreducible,

$$\mathcal{S} \;\hookrightarrow\; j_{o+}(\mathcal{S}|_{X\setminus\partial Q}) \;=\; j_{o+}\circ j_{c+}\mathcal{O}_Q(\mathcal{L}_\lambda|_Q) \;=\; j_+\mathcal{O}_Q(\mathcal{L}_\lambda|_Q)\,. \tag{61}$$

In general, constructing the $\mathcal{D}$-module direct image under an open embedding requires passage to the derived category. In our situation, because of (58), the direct image exists as a bona fide $\mathcal{D}_\lambda$-module. One calls $j_+\mathcal{O}_Q(\mathcal{L}_\lambda|_Q)$ the *standard sheaf* corresponding to the orbit Q, the parameter $\lambda \in \mathfrak{h}^*$, and one other simple datum that is necessary to pin down the meaning of $\mathcal{L}_\lambda|_Q$. The steps we outlined exhibit the irreducible $(\mathcal{D}_\lambda, K)$-module $\mathcal{S}$ as the *unique irreducible subsheaf* of the standard sheaf $j_+\mathcal{O}_Q(\mathcal{L}_\lambda|_Q)$.

At one extreme, the irreducible subsheaf $\mathcal{S}$ may coincide with the standard sheaf $j_+\mathcal{O}_Q(\mathcal{L}_\lambda|_Q)$ in which it lies, and at the opposite extreme, it may be much smaller. This phenomenon is governed by the behavior of sections near ∂Q. Very roughly, if $j_+\mathcal{O}_Q(\mathcal{L}_\lambda|_Q)$ has sections with various degrees of regularity along ∂Q, the unique irreducible subsheaf $\mathcal{S}$ consists of the "most regular" sections; when all sections have the same degrees of regularity along ∂Q, the standard sheaf $j_+\mathcal{O}_Q(\mathcal{L}_\lambda|_Q)$ is irreducible, hence equal to its unique irreducible subsheaf $\mathcal{S}$. The standard sheaf is more tractable than $\mathcal{S}$. In the crucial situation, when $\lambda+\rho$ is integrally dominant, one understands $H^0(X, j_+\mathcal{O}_Q(\mathcal{L}_\lambda|_Q))$, the *standard module* corresponding to the given set of data, quite well [17]. The space of sections $H^0(X,\mathcal{S})$ is the unique irreducible submodule of the standard module – or zero, which can happen only when $\lambda+\rho$ is singular. In view of (4.9), these results constitute a classification of the irreducible Harish-Chandra modules with infinitesimal character $\chi_{\lambda+\rho}$ – the *Beilinson-Bernstein classification.*

Two other classification schemes, due to Landglands [25] and Vogan-Zuckerman [36], predate Beilinson-Bernstein's. They, too, exhibit the irreducible Harish-Chandra modules as unique irreducible submodules, or dually as unique irreducible quotients, of certain standard modules. All three classifications obviously describe the same class of objects, but it is not clear a priori that the three types of standard modules agree or are dual to each other. That can be shown most transparently by geometric arguments [18]. Depending on the interplay between the orbit Q and the parameter λ, the higher cohomology groups of the standard sheaf $j_+\mathcal{O}_Q(\mathcal{L}_\lambda|_Q)$ may vanish even when part B) of Theorem (4.6) does not apply directly. One extreme case, with "Q as affine as possible", leads to the Langlands classification, the other, with "Q as close to projective as possible", to Vogan-Zuckerman's. The Beilinson-Bernstein situation lies between these two, and all three can be related via the intertwining functors we mentioned earlier.

4.4 Construction of $G_{\mathbb{R}}$-representations

Just as one can attach Harish-Chandra modules to K-orbits in the flag variety, $G_{\mathbb{R}}$-representations arise from $G_{\mathbb{R}}$-orbits. There are finitely many such orbits, and they are real algebraic subvarieties. We now equip X with the usual Hausdorff topology – not the Zariski topology, as in the previous section.

To motivate the discussion, we first look at two special cases. At one extreme, let us consider a group $G_{\mathbb{R}}$, subject to the usual hypotheses and the condition (45). We suppose that $G_{\mathbb{R}}$ contains a compact Cartan subgroup, and we fix an open $G_{\mathbb{R}}$-orbit $S \subset X$. As subgroup of G, $G_{\mathbb{R}}$ acts on $\mathcal{O}(\mathcal{L}_\lambda)$, the sheaf of holomorphic sections of a G-equivariant line bundle $\mathcal{L}_\lambda$. This action induces a natural linear action on the cohomology groups $H^p(S, \mathcal{O}(\mathcal{L}_\lambda))$ over the open $G_{\mathbb{R}}$-orbit S. The cohomology can be computed from the complex of $\mathcal{L}_\lambda$-valued Dolbeault forms. It is far from obvious, but the coboundary operator $\bar{\partial}$ has closed range, giving the cohomology groups natural Fréchet topologies, with respect to which $G_{\mathbb{R}}$ acts continuously. The resulting representations are admissible, of finite length, with infinitesimal character $\chi_{\lambda+\rho}$. When $\lambda+\rho$ is antidominant regular[3], the cohomology vanishes except in a single degree $s > 0$, and in that degree $p = s$, $H^s(S, \mathcal{O}(\mathcal{L}_\lambda))$ is a discrete series representation – more precisely, it is the maximal globalization of the Harish-Chandra module $\mathrm{HC}(V_{\lambda+\rho})$ of a discrete series representation $V_{\lambda+\rho}$. This construction provides a geometric realization, analogous to the Borel-Weyl-Bott theorem, of the entire discrete series. The reader may consult [7] for references and further details.

[3] i.e., $(\lambda+\rho, \alpha) < 0$ for all $\alpha \in \Phi^+$.

At the other extreme, we suppose that $S \subset X$ is a closed orbit of a split group $G_{\mathbb{R}}$; here "split" means that $G_{\mathbb{R}}$ contains a Cartan subgroup $A_{\mathbb{R}}$ such that every $a \in A^0_{\mathbb{R}}$ acts with real eigenvalues. Parenthetically we should remark that noncompact reductive groups $G_{\mathbb{R}}$ have finitely many Cartan subgroups, and may contain both a compact and a split Cartan subgroup. In the split case, the closed $G_{\mathbb{R}}$-orbit S is necessarily a real form of the flag variety, i.e., a submanifold such that the holomorphic tangent space T_xX at any $x \in S$ contains the tangent space T_xS of S as a real form: $T_xX = \mathbb{C} \otimes_{\mathbb{R}} T_xS$. For any $\lambda \in \Lambda$, $G_{\mathbb{R}}$ acts on $C^{-\omega}(S, \mathcal{L}_\lambda)$, the space of hyperfunction sections of $\mathcal{L}_\lambda$ over the real analytic manifold S. The hyperfunctions on a compact real analytic manifold carry a natural Fréchet topology. With respect to this topology, $G_{\mathbb{R}}$ acts continuously on $C^{-\omega}(S, \mathcal{L}_\lambda)$. The resulting representation is admissible, of finite length, with infinitesimal character $\chi_{\lambda+\rho}$; it belongs to the – non-unitary, in general – principal series of $G_{\mathbb{R}}$. Since λ has been confined to the lattice Λ, only some principal series representations can be obtained this way. One gets the others by letting λ range over $\mathfrak{h}^*$, the dual space of the universal Cartan, in which case $\mathcal{L}_\lambda$ still exists as $G_{\mathbb{R}}$-equivariant real analytic line bundle over the closed orbit S. However, for the moment we still want to suppose $\lambda \in \Lambda$, so that $\mathcal{L}_\lambda$ is well defined even as G-equivariant holomorphic line bundle on X.

At first glance, the cohomology groups $H^p(S, \mathcal{O}(\mathcal{L}_\lambda))$ over an open $G_{\mathbb{R}}$-orbit S and the space of hyperfunction sections $C^{-\omega}(S, \mathcal{L}_\lambda)$ over a closed orbit of a split group $G_{\mathbb{R}}$ might not seem to fit easily into a common framework. However, both can be expressed as Ext groups,

$$\begin{aligned} H^p(S, \mathcal{O}(\mathcal{L}_\lambda)) &= \operatorname{Ext}^p(j_!\mathbb{C}_S, \mathcal{O}(\mathcal{L}_\lambda)) \text{ if } S \text{ is an open } G_{\mathbb{R}}\text{-orbit,} \\ C^{-\omega}(S, \mathcal{L}_\lambda) &= \operatorname{Ext}^n(j_!\mathbb{C}_S, \mathcal{O}(\mathcal{L}_\lambda)) \text{ if } S \text{ is closed, } G_{\mathbb{R}} \text{ split,} \end{aligned} \tag{62}$$

and $n = \dim_{\mathbb{C}} X$; in both cases j denotes the inclusion $S \hookrightarrow X$, $\mathbb{C}_S$ the constant sheaf on S with fiber $\mathbb{C}$, and $j_!\mathbb{C}_S$ the sheaf on X obtained by taking the direct image with proper supports. One can describe the Ext groups equivalently as the right derived functors of $\operatorname{Hom}(\cdot,\cdot)$ in the second variable, or the left derived functors in the first variable. Properly interpreted, (62) represents only the extreme cases of a general construction [31], which attaches $G_{\mathbb{R}}$-representations to all $G_{\mathbb{R}}$-orbits. In this way one obtains the maximal globalizations of all standard modules in the Beilinson-Bernstein construction.

Kashiwara [20] observed that the results of [31] could be stated in more functorial language, conjecturally at least, which would then produce not just maximal globalizations of standard modules, but of all

induces a bijection

$$\{G_{\mathbb{R}}\text{-orbits on } X\} \quad \longleftrightarrow \quad \{K\text{-orbits on } X\},$$

which reverses the closure relationships.

As in (40), let us consider the example of $G_{\mathbb{R}} = SU(1,1)$, $G = SL(2,\mathbb{C})$, $K \simeq \mathbb{C}^*$. The flag variety of $\mathfrak{g}$ is $X = \mathbb{CP}^1 \simeq \mathbb{C} \cup \{\infty\}$, on which $a \in \mathbb{C}^* \simeq K$ acts as multiplication by a^2. Thus $\{0\}$, $\{\infty\}$, and $\mathbb{C}^*$ are the K-orbits. The group $SU(1,1)$ acts transitively on the unit disc D, the complement $X - D^{\mathrm{cl}}$ of the closure of D, and their common boundary $\partial D \simeq S^1$.

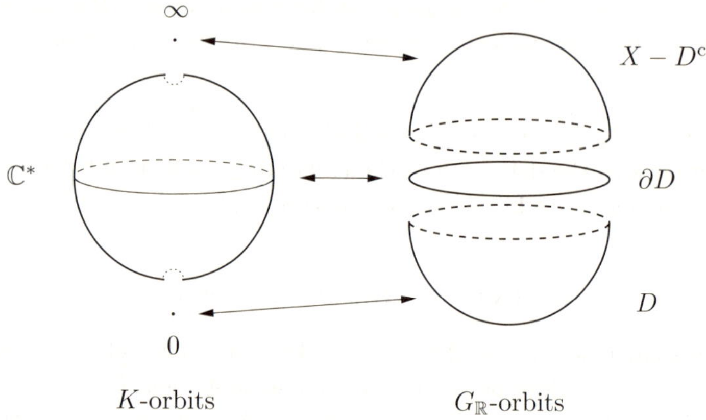

Figure 4.1 Matsuki correspondence for $SU(1,1)$.

In this particular situation, duality means that one of the two orbits contains the other, but that is not the case for a general group $G_{\mathbb{R}} = SU(1,1)$.

Matsuki's proof shows that K-equivariant twisted local systems on a K-orbit Q correspond bijectively to the $G_{\mathbb{R}}$-equivariant twisted local systems on the dual $G_{\mathbb{R}}$-orbit S, in both cases with the same twist $\lambda \in \mathfrak{h}^*$. As was mentioned, the $G_{\mathbb{R}}$-equivariant twisted local systems with twist λ may be regarded as the basic building blocks of the twisted $G_{\mathbb{R}}$-equivariant derived category $\mathbf{D}_{G_{\mathbb{R}}}(X)_\lambda$. Quite analogously, the K-equivariant derived category $\mathbf{D}_K(X)_\lambda$ is built up from K-equivariant twisted local systems with twist λ. But it is not at all obvious that the bijection between equivariant twisted local systems carries over to the extensions between them in the two categories. That was proved by Mirković, Uzawa, and Vilonen:

Theorem 4.12 (Matsuki correspondence of sheaves [28]) *Matsuki duality induces an equivalence of categories*

$$\Psi \,:\, \mathbf{D}_{G_{\mathbb{R}}}(X)_\lambda \ \xrightarrow{\ \sim\ }\ \mathbf{D}_K(X)_\lambda\,.$$

The covariant deRham functor [19, 27] establishes an equivalence of categories between the categories of, respectively, regular holonomic $\mathcal{D}$-modules and perverse sheaves constructible with respect to algebraic stratifications. This is the so-called *Riemann-Hilbert correspondence*[4]. In the context of an algebraic group action with finitely many orbits, all coherent equivariant $\mathcal{D}$-modules are automatically regular holonomic. The deRham functor therefore restricts to a well defined functor

$$\mathrm{dR}\ :\ \left\{\begin{matrix}\text{category of coherent}\\ (\mathcal{D},K)\text{-modules}\end{matrix}\right\}\ \longrightarrow\ \mathbf{D}_K(X)\,. \tag{66}$$

For our purposes, it will not matter that (66) defines an equivalence of categories to the subcategory of perverse objects in $\mathbf{D}_K(X)$. However, we do need the twisted version of the deRham operator,

$$\mathrm{dR}\ :\ \left\{\begin{matrix}\text{category of coherent}\\ (\mathcal{D}_\lambda,K)\text{-modules}\end{matrix}\right\}\ \longrightarrow\ \mathbf{D}_K(X)_{-\lambda}\,; \tag{67}$$

it is contravariant with respect to the twisting, hence the appearance of λ on the left and $-\lambda$ on the right. Our final statement relates the Beilinson-Bernstein construction to the construction of $G_{\mathbb{R}}$-representations in section 4.4. Like Theorem 4.10, it was conjectured by Kashiwara [20] and proved in [22].

Theorem 4.13 ([22]) *If $\mathcal{S}$ is a coherent $(\mathcal{D}_\lambda,K)$-module, and $\dim_{\mathbb{C}} X = n$, then the minimal globalization of the Harish-Chandra module $H^p(X,\mathcal{S})$ is isomorphic as $G_{\mathbb{R}}$-representation to the strong dual of $\mathrm{Ext}^{n-p}(\Psi^{-1}\circ \mathrm{dR}\,\mathcal{S}\,,\,\mathcal{O}_{-\lambda-2\rho})$, for any $p\in\mathbb{Z}$.*

The sheaf $\Psi^{-1}\circ\mathrm{dR}\,\mathcal{S}$ has twist $-\lambda$, but $\mathbf{D}_K(X)_{-\lambda}\simeq\mathbf{D}_K(X)_{-\lambda-2\rho}$ by the K-analogue of (64), so it does make sense to consider extensions between $\Psi^{-1}\circ\mathrm{dR}\,\mathcal{S}$ and $\mathcal{O}_{-\lambda-2\rho}$.

Some properties of representations are easier to understand in terms of the Beilinson-Bernstein construction, and others easier in terms of the $G_{\mathbb{R}}$-construction. Theorems 4.10 and 4.13 make it possible to "play off"

[4] The reader may consult Borel's book [8] for a discussion of algebraic $\mathcal{D}$-modules in general and the Riemann-Hilbert correspondence in particular.

the two sides against each other. The two theorems also play a crucial role in the proof of the Barbasch-Vogan conjectures [30].

Bibliography

1. M. Atiyah, *Characters of Semisimple Lie Groups*, Lecture notes (1976), reprinted in Collected Works of Michael Atiyah, Vol. 4, Springer, 1988, pp. 491–557.
2. V. Bargmann, *Irreducible unitary representations of the Lorentz group*, Ann. of Math. (2) 48 (1947), 568–640.
3. A. Beilinson and J. Bernstein, *Localisation de $\mathfrak{g}$-modules*, C. R. Acad. Sci. Paris 292 (1981), 15–18.
4. A. Beilinson and J. Bernstein, *A generalization of Casselman's submodule theorem*, in: Representation Theory of Reductive Groups, Progress in Mathematics, Vol 40, Birkhäuser, Boston, 1983, pp. 35–52.
5. J. L. Brylinski and M. Kashiwara, *Kazhdan-Lusztig conjecture and holonomic systems*, Inventiones Math. 64 (1981), 387–410.
6. J. Bernstein and V. Lunts, *Equivariant sheaves and functors*, Lecture Notes in Mathematics 1578, Springer, 1994.
7. V. Bolton and W. Schmid, *Discrete Series*, Proc. of Symposia in Pure Math. 61 (1997), 83–113.
8. A. Borel et al., *Algebraic $\mathcal{D}$-modules*, Perspectives in Mathematics, Academic Press, 1987.
9. R. Bott, *Homogeneous vector bundles*, Annals of Math. 66 (1957), 203–248.
10. W. Casselman, *Jacquet modules for real reductive groups*, Proc. of the International Congress of Mathematics, Helsinki, 1980, 557–563.
11. W. Casselman, *Canonical Extensions of Harish-Chandra modules to representations of G*, Can. Jour. Math. 41 (1989), 385–438.
12. R. Godement, *Sur la théorie des représentations unitaires*, Ann. of Math. (2) 53 (1951), 68–124.

13. R. Gunning and H. Rossi, *Analytic Functions of Several Complex Variables*, Prentice-Hall, Inc., Englewood Cliffs, N.J., 1965.
14. Harish-Chandra, *On some applications of the universal enveloping algebra of a semisimple Lie algebra*, Trans. Amer. Math. Soc. 70 (1951), 28–96.
15. Harish-Chandra, *Representations of a semisimple Lie group on a Banach space I*, Trans. Amer. Math. Soc. 75 (1953), 185–243.
16. S. Helgason, *Differential Geometry, Lie Groups, and Symmetric Spaces*, Academic Press, New York, 1978.
17. H. Hecht, D. Miličić, W. Schmid and J. Wolf, *Localization and standard modules for real semisimple Lie groups I: The duality theorem*, Invent. Math. 90 (1987), 297–332.
18. H. Hecht, D. Miličić, W. Schmid and J. Wolf, *Localization and standard modules for real semisimple Lie groups II: Irreducibility, vanishing theorems and classification*, preprint.
19. M. Kashiwara, *The Riemann-Hilbert problem for holonomic systems*, Publ. RIMS, Kyoto University 20 (1984), 319–365.
20. M. Kashiwara, *Open problems in group representations*, Proceedings of the Taniguchi Symposium 1986, RIMS preprint 569, Kyoto University 1987.
21. M. Kashiwara and P. Schapira, *Sheaves on Manifolds*, Springer, 1990.
22. M. Kashiwara and W. Schmid, *Quasi-equivariant $\mathcal{D}$-modules, equivariant derived category, and representations of reductive Lie groups*, Lie Theory and Geometry, in Honor of Bertram Kostant, Progress in Mathematics, vol. 123, Birkhäuser, Boston, 1994, pp. 457–488.
23. A. Knapp, *Lie Groups Beyond an Introduction*, Progress in Mathematics, vol. 140, Birkhäuser, 2002.
24. S. Lang, $SL(2,\mathbb{R})$, Graduate Texts in Mathematics, vol. 105, Springer-Verlag, New York, 1985.
25. R. P. Langlands, *On the classification of irreducible representations of real algebraic groups*, in: Representation Theory and Harmonic Analysis on Semisimple Lie Groups, Mathematical Surveys and Monographs No. 31, Amer. Math. Soc. 1989, pp. 101–170.
26. T. Matsuki, *Orbits on affine symmetric spaces under the action of parabolic subgroups*, Hiroshima Math. Jour. 12 (1982), 307–320.
27. Z. Mebkhout, *Une équivalance de catégories - Une autre équivalence de catégories*, Comp. Math. 51 (1984), 55–64.
28. I. Mirković, T. Uzawa and K. Vilonen, *Matsuki correspondence for sheaves*, Inventiones Math. 109 (1992), 231–245.
29. W. Schmid, *Boundary value problems for group invariant differen-*

tial equations, in: Élie Cartan et les Mathématiques d'Aujourd'hui, Astérisque 1985, numéro hors séries (1985), 311–321.

30. W. Schmid and K. Vilonen, *Characteristic cycles and wave front cycles of representations of reductive groups*, Annals of Math. (2), 151 (2000), 1071–1118.
31. W. Schmid and J. A. Wolf, *Geometric quantization and derived functor modules for semisimple Lie groups*, Jour. Funct. Anal. 90 (1990), 48–112.
32. J.-P. Serre, *Représentations linéaires et espaces homogènes Kählériennes des groupes de Lie compacts*, Séminaire Bourbaki, 6e année, 1953/54, Exposé 100, Inst. Henri Poincaré, Paris, 1954; reprinted with corrections, 1965.
33. W. Soergel, *An irreducible not admissible Banach representation of* $SL(2,\mathbb{R})$, Proc. Amer. Math. Soc, 104 (1988), 1322–1324.
34. F. Treves, *Topological Vector Spaces, Distributions and Kernels*, Academic Press, 1967.
35. N. R. Wallach, *Asymptotic expansion of generalized matrix entries of real reductive groups* in: Lie Group Representations I, Lecture Notes in Mathematics, vol. 1024, Springer, 1983, pp. 287–369.
36. D. A. Vogan, *Representations of Real Reductive groups*, Progress in Mathematics, vol. 15, Birkhäuser, Boston, 1981.
37. G. Warner, *Harmonic Analysis on Semi-Simple Lie Groups*, vols. I and II, Springer-Verlag, 1972.

PART FIVE

Deformation theory: A powerful tool in physics modelling

Daniel Sternheimer
Daniel.Sternheimer@u-bourgogne.fr
Institut de Mathématiques de Bourgogne - CNRS UMR 5584
Université de Bourgogne
9 avenue Alain Savary, BP 47870
F-21078 Dijon Cedex, France

ABSTRACT
Deformation theory is a powerful tool – so far, a posteriori – in modelling physical reality. We start with a short historical and philosophical review of the context and concentrate this rapid presentation on three interrelated directions where deformation theory is essential in bringing a new framework – which has then to be developed using adapted tools, some of which come from the deformation aspect. Minkowskian space-time can be deformed into Anti de Sitter, where representation theory shows us that massless particles become composite (also dynamically). Nonlinear group representations and covariant field equations, coming from interactions, can be viewed as some deformation of their linear (free) part, which provides a good framework for treating problems in that area. Last but not least, (algebras associated with) classical mechanics (and field theory) on a Poisson phase space can be deformed to (algebras associated with) quantum mechanics (and quantum field theory). That is now a frontier domain in mathematics and theoretical physics called deformation quantization, with multiple ramifications, avatars and connections. These include representation theory, quantum groups (when considering Hopf algebras instead of associative or Lie algebras), noncommutative geometry and manifolds, algebraic geometry (even algebraic curves à la Zagier), number theory, and of course what is regrouped under the name of M-theory. We shall here look at these from the unifying point of view of deformation theory and refer to a limited number of papers as a starting point for further study.

1 Introduction

1.1 It ain't necessarily so

Mathematics proceeds by logical deduction: *If A, and A implies B, then B.* In other words, A is a sufficient condition for B to hold. As simple as that sentence may seem, it is often distorted in ordinary life where (due to external reasons) one is tempted to take for necessary a sufficient condition. Schematically, it can be expressed as follows: *Given that A implies B, if I find B nice (thus want A because it will give me B), then A.* The subtle logical mistake is perpetrated by almost all in experimental sciences when building models.

The need for modelling is as old as Science: more and more data are being collected and it is natural to try and put some order there. So from experimental data E one imagines a model M that can explain them. Eventually (with deeper intuition) one can sometimes show that the model M can be derived from more fundamental principles, from a theory T. That is the implicit part, taken for granted by experimental scientists (A implies B).

Now if new data $E_1 \supsetneq E$ are found that can also be derived from T, i.e. B becomes nice, the model or theory receives experimental confirmation (then A). One does not argue with success. The confusion between necessary and sufficient conditions may go as far as saying that abstract entities involved in T or M were "directly observed" with the new data: in fact, what has been observed is only a consequence of these entities in some model. [See e.g. in the Press Release for the 1999 Nobel prizes the remark that "This [the top] quark was observed directly for the first time in 1995 at the Fermilab in the USA," somewhat strange for a "particle" supposed to be confined and thus not directly observable.] The confusion is enhanced by the fact that our interpretation of the raw experimental data is made within existing models or theories, so that what we call an experimental result may, in fact, be theory-dependent.

But it often happens that with a larger data set E', the new data will not be easily cast in the existing model. Then there will be a need to develop a new model M', if possible deriving from a new theory T', that can explain everything observed *so far* (one should not hope for a definitive theory of everything). A scientist should therefore, even (especially) when everything seems for the best in the best of possible worlds and some are sure that we can now explain everything, be always prepared for surprises and have, in the back of his mind, a tune playing *it ain't necessarily so* in relation with the best accepted theories – even more so when trying to block some avenues with so called "no go" theorems, overlooking the hypotheses (sometimes hidden) on which they rely or the lack of rigour in their proofs.

Such was the case towards the end of the nineteenth century with classical Newtonian mechanics and electromagnetism. What happened then shows that deformation theory, developed in an appropriate context, can lead us to such "deformed" models and theories. In this review I shall give strong "experimental evidence" to the effectiveness of deformation theory, developed in an adequate context. The examples involve the three components of this School (P, Q and R).

1.2 Epistemological importance of deformation theory

A scientist should try and answer three questions: why, what and how. The bulk of the work is of course devoted to the last question. But if research is maybe 1% inspiration and 99% perspiration, the inspiration is an essential ingredient.

It is certainly better to know 'what' one is doing, and it helps a lot to know 'why.' The knowledge of 'why' can often be imprecise and implicit. But very few works turn out to be important if the answer to the question 'why?' is 'why not?' – or if the research is merely solving a problem posed by some adviser or a guru, without asking oneself why is the problem important to solve.

In answering the three questions the human mind uses two very different approaches: intuition and deduction. The distinction between both, with examples, plays a major role in the work of Daniel Kahneman, 2002 Nobel laureate in Economics "for having integrated insights from psychological research into economic science..." Intuition is an important – albeit hard to evaluate – factor in the evolution of markets; its effects may contradict (at least locally in time) what logical deduction tells us. In science these two factors are present, with intuition playing an important role. But in science there is an interaction between both approaches, since an extensive study of notions often makes them so familiar that they can be subject to intuition. For scientists whose works are seminal, that interaction is probably very intense.

Deformation theory provides a partial answer to the question *how*, at least in the mathematical formulation and study of fundamental physics, which is what we call physical mathematics. The knowledge of the category (in the mathematical sense) where one defines the deformation clarifies *what* is done. We shall start by explaining *why* it is a powerful tool.

One should never forget that physical theories have their domain of applicability defined by the relevant distances, velocities, energies, etc. involved. The passage from one domain (of distances, etc.) to another does not happen in an uncontrolled way. Rather, experimental phenomena appear that cause a paradox and contradict accepted theories. Eventually a new fundamental constant enters and the formalism is modified. Then the attached structures (symmetries, observables, states, etc.) *deform* the initial structure. Namely, we have a new structure which in the limit, when the new parameter goes to zero, coincides with the previous formalism. The question is therefore, in which category do we seek for deformations? Usually physics is rather conservative and if we start e.g. with the category of associative or Lie algebras, we tend to deform in the same category. But there are important examples of generalizations of this principle: e.g. quantum groups are deformations of Hopf algebras.

The discovery of the non-flat nature of Earth may be the first example of this phenomenon. Closer to us, the paradox coming from the Michelson and Morley experiment (1887) was resolved in 1905 by Einstein with the special theory of relativity: in our context, one can express that by saying that the Galilean geometrical symmetry group of Newtonian mechanics is deformed to the Poincaré group, the new fundamental constant being c^{-1} where c is the velocity of light in vacuum.

It is interesting to note that a first mathematical example of deformations was introduced at around the same time with the Riemann surface theory, though deformations became systematically studied in the mathematical literature only at the end of the fifties with the profound works of Kodaira and Spencer [45] on deformations of complex analytic structures. Now, when one has an action on a geometrical structure, it is natural to try and "linearise" it by inducing from it an action on an algebra of functions on that structure. This is implicitly what Gerstenhaber did shortly afterwards [40] with his definition and thorough study of deformations of rings and algebras.

It is in the Gerstenhaber sense that the Galileo group is deformed to the Poincaré group; that operation is the inverse of the notion of group contraction introduced ten years before, empirically, by Ínönü and Wigner [44], an earlier example of which can be found in [57]. This fact triggered strong interest for deformation theory in France among a number of theoretical physicists, including Flato who had just arrived from the Racah school and knew well the effectiveness of symmetry in physical problems. He was soon to realize that, however important symmetry is as a notion and a tool in a mathematical treatment of physical problems, it is not the only one and should be complemented with other (often related) concepts: The notion of deformation can be applied to a variety of categories that are used to express mathematically the physical reality.

One should not forget that mathematics arose as an abstraction of the physical world. Until the 19th century, most leading mathematicians were also physicists, and vice-versa (sometimes also philosophers): Archimedes, Newton, Pascal, Laplace, Gauss, to mention just a few in the "Western World". In the middle of the 20th century the two communities became so widely separated by a kind of "Babel tower effect" that Wigner could marvel about the "unreasonable effectiveness of mathematics in theoretical physics". As Sir Michael Atiyah put it in his closing talk at ICMP 2000 (paraphrasing Oscar Wilde about the US and UK), "mathematics and physics became two communities separated by a common language." Many mathematicians were proud of knowing nothing about physics, and many physicists despise papers that are too mathematical for them; every senior scientist can put famous names behind those attitudes, so I shall refrain from doing so.

In recent years the trend has been reversed, at least in mathematics. Somehow, among the variety of mathematical problems that the human brain can imagine, those having a physical origin tend to be most seminal – and mathematicians again realise that. The number of "quantum Fields medals", from Connes in 1982 to Kontsevich in 1998, is a clear proof of it. In physics the two trends coexist, not always peacefully, sometimes with exaggeration. Many physicists tend to agree with Goethe who said that "mathematicians are like Frenchmen, they translate everything into their own language and henceforth it is something entirely different." But more and more now realise the importance of being bilingual; as with languages, it is of consequence to learn both languages at a young age, because knowing facts (by learning and deduction) is one thing and feeling them (intuitively) is another question. There are at present a few living examples of truly bilingual scientists, in all generations, but not enough.

In addition to, but not unrelated with, the importance of deformation theory in model building, to which the bulk of this paper is devoted, another fact played a major role in Flato's philosophy: space-time cannot be disconnected from truly fundamental models. Nowadays this seems obvious, albeit with space-times that may have more than the traditional four dimensions, but insisting in keeping that direction of research open (see e.g. [37]) was considered heretical in the mid 60's when the mainstream would not tolerate anything but particle spectroscopy based on (phenomenological) unitary groups commuting with the Poincaré group. In this spirit we were at that time led to look at the conformal group and to another group [32] involving two kinds of translations, vectorial and spinorial, applications of which to neutrino physics gave a prototype of the Wess-Zumino [60] Poincaré supersymmetry.

On the epistemological side, that approach is in line with the philosophy of Kant – and with Spinoza's pantheistic views, according to which (in mathematical terms) our Universe is a representation of an abstract structure named God. The representation is possibly unfaithful, but the question is metaphysical since we know no other representation. As pointed out in the Summer of 2002 in a weekly magazine by the present French Minister of Education, the philosopher Luc Ferry, Kant was the first to invert the traditional concept that God created man from his own image. Kant, a deeply religious person, did not go as far as the (blaspheming) opposite, later reached e.g. by marxists, but his view was that one has to start with man's imperfection and try and get from it closer to God's perfection. Deformation theory provides us with a tool to do just that, starting with an imperfect description and deforming it into a less imperfect one.

Doing so one may discover that often "complicated is simpler:" The richer (deformed) structures carry more information, and may exhibit a variety of properties that make them easier to tackle mathematically than their more degenerate special cases.

The presentation that follows deals with a "trilogy" of interrelated subjects where the deformation insight permits significant progress. In his closing lecture of TH2002 in Paris, C.N. Yang described what he called the "three melodies" that dominated the development of physics in the twentieth century: quantisation, symmetries and "the phase factor" – the latter being of course, dealing with gauge theories, his favorite. Interestingly those three tunes were already present – the latter in embryonic form – in Hermann Weyl's seminal book [61] from 1928, i.e. quantisation, group theory and covariant field equations (in particular Maxwell-Dirac, with their Abelian gauge). Missing there is the concept of deformation which, as we have shown already in the 70's, explains the essence of quantisation. It also explains the passage from one kind of symmetry to another, e.g. from the symmetry of classical (Galilean) mechanics to that of special relativity and from the latter to Anti de

Sitter when a tiny negative curvature is permitted. Then one cannot do physics without looking at field equations, and the latter are usually covariant under a symmetry group; but fields interact, in particular (but not only) when operating a measurement, which brings in nonlinear field equations, a kind of deformation of the free equations when a coupling constant appears. Finally one needs, there also, to take into account quantum effects, which brings us back to the problem of quantisation. That is the basis of the trilogy pushed forward by Flato and coworkers since the 70's, which I survey here, devoting a little more space to the quantisation part of the trilogy.

2 Composite elementary particles in AdS microworld

It follows from our deformation philosophy that, in order to *anticipate* new formalisms, we have to study deformations of the algebraic structures attached to a given formalism – and of course their representations, essential in physical applications. The only question is, in which category do we perform this search for deformations. Usually physics is rather conservative and if we start e.g. with the category of associative or Lie algebras, we tend to deform in this category. In the passage from Galilean physics to special relativity (new parameter c^{-1}, where c is the speed of light), we deform the symmetry of the theory. In the Lie group (or algebra) category, there is a further deformation, giving rise to physics in (anti) de Sitter space-time (the new parameter being the curvature). It is this last aspect which we shall present here.

Recent experimental data indicate that the cosmological constant is most likely positive, suggesting (assuming a space-time of constant curvature, at least in first approximation) a de Sitter universe at cosmological distances. At our level, for (almost) all practical purposes, space-time is Minkowskian (flat). We shall assume that, at a much smaller scale, a tiny constant negative curvature is present, i.e. an anti de Sitter (AdS) microworld. This Ansatz might be a consequence of the hidden presence, at that level, of compactified extra dimensions and should be related with 't Hooft ideas on holography [43], ideas that go beyond the AdS/CFT Maldacena conjecture [52]. As we shall see, that hypothesis has far reaching consequences that among others could permit to go beyond the Standard Model and in particular explain neutrino oscillations and *PC* violation, and predict new mesons associated with multiple Higgs.

In the Lie group category the deformation chain stops at AdS but one can deform further the symmetry in the Hopf algebra category, to an AdS quantum group, where new and interesting features appear, including some very surprising (finite dimensional unitary representations) at root of unity [31]. We shall here be conservative and remain within usual AdS. But one should be open to the possibility to deform further, quantising the symmetry group of AdS space-time and/or even space-time itself, in a "stringy" way (e.g. adding 6 extra dimensions, compactified, so as to give some "fuzziness" to points in AdS_4, which might explain the local negative curvature there) or with noncommutative geometry as in [12].

2.1 A qualitative overview

The strategy is the following. AdS_4 group representation theory shows us that the UIR (unitary irreducible representations) which, for many good reasons (see e.g. [2]), should be called massless, are (in contradistinction with the flat space limit) composed of two more degenerate UIR of (the covering of) the AdS_4 group $SO(3,2)$. The latter were discovered by Dirac [17] and called singletons because the states appear on a single line and not on a lattice. They are naturally confined because their energy is proportional to angular momentum times the tiny curvature, which would require a laboratory of cosmic dimensions to get a measurable energy. We have called them *Di* and *Rac*, on the pattern of "bra" and "ket". They are the massless representations of the Poincaré group in 2+1 dimensional space, where $SO(3,2)$ is the conformal group (AdS_4/CFT_3 correspondence).

So far that compositeness is kinematical. Dynamics require in particular the consideration of field equations, initially at the first quantised level, in particular the analogue of the Klein-Gordon equation in AdS_4 for the *Rac*. There, as can be expected of massless (in 2+1 space-time) representations, gauges appear. We thus have to deal with indecomposable representations, triple extensions of UIR, as in the Gupta-Bleuler (GB) theory, and their tensor products. It is also desirable to take into account conformal covariance at these GB-triplets level. The situation gets therefore much more involved, quite different from the flat space limit, which makes the theory even more interesting.

One can then attempt to "plug into" conventional QED by considering a massless photon composed of two scalar singletons. The idea is to take creation and annihilation operators for the *Rac* that satisfy unusual commutation relations (which is fine for confined entities) in such a way that for the 2-*Rac* states (photons), the creation and annihilation operators satisfy the usual canonical commutation relations (CCR). We thus get a new and interesting infinite-dimensional Lie algebra, a kind of "square root" of the CCR. The theory can be completed, including taking into account conformal covariance of triplets, and composite QED was established [28].

After QED the natural step is to introduce compositeness in electroweak theory. Along the lines described above, that would require finding a kind of "square root of superalgebra", with both CAR and CCR included, obtained from creation and annihilation operators for $Di \oplus Rac$. That has yet to be done. Some steps in that direction have been initiated but the mathematical problems are extremely complicated, even more so since now the three flavours of leptons have to be considered.

But here a more pragmatic approach can be envisaged [38], triggered by recent experimental data which indicate that there are oscillations between various flavours of neutrinos. The latter would thus not be massless. This is not as surprising as it seems from the AdS point of view, because one of the attributes of masslessness is the presence of gauges. These are group theoretically associated with the limit of unitarity in the representations diagram, and the neutrino is above that limit in AdS: the *Di* is at the limit. Thus, all 9 leptons (electrons, muons, tau and their neutrinos) can be treated on an equal footing. One is then tempted to write them in a square table and consider them as composites $L^A_\beta = R^A D_\beta$. In this empirical approach, the vector mesons of the electroweak model are $Rac - Rac$ composites and the model predicts a new set of vector mesons that are $Di - Di$ composites and that play exactly the same role for the flavour symmetry $U_F(2)$ as the weak vector bosons do for the weak group $U_W(2)$. A set (maybe five pairs) of Higgs fields would have Yukawa couplings to the leptons currents and massify the leptons (and the vector mesons and the new mesons). This attempt has been developed in part in [38] (Frønsdal and I are still pursuing that direction) and is qualitatively promising. In addition to the neutrino masses it could explain why the Higgs has so far escaped detection: instead of one "potato" one has a gross purée of five, far more difficult to isolate from background.

Quantitatively however its predictive power is limited by the presence of too many free parameters. Maybe the addition to the picture of a deformation induced by the strong force and of the 18 quarks (which could be written in a cube and also considered composite) using fully the orthosymplectic AdS_4 supersymmetry and conformal covariance, and possibly the power of noncommutative geometry [12], would make this new "composite standard model" more predictive.

Intuitively the picture could be the following. Around the Planck length we would have AdS microworlds, kind of black holes with which we can communicate (in a way reminding 't Hooft's holography [43]) only by interaction at the surface, which in this case would be the "cone at infinity" of these AdS microworlds (at this level, Planck length would be treated mathematically as infinity for these microworlds) where the singleton states live. The interaction of 2-singleton states with ambiant Higgs fields (their presence might explain the "missing mass" of the Universe) would create the flavourless massless photons, the flavoured massive leptons that we know, and shorter lived quarks that are confined near the surface and, in turn, give the hadrons as in conventional theory. At present that picture is Science Fiction and it will remain so until more precise models can be built along similar lines.

That is a challenge for the next generations, in terms both of the mathematical tools that will have to be developed and of the physical ideas and calculations required. Moshe died in sight of the Promised Land; we are witnessing the crumbling of the walls of Jericho (the Standard Model); it is still a long way up to Jerusalem.

2.2 A brief more precise overview of the present state of singleton symmetry and field theory

References and a short account can be found in [30] of which we shall now, so as to give more background to the previous discussion, present some highlights. We denote by $D(E_0, s)$ the minimal weight representations of the twofold covering of the connected component of the identity of $SO(2,3)$. Here E_0 is the minimal $SO(2)$ eigenvalue and the half-integer s is the spin. These irreducible representations are unitary provided $E_0 \geq s+1$ for $s \geq 1$ and $E_0 \geq s + \frac{1}{2}$ for $s = 0$ and $s = \frac{1}{2}$. The *massless representations* of $SO(2,3)$ are defined (for $s \geq \frac{1}{2}$) as $D(s+1, s)$ and (for helicity zero) $D(1,0) \oplus D(2,0)$. At the limit of unitarity the Harish Chandra module $D(E_0, s)$ becomes indecomposable and the physical UIR appears as a quotient, a hall-mark of gauge theories. For $s \geq 1$ we get in the limit an indecomposable representation

$D(s+1,s) \to D(s+2,s-1)$, a shorthand notation [28] for what mathematicians would write as a short exact sequence of modules.

In gauge theories one needs extensions involving more than two UIRs. A typical situation is the case of flat space electromagnetism where one has the classical Gupta-Bleuler triplet which, in our shorthand notations, can be written $Sc \to Ph \to Ga$. Here Sc (scalar modes) and Ga (gauge modes) are massless zero-helicity UIRs of the Poincaré (inhomogeneous Lorentz) group while Ph is the module of physical modes, transforming under a sum of two UIRs of the Poincaré group with mass 0 and helicity $s = \pm 1$. The scalar modes can be suppressed by a gauge fixing condition (e.g. the Lorentz condition) and one is left with a nontrivial extension $Ph \to Ga$ on the vector space $Ph \dot{+} Ga$ which has no invariant nondegenerate metric and cannot be quantised covariantly. However the above Gupta-Bleuler triplet is an indecomposable representation (a nontrivial successive extension $Sc \to (Ph \to Ga)$) on a space which admits an invariant nondegenerate (but indefinite) Hermitian form and it must be used in order to obtain a covariant quantisation of this gauge theory. We shall meet here a similar situation, which in fact cannot be avoided.

For $s = 0$ and $s = \frac{1}{2}$, the above mentioned gauge theory appears not at the level of the massless representations $D(1,0) \oplus D(2,0)$ and $D(\frac{3}{2},\frac{1}{2})$ but at the limit of unitarity, the singletons $Rac = D(\frac{1}{2},0)$ and $Di = D(1,\frac{1}{2})$. These UIRs remain irreducible on the Lorentz subgroup SO(1,3) and on the (1+2) dimensional Poincaré group, of which $SO(2,3)$ is the conformal group. The singleton representations have a fundamental property: $(Di \oplus Rac) \otimes (Di \oplus Rac) = (D(1,0) \oplus D(2,0)) \oplus 2\bigoplus_{s=\frac{1}{2}}^{\infty} D(s+1,s)$. Note that all the representations that appear in the decomposition are massless representations. Thus, in contradistinction with flat space, in AdS_4, massless states are "composed" of two singletons. The flat space limit of a singleton is a vacuum and, even in AdS_4, the singletons are very poor in states: their (E,j) diagram has a single trajectory (hence their name). In normal units a singleton with angular momentum j has energy $E = (j+\frac{1}{2})\rho$, where ρ is the curvature of the AdS_4 universe. This means that only a laboratory of cosmic dimensions can detect a j large enough for E to be measurable. Elementary particles would then be composed of two (possibly also three or more) singletons and/or anti singletons, the latter being associated with the contragredient representations. As with quarks, several (at present three) flavours of singletons (and anti singletons) should eventually be introduced to account for all elementary particles. In order to pursue this point further we need to give a little more details on how to develop a field theory of singletons and of particles composed of singletons.

For reasons explained in [26, 30] and references quoted therein, we consider for the *Rac*, the dipole equation $(\Box - \frac{5}{4}\rho)^2\phi = 0$ with the boundary conditions $r^{\frac{1}{2}}\phi < \infty$ as $r \to \infty$, which carries the non-decomposable representation $D(\frac{1}{2},0) \to D(\frac{5}{2},0)$. Quantisation needs a non-degenerate, invariant symplectic structure. This requires the introduction of additional modes, canonically conjugate to the gauge modes (compare the situation in electrodynamics where Maxwell theory has no momentum conjugate to gauge modes), to give to the total space the symmetric form $D(\frac{5}{2},0) \to D(\frac{1}{2},0) \to D(\frac{5}{2},0)$ or " scalar → transverse → gauge". A remarkable fact is that this theory is a *topological field theory*; that is [25], the physical solutions manifest themselves only by their boundary values at $r \to \infty$: $\lim r^{\frac{1}{2}}\phi$ defines a field on the 3-dimensional boundary at infinity. There, on the boundary, gauge invariant interactions are possible and make a 3-dimensional conformal field theory (CFT). A 5-dimensional analogue of this 4-dimensional theory is the 5-dimensional Anti de Sitter/4-dimensional conformal field theory (AdS_5/CFT_4) duality which has found an interesting interpretation by Maldacena [52] in the context of strings and branes.

However, if massless fields (in 4 dimensions) are singleton composites, then singletons must come to life as four dimensional objects, and this requires the introduction of unconventional statistics. The requirement that the bilinears have the properties of ordinary (massless) bosons also tells us that the statistics of singletons must be of another sort. The basic idea is [28] that we can decompose the singleton field operator as $\phi(x) = \sum_{-\infty}^{\infty} \phi^j(x)a_j$ in terms of positive energy creation operators $a^{*j} = a_{-j}$ and annihilation operators a_j (with $j > 0$) without so far making any assumptions about their commutation relations. The choice of commutation relations comes later, when requiring that photons, considered as $2 - Rac$ fields (using the full tensor product of the two singleton triplets) be Bose-Einstein quanta. The singletons are then subject to unconventional statistics (which is perfectly admissible since they are naturally confined) and an appropriate Fock space can be constructed. Based on these principles, a (conformally covariant) composite QED theory was constructed [28], with all the good features of the usual theory. In addition the BRST structure of singleton gauge theory induces [27] the BRST structure of the electromagnetic potential.

A more recent contribution [29] to this interpretation of massless fields as singleton composites deals with gravitons, giving an explicit expression for the weak gravitational potential in terms of singleton bilinears. If this idea is introduced in the context of bulk/boundary duality, it is natural to relate massless fields on the bulk to conserved currents on the boundary. But we are interested in the composite nature of massless fields on space time (the bulk), and a direct current-field identity is then inappropriate. It was shown [29] that the dipole formulation provides a natural construction of all massless fields in terms of bilinears that are conserved only by virtue of the gauge fixing condition on constituent singleton fields.

3 Nonlinear covariant field equations

A cohomological (formal), then analytical, study of nonlinear Lie group representations was started by us about 27 years ago [34]. Nonlinear representations can be viewed as successive extensions of their linear part S^1 by its (symmetric) tensorial powers $\otimes^n S^1$, $n \geq 2$: first S^1 by $S^1 \otimes S^1$, then the result by $\otimes^3 S^1$ and so on. Cohomology plays thus a natural role. E.g. it is sufficient to have at least one invertible operator in the representation of the centre of the enveloping algebra for the corresponding 1-cohomology to vanish, rendering trivial an associated extension.

That theory has given spectacular applications to covariant nonlinear partial differential equations, in particular nonlinear and especially the coupled Maxwell-Dirac equations (first-quantised electrodynamics) [35, 36]. In such equations the nonlinearity appears as coupled to the linear (free) equations, with a coupling constant that plays the role of deformation parameter. Once the classical covariant field equations are studied enough in details one can think [18] of studying their quantisation along the lines of deformation quantisation, e.g. by considering the quantised fields as functionals over the initial data of the classical equations. This part is thus a natural component of our trilogy. We shall not enter here into technical details and shall be satisfied with a qualitative presentation of some consequences.

NonLinear Klein-Gordon equation The nonlinear Klein-Gordon equation (NLKG) can be written as:

$$(\Box + m^2)\ \varphi(t,x) = P(\varphi(t,x), \frac{\partial}{\partial t}\varphi(t,x), \nabla\varphi(t,x))$$

where $m^2 > 0$, $x \in \mathbb{R}^n$, $n \geq 2$ and P is analytic (or only C^∞) with

no constant and no linear term ($P(0) = 0 = dP(0)$). We transform it by standard methods into an *evolution equation.* We introduce appropriate Banach spaces which are completions of the *differentiable vectors* space E_∞ for the associated linear representation of the Poincaré Lie algebra. Then *local solutions* are obtained by Lie theory [34]. *Global solutions* [58] follow from the linearisability of the time translations which (together with asymptotic freedom) will be a consequence of the existence of a solution to a related integral Yang-Feldman-Källén equation. In this way one obtains global nonlinear representations, analytic linearisability, global solutions and asymptotic completeness. For precise statements, see [58, 35] and references therein. These methods permit to include quadratic interactions in the equation in physical 1+3 dimensions, that had not been treated before. All these are scalar field equations. Equations involving massless particles are more difficult to treat, in particular due to infrared divergencies. Nevertheless the general framework presented here is powerful enough to permit their treatment.
Asymptotic completeness, global existence and the infrared problem for the Maxwell-Dirac equations. We refer here to the extensive monograph [36] and especially to its introduction where the main results are sketched.

The classical Maxwell-Dirac (MD) equations read, in the usual notations of 3+1 dimensional space-time, $\Box A_\mu = \overline{\psi}\gamma_\mu\psi$, $(i\gamma^\mu\partial_\mu + m)\psi = A_\mu\gamma^\mu\psi$, $\partial_\mu A^\mu = 0$, where A_μ is the electromagnetic potential, $m > 0$, $0 \leq \mu \leq 3$, $\overline{\psi} = \psi^+\gamma_0$, ψ^+ being the Hermitian conjugate of the Dirac spinor ψ.
The Infrared Problem. On the classical level the infrared problem consists of determining to which extent the long-range interaction created by the coupling $A^\mu j_\mu$ between the electromagnetic potential A_μ and the current $j_\mu = \overline{\psi}\gamma_\mu\psi$ is an obstruction for the separation, when $|t| \to \infty$, of the nonlinear relativistic system into two asymptotic isolated relativistic systems, one for the electromagnetic potential A_μ and one for the Dirac field ψ. It has been proved in [36] that there is such an obstruction, which in particular implies that *asymptotic in and out states do not transform according to a linear representation of the Poincaré group.* This constitutes a serious problem for the second quantisation of the asymptotic (in and out going) fields. The particle interpretation usually requires free relativistic fields, i.e. at least a linear representation of the Poincaré group $\mathcal{P}_0$. Here we introduce nonlinear representations $U^{(-)}$ and $U^{(+)}$ of the Poincaré group which give the Poincaré transformation of the

asymptotic in and out states and permit a particle interpretation. In mathematical terms the infrared problem of the MD equations consists of determining diffeomorphisms (*modified wave operators*) Ω_ε satisfying $U_g^{(\varepsilon)} = \Omega_\varepsilon^{-1} \circ U_g \circ \Omega_\varepsilon$ with $g \in \mathcal{P}_0$, $\varepsilon = \pm$, the asymptotic representations $U^{(\varepsilon)}$ being differentiable.

The same methods can be used for nonabelian gauge theories (of the Yang-Mills type) coupled with fermions. The aim here is to separate asymptotically the linear (modulo an infrared problem that can be a lot worse in the nonabelian case) equation for the spinors from the pure Yang-Mills equation (the A_μ part). The next step would be to linearise analytically the pure Yang-Mills equation (that is known to be formally linearisable), and then to combine all this with the deformation quantisation approach to deal rigorously with the corresponding quantum field theories.

The results on the Maxwell-Dirac equations give indications how a true quantum field theory (i.e. not based on perturbative theory) can be developed on the basis of this first quantised (classical) field theory, dealing in particular with the infrared problem and the definition of observables. The quantisation should be based on the mathematical facts found here and not on a nonrigourous perturbation theory developed from the free field by canonical quantisation or using some algebraic postulates which (however interesting they may seem) reflect sometimes a "wishful thinking". In other words the path to follow should be based on "quantum deformations" (in the sense of star products) of the "classical" theory presented here. In this context it is important to get existence theorems for large initial data and to be able to localise specific solutions corresponding to large initial data, such as of the soliton or instanton type. In 4-dimensional space-time these are very hard problems, which is no surprise: Problems worthy of attack prove their worth by hitting back!

4 Quantisation is a deformation

4.1 The Gerstenhaber theory of deformations of algebras

A concise formulation of a Gerstenhaber deformation of an algebra (associative, Lie, bialgebra, etc.) is [40, 7]:

DEFINITION. A deformation of an algebra A over a field $\mathbb{K}$ is a $\mathbb{K}[[\nu]]$-

algebra $\tilde{A}$ such that $\tilde{A}/\nu\tilde{A} \approx A$. Two deformations $\tilde{A}$ and $\tilde{A}'$ are said equivalent if they are isomorphic over $\mathbb{K}[[\nu]]$ and $\tilde{A}$ is said trivial if it is isomorphic to the original algebra A considered by base field extension as a $\mathbb{K}[[\nu]]$-algebra.

Whenever we consider a topology on A, $\tilde{A}$ is supposed to be topologically free. For associative (resp. Lie) algebras, the above definition tells us that there exists a new product $*$ (resp. bracket $[\cdot,\cdot]$) such that the new (deformed) algebra is again associative (resp. Lie). Denoting the original composition laws by ordinary product (resp. $\{\cdot,\cdot\}$) this means that, for $u, v \in A$ (we can extend this to $A[[\nu]]$ by $\mathbb{K}[[\nu]]$-linearity) we have:

$$u * v = uv + \sum_{r=1}^{\infty} \nu^r C_r(u,v) \tag{1}$$

$$[u,v] = \{u,v\} + \sum_{r=1}^{\infty} \nu^r B_r(u,v) \tag{2}$$

where the C_r are Hochschild 2-cochains and the B_r (skew-symmetric) Chevalley 2-cochains, such that for $u, v, w \in A$ we have $(u * v) * w = u * (v * w)$ and $\mathcal{S}[[u,v],w] = 0$, where $\mathcal{S}$ denotes summation over cyclic permutations.

For a (topological) *bialgebra* (an associative algebra A where we have in addition a coproduct $\Delta : A \longrightarrow A \otimes A$ and the obvious compatibility relations), denoting by $\otimes_\nu$ the tensor product of $\mathbb{K}[[\nu]]$-modules, we can identify $\tilde{A}\hat{\otimes}_\nu\tilde{A}$ with $(A\hat{\otimes}A)[[\nu]]$, where $\hat{\otimes}$ denotes the algebraic tensor product completed with respect to some topology (e.g. projective for Fréchet nuclear topology on A), we similarly have a deformed coproduct $\tilde{\Delta} = \Delta + \sum_{r=1}^{\infty} \nu^r D_r$, $D_r \in \mathcal{L}(A, A\hat{\otimes}A)$, satisfying $\tilde{\Delta}(u*v) = \tilde{\Delta}(u)*\tilde{\Delta}(v)$. In this context appropriate cohomologies can be introduced [41, 6]. There are natural additional requirements for Hopf algebras.

Equivalence means that there is an isomorphism $T_\nu = I + \sum_{r=1}^{\infty} \nu^r T_r$, $T_r \in \mathcal{L}(A, A)$ so that $T_\nu(u *' v) = (T_\nu u * T_\nu v)$ in the associative case, denoting by $*$ (resp. $*'$) the deformed laws in $\tilde{A}$ (resp. $\tilde{A}'$); and similarly in the Lie, bialgebra and Hopf cases. In particular we see (for $r = 1$) that a deformation is trivial at order 1 if it starts with a 2-cocycle which is a 2-coboundary. More generally, exactly as above, we can show [3] ([41, 6] in the Hopf case) that if two deformations are equivalent up to some order t, the condition to extend the equivalence one step further is that a 2-cocycle (defined using the T_k, $k \leq t$) is the coboundary of the required T_{t+1} and therefore *the obstructions to equivalence lie in the*

2-cohomology. In particular, if that space is null, all deformations are trivial.

Unit. An important property is that a *deformation of an associative algebra with unit* (what is called a unital algebra) is again unital, and *equivalent to a deformation with the same unit.* This follows from a more general result of Gerstenhaber (for deformations leaving unchanged a subalgebra) and a proof can be found in [41].
REMARK. 1) In the case of (topological) *bialgebras* or *Hopf* algebras, *equivalence* of deformations has to be understood as an isomorphism of (topological) $\mathbb{K}[[\nu]]$-algebras, the isomorphism starting with the identity for the degree 0 in ν. A deformation is again said *trivial* if it is equivalent to that obtained by base field extension. For Hopf algebras the deformed algebras may be taken (by equivalence) to have the same unit and counit, but in general not the same antipode.
2) Deformations that are more general than those of Gerstenhaber can (and have been [19, 56, 54]) introduced, where e.g. the deformation "parameter" may act on the algebra.

4.2 From quantisation to the invention of deformation quantisation

The need for quantisation appeared for the first time in 1900 when, faced with the impossibility to explain otherwise the black body radiation, Planck proposed the quantum hypothesis: the energy of light is not emitted continuously but in quanta proportional to its frequency. He wrote h for the proportionality constant which bears his name. This paradoxical situation got a beginning of a theoretical basis when, in 1905, Einstein came with the theory of the photoelectric effect – for which he was awarded the Nobel prize in 1922 (for 1921). Around 1920, Prince Louis de Broglie was introduced to the photoelectric effect, together with the Planck–Einstein relations and the theory of relativity, in the laboratory of his much older brother, Maurice duc de Broglie. This led him, in 1923, to his discovery of the duality of waves and particles, which he described in his celebrated Thesis published in 1925, and to what he called 'mécanique ondulatoire'. German and Austrian physicists, in particular, Hermann Weyl, Werner Heisenberg and Erwin Schrödinger, followed by Niels Bohr, transformed it into the quantum mechanics that we know, where the observables are operators in Hilbert spaces of wave functions – and were lead to its probabilistic interpretation that neither Einstein nor de Broglie were at ease with.

Intuitively, classical mechanics is the limit of quantum mechanics when $\hbar = \frac{h}{2\pi}$ goes to zero. But how can this be realised when in classical mechanics the observables are functions over phase space (a Poisson manifold) and not operators? The deformation philosophy promoted by Flato shows the way: one has to look for deformations of algebras of classical observables, functions over Poisson manifolds, and realise there quantum mechanics in an *autonomous* manner.

What we call "deformation quantisation" relates to (and generalizes) what in the conventional (operator) formulation are the Heisenberg picture and Weyl's quantisation procedure. In the latter [61], starting with a classical observable $u(p, q)$, some function on phase space $\mathbb{R}^{2\ell}$ (with $p, q \in \mathbb{R}^\ell$), one associates an operator (the corresponding quantum observable) $\Omega(u)$ in the Hilbert space $L^2(\mathbb{R}^\ell)$ by the following general recipe:

$$u \mapsto \Omega_w(u) = \int_{\mathbb{R}^{2\ell}} \tilde{u}(\xi, \eta)\exp(i(P.\xi + Q.\eta)/\hbar)w(\xi, \eta)\, d^\ell\xi d^\ell\eta \qquad (3)$$

where $\tilde{u}$ is the inverse Fourier transform of u, P_α and Q_α are operators satisfying the canonical commutation relations $[P_\alpha, Q_\beta] = i\hbar\delta_{\alpha\beta}$ $(\alpha, \beta = 1, ..., \ell)$, w is a weight function and the integral is taken in the weak operator topology. What is now called normal ordering corresponds to choosing the weight $w(\xi, \eta) = \exp(-\frac{1}{4}(\xi^2 \pm \eta^2))$, standard ordering (the case of the usual pseudodifferential operators in mathematics) to $w(\xi, \eta) = \exp(-\frac{i}{2}\xi\eta)$ and the original Weyl (symmetric) ordering to $w = 1$. An inverse formula was found shortly afterwards by Eugene Wigner [62] and maps an operator into what mathematicians call its symbol by a kind of trace formula. For example Ω_1 defines an isomorphism of Hilbert spaces between $L^2(\mathbb{R}^{2\ell})$ and Hilbert-Schmidt operators on $L^2(\mathbb{R}^\ell)$ with inverse given by

$$u = (2\pi\hbar)^{-\ell}\, \mathrm{Tr}[\Omega_1(u)\exp((\xi.P + \eta.Q)/i\hbar)] \qquad (4)$$

and if $\Omega_1(u)$ is of trace class one has $\mathrm{Tr}(\Omega_1(u)) = (2\pi\hbar)^{-\ell} \int u\,\omega^\ell \equiv \mathrm{Tr}_{\mathrm{M}}(u)$, the "Moyal trace", where ω^ℓ is the (symplectic) volume dx on $\mathbb{R}^{2\ell}$. Numerous developments followed in the direction of phase-space methods, many of which are described in [1]. Of particular interest to us here is the question of finding an interpretation to the classical function u, symbol of the quantum operator $\Omega_1(u)$; this was the problem posed (around 15 years after [62]) by Blackett to his student Moyal. The (somewhat naive) idea to interpret it as a probability density had of course to be rejected (because u has no reason to be positive) but, looking for a direct expression for the symbol of a quantum commutator, Moyal found [53] what is now called the Moyal bracket:

$$\begin{aligned} M(u_1, u_2) &= \nu^{-1} \sinh(\nu P)(u_1, u_2) \\ &= P(u_1, u_2) + \sum_{r=1}^{\infty} \frac{\nu^{2r}}{(2r+1)!} P^{2r+1}(u_1, u_2) \end{aligned} \tag{5}$$

where $2\nu = i\hbar$, $P^r(u_1, u_2) = \Lambda^{i_1 j_1} \ldots \Lambda^{i_r j_r}(\partial_{i_1 \ldots i_r} u_1)(\partial_{j_1 \ldots j_r} u_2)$ is the rth power ($r \geq 1$) of the Poisson bracket bidifferential operator P, $i_k, j_k = 1, \ldots, 2\ell$, $k = 1, \ldots, r$ and $(\Lambda^{i_k j_k}) = \begin{pmatrix} 0 & -I \\ I & 0 \end{pmatrix}$. To fix ideas we may assume here $u_1, u_2 \in C^\infty(\mathbb{R}^{2\ell})$ and the sum is taken as a formal series (the definition and convergence for various families of functions u_1 and u_2 was also studied, including in [3]). A similar formula for the symbol of a product $\Omega_1(u)\Omega_1(v)$ had been found a little earlier [42] and can now be written more clearly as a (Moyal) *star product*:

$$u_1 *_M u_2 = \exp(\nu P)(u_1, u_2) = u_1 u_2 + \sum_{r=1}^{\infty} \frac{\nu^r}{r!} P^r(u_1, u_2). \tag{6}$$

One recognizes in (6) a special case of (1), and similarly for the bracket. So, via a Weyl quantisation map, the algebra of quantised observables can be viewed as a deformation of that of classical observables.

Several integral formulas for the star product have been introduced and the Wigner image of various families of operators (including bounded operators on $L^2(\mathbb{R}^\ell)$) were studied. The formal series may be deduced (see e.g. [5]) from an integral formula of the type:

$$(u_1 * u_2)(x) = c_\hbar \int_{\mathbb{R}^{2\ell} \times \mathbb{R}^{2\ell}} u_1(x+y) u_2(x+z) e^{-\frac{i}{\hbar} \Lambda^{-1}(y,z)} dy dz. \tag{7}$$

It was noticed, however after deformation quantisation was introduced, that the composition of symbols of pseudodifferential operators (ordered, like differential operators, "first q, then p") used e.g. in index theorems, is a star product. Starting from field theory, where normal (Wick) ordering is essential (the role of q and p above is played by $q \pm ip$), Berezin [4] developed in the mid-seventies an extensive study of what he called "quantisation", based on the correspondence principle and Wick symbols. It is essentially based on Kähler manifolds and related to pseudodifferential operators in the complex domain [9]. However in his theory (which we noticed rather late), as in the studies of various orderings [1], the important concepts of *deformation* and *autonomous* formulation of quantum mechanics in general phase space are absent.

Quantisation involving more general phase spaces was treated, in a somewhat systematic manner, only with Dirac constraints [16]: second class Dirac constraints restrict phase space from some $\mathbb{R}^{2\ell}$ to a symplectic manifold W imbedded in it (with induced symplectic form), while first class constraints further restrict to a Poisson manifold with symplectic foliation (see e.g. [33]). The question of quantisation on such manifolds was certainly treated by many authors (including [16]) but did not go beyond giving some (often useful) recipes and hoping for the best.

A first systematic attempt started around 1970 with what was called soon afterwards *geometric quantisation* [51], a by-product of Lie group representations theory where it gave significant results. It turns out that it is geometric all right, but its scope as far as quantisation is concerned has been rather limited since few classical observables could be quantised, except in situations which amount essentially to the Weyl case considered above. In a nutshell one considers phase-spaces W which are coadjoint orbits of some Lie groups (the Weyl case corresponds to the Heisenberg group with the canonical commutation relations); there one defines a "prequantisation" on the Hilbert space $L^2(W)$ and tries to halve the number of degrees of freedom by using polarizations (often complex ones, which is not an innocent operation as far as physics is concerned) to get a Lagrangean submanifold $\mathcal{L}$ of dimension half of that of W and quantised observables as operators in $L^2(\mathcal{L})$. A recent exposition can be found in [63]. Since physicists have no problem with quantising classical observables (at least in flat space), there was clearly a practical gap that needed to be filled. From the conceptual point of view, the "quantum jump" in the nature of observables required also an explanation. The answer to both questions was given by deformation quantisation, reviewed recently more in details in [20, 59].

4.3 Deformation quantisation and its developments

We want to stress that deformation quantisation is not merely "a reformulation of quantising a mechanical system" [22], e.g. in the framework of Weyl quantisation: *The process of quantisation itself is a deformation.* In order to show that explicitly it was necessary to treat in an *autonomous* manner significant physical examples, without recourse to the traditional operator formulation of quantum mechanics. That was achieved in [3] with the paradigm of the harmonic oscillator and more, including the angular momentum and the hydrogen atom. In particular what plays here the role of the unitary time evolution operator of a quantised system is the "star exponential" of its classical Hamiltonian H (expressed as a usual exponential series but with "star powers" of $tH/i\hbar$, t being the time, and computed as a distribution both in phase space variables and in time); in a very natural manner, the spectrum of the quantum operator corresponding to H is the support of the Fourier-Stieltjes transform (in t) of the star exponential (what Laurent Schwartz had called the spectrum of that distribution). We thus get the discrete spectrum $(n+\frac{\ell}{2})\hbar$ of the *harmonic oscillator* $H=\frac{1}{2}(p^2+q^2)$ and the continuous spectrum $\mathbb{R}$ for the dilation generator pq. The eigenprojectors are given [3] by known special functions on phase-space (generalized Laguerre and hypergeometric, multiplied by some exponential). Other examples can be brought to this case by functional manipulations [3]. For instance the Casimir element of $\mathfrak{so}(\ell)$ representing *angular momentum* has $n(n+(\ell-2))\hbar^2$ for spectrum. For the *hydrogen atom*, with Hamiltonian $H=\frac{1}{2}p^2-|q|^{-1}$, the Moyal product on $\mathbb{R}^8$ induces a star product on $X=T^*S^3$; the energy levels, solutions of $(H-E)*\phi=0$, are found to be (as they should) $E=\frac{1}{2}(n+1)^{-2}\hbar^{-2}$ for the discrete spectrum, and $E\in\mathbb{R}^+$ for the continuous spectrum. We thus have recovered, in a completely autonomous manner entirely within deformation quantisation, the results of "conventional" quantum mechanics in these typical examples. Further examples were (and are still being) developed, in particular in the direction of field theory.

That aspect of deformation theory has in the past 27 years or so been extended considerably. It now includes general symplectic [15, 23, 24, 55] and Poisson (finite dimensional) manifolds [46, 47, 10], with further results for infinite dimensional manifolds, for "manifolds with singularities" and for algebraic varieties, and has many far reaching ramifications in both mathematics and physics (see e.g. a brief overview in [20]). As in quantisation itself [61], symmetries (group theory) play a special role and an autonomous theory of star representations of Lie groups was developed, in the nilpotent and solvable cases of course (due to the importance of the orbit method there), but also in other significant examples. Among the latter are those where one makes full use of the Hopf algebra structures and of the "duality" between the group structure and the set of its irreducible representations, recently reviewed in [8]. Deformation theory (and Hopf algebras) are seminal in a variety of problems ranging from theoretical physics (see e.g. [14, 20]), including renormalisation and Feynman integrals and diagrams, to algebraic geometry and number theory (see e.g. [48, 50]), including algebraic curves à la Zagier (cf. Connes' lectures at Collège de France [13] and his lecture in PQR2003.)

We shall not here go further in the details of the developments of deformation quantisation, for which we refer e.g. to the latest reviews [20, 8]. More can be found on the deformation quantisation web site:

`http://idefix.physik.uni-freiburg.de/~star/`

To conclude this short presentation, we point out that deformation quantisation is intimately related to the other topics of the present school. As can be seen from its origin and from the many developments of "star representations" of Lie groups, there could be mutually beneficial interaction between deformation quantisation and the representation theory of Lie groups (especially reductive, where there is still important work to be done). The metamorphoses of deformation quantisation, in particular for Poisson manifolds [46, 47, 49] and algebraic varieties [48], are related to some of the deepest recent mathematical developments. Poisson geometry and groupoids are important tools there. Furthermore, quantum groups can be viewed as an avatar of deformation quantisation [8], when the category in question is that of Hopf algebras and natural topological vector space topologies, associated with Lie group representation theory, are introduced. The more algebraic theory of polynomials of noncommutative variables developed recently by Gelfand (see e.g. [39]), and especially noncommutative geometry, are very much related to deformation quantisation in several respects. Some are presented in Connes' book [11]; a very elaborate beginning of the theory of noncommutative manifolds (especially in dimension 4) can be found in [12]; they play an increasing role in modern theoretical physics, including string and M-theory, where star products play a role [21]. Exciting times ...

Bibliography

1. Agarwal G.S. and Wolf E. "Calculus for functions of noncommuting operators and general phase-space methods in quantum mechanics I, II, III," *Phys. Rev.* **D2** (1970), 2161–2186, 2187–2205, 2206–2225.
2. Angelopoulos E., Flato M., Fronsdal, C. and Sternheimer D. "Massless particles, conformal group and De Sitter universe," *Phys. Rev.* **D23** (1981), 1278–1289.
3. Bayen F., Flato M., Fronsdal C., Lichnerowicz A. and Sternheimer D. "Deformation theory and quantisation I, II," Ann. Phys. (NY) (1978) **111**, 61–110, 111–151.
4. Berezin F. A. "General concept of quantisation," Comm. Math. Phys. **40** (1975), 153–174; "Quantisation," Izv. Akad. Nauk SSSR Ser. Mat. **38** (1974), 1116–1175; "Quantisation in complex symmetric spaces," Izv. Akad. Nauk SSSR Ser. Mat. **39** (1975), 363–402, 472. [English translations: Math. USSR-Izv. **38** 1109–1165 (1975) and **39**, 341–379 (1976)].
5. Bieliavsky P. "Strict quantisation of solvable symmetric spaces," `math.QA/0011144` (2000).
   Bieliavsky P., Bonneau P. and Maeda Y. "Universal Deformation Formulae, Symplectic Lie groups and Symmetric Spaces," `math.QA/0308189`. "Universal Deformation Formulae for Three-Dimensional Solvable Lie groups," `math.QA/0308188`.
6. Bonneau P. "Cohomology and associated deformations for not necessarily co-associative bialgebras." *Lett. Math. Phys.*, **26** (1992), 277–283.
7. Bonneau P., Flato M., Gerstenhaber M. and Pinczon G. "The hidden group structure of quantum groups: strong duality, rigidity and preferred deformations". *Comm. Math. Phys.* **161** (1994), 125–156.

Bidegain F. and Pinczon G. "Quantisation of Poisson-Lie Groups and Applications," *Comm. Math. Phys.* **179** (1996), 295–332.

8. Bonneau P. and Sternheimer D. "Topological Hopf algebras, quantum groups and deformation quantisation," pp. 55–70 in *Hopf algebras in non-commutative geometry and physics*, S. Caenepeel and F. Van Oystaeyen, eds., *Lecture Notes Pure Appl. Math.*, Dekker, New York, 2004 (`math.QA/0307277`).
9. Boutet de Monvel L. and Guillemin V. *The spectral theory of Toeplitz operators.* Annals of Mathematics Studies **99**, Princeton University Press (1981).
10. Cattaneo A.S. and Felder G. "A path integral approach to the Kontsevich quantisation formula," *Comm. Math. Phys.* **212** (2000) 591–611; "On the Globalization of Kontsevich's Star Product and the Perturbative Poisson Sigma Model," *Prog.Theor.Phys.Suppl.* **144** (2001) 38–53 (`hep-th/0111028`).
    Cattaneo A.S. "On the integration of Poisson manifolds, Lie algebroids, and coisotropic submanifolds," `math.SG/0308180`.
11. Connes A. *Noncommutative Geometry*, Academic Press, San Diego 1994; "Noncommutative differential geometry," Inst. Hautes Études Sci. Publ. Math. No. 62 (1985), 257–360.
    "Noncommutative Geometry Year 2000," pp. 49–110 in *Highlights of mathematical physics (London, 2000)*, Amer. Math. Soc., Providence RI, 2002. (`math.QA/0011193`).
12. Connes A. and Dubois-Violette M. "Noncommutative finite-dimensional manifolds. I. Spherical manifolds and related examples." *Comm. Math. Phys.* **230** (2002), 539–579. "Moduli space and structure of noncommutative 3-spheres", *Lett. Math. Phys.* **66** (2003), 91–121.
13. Connes A. and Moscovici A. "Modular Hecke Algebras and their Hopf Symmetry," `math.QA/0301089`; and "Rankin-Cohen Brackets and the Hopf Algebra of Transverse Geometry," `math.QA/0304316` (to appear in the Moscow Mathematical Journal, volume dedicated to Pierre Cartier.)
14. Connes A. and Kreimer D. "Renormalization in quantum field theory and the Riemann-Hilbert problem. I: The Hopf algebra structure of graphs and the main theorem"; II: The β-function, diffemorphisms and the renormalization group," *Comm. Math. Phys.* **210** (2000) 249–273; **216** (2001) 215–241.
15. De Wilde M. and Lecomte P.B.A. "Existence of star-products and of formal deformations of the Poisson Lie algebra of arbitrary symplec-

tic manifolds," *Lett. Math. Phys.* **7** (1983), 487–496.

16. Dirac P.A.M. *Lectures on Quantum Mechanics*, Belfer Graduate School of Sciences Monograph Series No. 2, Yeshiva University, New York 1964. "Generalized Hamiltonian dynamics," *Canad. J. Math.* **2** (1950), 129–148.
17. Dirac P.A.M. "A remarkable representation of the $3 + 2$ de Sitter group", *J. Math. Phys.* **4** (1963), 901–909.
18. Dito J. "Star-product approach to quantum field theory: the free scalar field," *Lett. Math. Phys.* **20** (1990), 125–134; "Star-products and nonstandard quantisation for K-G equation," *J. Math. Phys.* **33** (1992), 791–801; "An example of cancellation of infinities in star-quantisation of fields," *Lett. Math. Phys.* **27** (1993), 73–80; "Deformation quantisation of covariant fields," pp. 55–66 in: *Deformation quantisation* (G.Halbout ed.), IRMA Lectures in Math. Theoret. Phys. **1**, Walter de Gruyter, Berlin 2002 (`math.QA/0202271`).
19. Dito G., Flato M., Sternheimer D. and Takhtajan L. "Deformation Quantisation and Nambu Mechanics," *Comm. Math. Phys.* **183** (1997) 1–22 (`hep-th/9602016`).
20. Dito G. and Sternheimer D., "Deformation Quantisation: Genesis, Developments and Metamorphoses," pp. 9–54 in: *Deformation quantisation* (G.Halbout ed.), IRMA Lectures in Math. Theoret. Phys. **1**, Walter de Gruyter, Berlin 2002 (`math.QA/0202168`).
21. Douglas M.R., Liu H., Moore, G., Zwiebach B. "Open string star as a continuous Moyal product." *J. High Energy Phys.* 0204 (2002), No. 22, 27 pp. (`hep-th/0202087`)
22. Douglas M.R. and Nekrasov N.A. "Noncommutative field theory," *Rev. Mod. Phys.* **73** (2001) 977–1029 (`hep-th/0106048`).
23. Fedosov B.V. "A simple geometrical construction of deformation quantisation," *J. Diff. Geom.* **40** (1994), 213–238.
24. Fedosov B.V. *Deformation quantisation and index theory*. Mathematical Topics **9**, Akademie Verlag, Berlin (1996).
25. Flato M. and Fronsdal C. "Quantum field theory of singletons. The Rac," *J. Math. Phys.* **22** (1981), 1100–1105.
26. Flato M. and Fronsdal C. "The singleton dipole," *Comm. Math. Phys.* **108** (1987), 469–482.
27. Flato M. and Fronsdal C. "How BRST invariance of QED is induced from the underlying field theory," *Physics Letters B* **189** (1987), 145–148.
28. Flato M. and Fronsdal C. "Composite Electrodynamics". *J. Geom.*

Phys. **5** (1988), 37–61.
Flato M., Fronsdal C. and Sternheimer D. “Singletons as a basis for composite conformal quantum electrodynamics” pp. 65–76, in *Quantum Theories and Geometry*, Mathematical Physics Studies **10**, Kluwer Acad. Publ. Dordrecht (1988).

29. Flato M. and Frønsdal C. “Interacting Singletons,” *Lett. Math. Phys.* **44** (1998), 249–259 (hep-th/9803013).
30. Flato M., Frønsdal C. and Sternheimer D. “Singletons, physics in AdS universe and oscillations of composite neutrinos”, *Lett. Math. Phys.* **48** (1999), 109–119.
31. Flato M., Hadjiivanov L.K. and Todorov I.T. “Quantum Deformations of Singletons and of Free Zero-Mass Fields,” *Foundations of Physics* **23** (1993), 571–586.
32. Flato M., Hillion P. and Sternheimer D. “Un groupe de type Poincaré associé à l’équation de Dirac-Weyl, *C.R. Acad. Sci. Paris* **264** (1967), 85–88.
Flato M. and Hillion P. “Poincaré-like group associated with neutrino physics, and some applications, *Physical Review* **D1** (1970), 1667–1673.
33. Flato M., Lichnerowicz A. and Sternheimer D. “Deformations of Poisson brackets, Dirac brackets and applications,” *J. Math. Phys.* **17** (1976), 1754–1762.
34. Flato M., Pinczon G. and Simon J. “Non-linear representations of Lie groups,” *Ann. Sci. Éc. Norm. Sup. (4)* **10** (1977), 405–418.
35. Flato M., Simon J.C.H., Sternheimer D. and Taflin E. “Nonlinear relativistic wave equations in general dimensions.” pp. 53–70 in: *Collection of papers on geometry, analysis and mathematical physics*, World Sci. Publishing, River Edge, NJ, 1997.
36. Flato M., Simon J.C.H. and Taflin E. “Asymptotic completeness, global existence and the infrared problem for the Maxwell-Dirac equations.” *Mem. Amer. Math. Soc.* **127** (1997), no. 606, x+311 pp (hep-th/9502061)
37. Flato M. and Sternheimer D. “Remarks on the connection between external and internal symmetries,” *Phys. Rev. Lett.* **15** (1965), 934–936. “On the connection between external and internal symmetries of strongly interacting particles,” *J. Math. Phys.* **7** (1966), 1932–1958. “Local representations and mass-spectrum,” *Phys. Rev. Lett.* **16** (1966), 1185–1186. “Poincaré partially integrable local representations and mass-spectrum,” *Comm. Math. Phys.* **12** (1969), 296–303.

38. Frønsdal C. "Singletons and neutrinos," Conférence Moshé Flato 1999 (Dijon). *Lett. Math. Phys.* **52** (2000) 51–59.
39. Gelfand I., Retakh V. and Wilson R.L. "Quadratic linear algebras associated with factorizations of noncommutative polynomials and noncommutative differential polynomials." *Selecta Math.* (N.S.) **7** (2001), 493–523.
40. Gerstenhaber M. "The cohomology structure of an associative ring," *Ann. Math.* **78** (1963), 267–288. "On the deformation of rings and algebras," *ibid.* **79** (1964), 59–103; and (IV), *ibid.* **99** (1974), 257–276.
41. Gerstenhaber M. and Schack S.D. "Algebraic cohomology and deformation theory," pp. 11–264 in: M. Hazewinkel and M. Gerstenhaber (eds.), *Deformation Theory of Algebras and Structures and Applications*, NATO ASI Ser. C **247**, Kluwer Acad. Publ., Dordrecht 1988; "Algebras, bialgebras, quantum groups and algebraic deformations," in *Deformation Theory and Quantum Groups with Applications to Mathematical Physics* (M. Gerstenhaber and J. Stasheff, eds.), Contemporary Mathematics **134**, 51–92, American Mathematical Society, Providence (1992).
42. Groenewold A. "On the principles of elementary quantum mechanics," Physica **12** (1946), 405–460.
43. 't Hooft Gerard. "Dimensional reduction in quantum gravity," pp. 284–296 in *Salamfestschrift: a collection of talks* (A. Ali, J. Ellis and S. Randjbar-Daemi eds.), World Scientific 1993 (`gr-qc/9310026`). "The holographic principle: opening lecture," pp. 72–86 in *Erice 1999, Basics and highlights in fundamental physics* (edited by Antonino Zichichi), World Scientific, Singapore 2001 (`hep-th/0003004`).
44. Inönü E. and Wigner E.P. "On the contraction of groups and their representations," *Proc. Nat. Acad. Sci. U. S. A.* **39** (1953), 510–524.
45. Kodaira K. and Spencer D.C. "On deformations of complex analytic structures," *Ann. Math.* **67** (1958), 328–466.
46. Kontsevich M. "Deformation quantisation of Poisson manifolds I". `q-alg/9709040` (1997) *Lett. Math. Phys.* bf 66 (3) (2003), 157–216.
47. Kontsevich M. "Operads and motives in deformation quantisation," *Lett. Math. Phys.* **48** (1999), 35–72.
48. Kontsevich M. "Deformation quantisation of algebraic varieties," *Lett. Math. Phys.* **56** (2001), 271–294.
49. Kontsevich M. and Soibelman Y. "Deformations of algebras over operads and the Deligne conjecture," pp. 255–307 in: G. Dito and D. Sternheimer (eds.), *Conférence Moshé Flato 1999*, Math. Phys.

Stud. **21**, Kluwer Acad. Publ., Dordrecht 2000.

50. Kontsevich M. and Zagier D. "Periods", in *Mathematics unlimited—2001 and beyond*, 771–808, Springer, Berlin 2001.
51. Kostant B. *Quantisation and unitary representations*, pp. 87–208 in: Lecture Notes in Math. **170**, Springer Verlag, Berlin 1970; "On the definition of quantisation," pp. 187–210 in: *Géométrie symplectique et physique mathématique* (Colloq. Int. CNRS, N° 237), Éds. CNRS, Paris 1975; "Graded manifolds, graded Lie theory, and prequantisation," pp. 177–306 in *Differential geometrical methods in mathematical physics*, Lecture Notes in Math. **570**, Springer, Berlin 1977.
52. Maldacena J.M. "The Large N Limit of Superconformal Field Theories and Supergravity," *Adv. Theor. Math. Phys.* **2** (1998), 231–252 (`hep-th/9711200`).
53. Moyal J.E. "Quantum mechanics as a statistical theory," *Proc. Cambridge Phil. Soc.* **45** (1949), 99–124.
54. Nadaud F. "Generalized deformations, Koszul resolutions, Moyal Products," *Reviews Math. Phys.* **10** (1998), 685–704; "Generalized Deformations and Hochschild Cohomology," *Lett. Math. Phys.* **58** (2001), 41–55.
55. Omori H., Maeda Y. and Yoshioka A. "Weyl manifolds and deformation quantisation," *Adv. Math.* **85** (1991), 225–255; "Existence of a closed star product," *Lett. Math. Phys.* **26** (1992), 285–294; "Deformation quantisations of Poisson algebras," *Contemp. Math.* **179** (1994), 213–240.
56. Pinczon G. "Non commutative deformation theory," *Lett. Math. Phys.* **41** (1997), 101–117.
57. Segal I.E. "A class of operator algebras which are determined by groups," Duke Math. J. **18** (1951), 221–265.
58. Simon, J.C.H. and Taflin, E. "The Cauchy problem for non-linear Klein-Gordon equations," Comm. Math. Phys. **152**, 433–478 (1993). "Nonlinear relativistic evolution equations: survey of a new approach," Conférence Moshé Flato 1999, Vol. I (Dijon), 389–402, Math. Phys. Stud. **21**, Kluwer Acad. Publ., Dordrecht, 2000.
59. Sternheimer D. "Deformation quantisation: Twenty years after," pp. 107–145 in: J. Rembieliński, (ed.), *Particles, Fields, and Gravitation (Łodz 1998)*, AIP Press, New York 1998 (`math.QA/9809056`).
60. Wess J. and Zumino B. "Supergauge transformations in four dimensions", *Nuclear Phys.* **B70** (1974), 39–50.
61. Weyl, H. *The theory of groups and quantum mechanics*, Dover, New-York (1931), translated from *Gruppentheorie und Quantenmechanik*,

Hirzel Verlag, Leipzig (1928); "Quantenmechanik und Gruppentheorie," Z. Physik **46** (1927), 1–46.

62. Wigner, E.P. "Quantum corrections for thermodynamic equilibrium," Phys. Rev. **40** (1932), 749–759.
63. Woodhouse N. M. J. *Geometric quantisation.* Oxford Science Publications, Oxford Mathematical Monographs. Oxford University Press, New York (second edition, 1992).

This brief review is an expanded version of the lecture given by the author at the Euroschool PQR2003 in Brussels. The main complements are: `math.QA/0202168` [20] for Section IV, [36] for Section III and [30] for Section II.

It is impossible not to associate this panorama with the ever present memory of Moshé Flato, the founder of, and a main player in, the field of deformation theory in view of physical applications, with whom I had the privilege to work as a team for thirty five years.

MSC (2000): 53D55, 81T10, 81R05, 81R20; 17B37, 19K56, 22E45, 22E70, 35Q53, 35Q55, 35Q75, 46L65, 46M20, 58B32, 58B34, 58J20, 58J22, 58J40, 81Q30, 81Q70, 81R10, 81R50, 81R60, 81S10, 81S30, 81T10, 81T60, 81T70.

Index